BIG IDEAS MATH®

Red

Resources by Chapter

- Family and Community Involvement
- Start Thinking! and Warm Up
- Extra Practice (A and B)
- Enrichment and Extension
- Puzzle Time
- Technology Connection
- Project with Rubric

Erie, Pennsylvania

ISBN 13: 978-1-60840-475-9
ISBN 10: 1-60840-475-7

456789-VLP-17 16 15

Contents

About the Resources by Chapter

Family and Community Involvement (English and Spanish)

The Family and Community Involvement letters provide a way to quickly communicate to family members how they can help their student with the material of the chapter. They make the mathematics less intimidating and provide suggestions for helping students see mathematical concepts in common activities.

Start Thinking! and Warm Up

Each Start Thinking! and Warm Up includes two options for getting the class started. The Start Thinking! questions provide students with an opportunity to discuss thought-provoking questions and analyze real-world situations. The Warm Up questions review prerequisite skills needed for the lesson.

Extra Practice

The Extra Practice exercises provide additional practice on the key concepts taught in the lesson. There are two levels of practice provided for each lesson: A (basic) and B (average). Chapters with Extensions also have Extra Practice.

Enrichment and Extension

Each Enrichment and Extension extends the lesson and provides a challenging application of the key concepts.

Puzzle Time

Each Puzzle Time provides additional practice in a fun format in which students use their mathematical knowledge to solve a riddle. This format allows students to self-check their work.

Technology Connection

Each Technology Connection provides opportunities for students to explore mathematical concepts using tools such as scientific and graphing calculators, spreadsheets, geometry software, and the Internet.

Project with Rubric

The Projects summarize key concepts. They require students to investigate a concept, gather and analyze data, and summarize the results. Scoring rubrics are provided.

Chapter 1

Name____ Date__________

Chapter 1 Integers

Dear Family,

Hiking can be good exercise and a nice opportunity to talk with your family. Whether you walk to a nearby park, or travel to a favorite hiking trail, you usually plan to end up where you start. How far have you traveled on your walk, then? In a sense, you have gone nowhere—you have traveled zero distance.

This isn't the whole story, though. What also happens is that you travel a certain distance there and then the same distance back. The two distances are in opposite directions, so they bring you back to your starting point. But if you want to know how far you have walked, you talk about the distance without regard to direction—the *absolute value*. If the park is a mile away, you walk one mile there and one mile back—two miles in total.

The same reasoning applies when you walk up a hill. If you climb a 300-foot hill, you are going up 300 feet. If you want to end up back at the bottom, you must eventually climb down 300 feet. How much have you climbed in total?

While you are walking, keep track of how far you have traveled. You might talk with your student about the following:

- When is it helpful to assign direction (positive or negative) to each part of the walk?
- When is it helpful to ignore direction and just use the absolute value of the distance?

Usually we express the up direction as a positive number and the down direction as a negative number. Talk with your student about why that might be the case. With your student, think of situations where the reverse might be more convenient.

Enjoy the sunshine while you walk and talk with your student!

Nombre ____ Fecha __________

Números enteros

Estimada Familia:

Salir de excursión puede ser un buen ejercicio y una buena oportunidad para conversar con su familia. Ya sea que caminen a un parque cercano o viajen a su sendero de excursión favorito, generalmente planean terminar donde empezaron. Entonces, ¿cuánto han recorrido en su caminata? En cierto sentido, no han ido a ningún lugar—han recorrido una distancia cero.

Sin embargo, aquí no acaba todo. Lo que también sucede es que recorrieron una cierta distancia hacia allá y luego recorrieron la misma distancia de regreso. Las dos distancias están en direcciones opuestas, así que los traen de vuelta a su punto de partida. Pero si desean saber cuánto han caminado, se habla sobre la distancia sin considerar la dirección—es decir, el *valor absoluto*. Si el parque queda a una milla de distancia, ustedes caminan una milla hacia allá y una milla de regreso—dos millas, en total.

El mismo razonamiento se aplica cuando se trepa por una colina. Si trepa una colina de 300 pies, está ascendiendo 300 pies. Si desea terminar de regreso en la parte inferior, eventualmente tendrá que bajar 300 pies. ¿Cuánto ha trepado en total?

Mientras está caminando, lleve un registro de la distancia que ha recorrido. Querrá hablar con su estudiante acerca de lo siguiente:

- ¿Cuándo es útil asignar una dirección (positiva o negativa) a cada parte de la caminata?
- ¿Cuándo es útil ignorar una dirección y sólo usar el valor absoluto de la distancia?

Generalmente expresamos la dirección ascendente con un número positivo y la descendente con un número negativo. Converse con su estudiante acerca de por qué sería ese el caso. Con su estudiante, piense en situaciones en donde lo inverso podría ser más conveniente.

¡Disfrute de la luz solar mientras camina y conversa con su estudiante!

Activity 1.1 Start Thinking!
For use before Activity 1.1

Explain how football is like solving a math problem.

Activity 1.1 Warm Up
For use before Activity 1.1

Copy and complete the statement using $<$ or $>$.

1. 12 __?__ 14

2. 36 __?__ 26

3. -2 __?__ -5

4. -15 __?__ -8

5. 13 __?__ -10

6. -20 __?__ 19

Lesson 1.1 Start Thinking!

For use before Lesson 1.1

When you go to school in the morning, you travel in one direction. Then returning home, you travel in the other direction. How does this compare to a number line where one direction is positive and the other is negative? Do you ever travel in a negative direction?

Lesson 1.1 Warm Up

For use before Lesson 1.1

Find the absolute value.

1. $|15|$ **2.** $|-23|$ **3.** $|7|$ **4.** $|-35|$

5. $|-43|$ **6.** $|0|$ **7.** $|39|$ **8.** $|-212|$

Name______________________________ Date__________

1.1 Practice A

Find the absolute value.

1. $|-7|$ **2.** $|12|$ **3.** $|-13|$ **4.** $|0|$

Copy and complete the statement using $<$, $>$, or $=$.

5. $|-4| \underline{\ ?\ } 2$ **6.** $7 \underline{\ ?\ } |-7|$ **7.** $|8| \underline{\ ?\ } 5$

8. While playing a game, you move back 5 spaces with your roll of the number cube. Your friend moves forward 3 spaces. Write each amount as an integer.

Order the values from least to greatest.

9. $-1, |5|, |4|, 8, |-1|$ **10.** $|2|, 0, |5|, 6, |3|$

Simplify the expression.

11. $|-19|$ **12.** $-|-8|$ **13.** $-|13|$

14. You are kite sailing on the ocean. The table gives your height at different times.

Time (seconds)	0	1	2	3
Height (feet)	2	4	6	8

a. How many feet do you move each second?

b. What is your speed? Give the units.

c. Is your velocity positive or negative?

d. What is your velocity? Give the units.

15. Use a number line.

a. Graph and label the following points on a number line: $T = 1$, $L = -8$, $E = 4$, $A = -5$. What word do the letters spell?

b. Graph and label the absolute value of each point in part (a). What word do the letters spell now?

16. Write an integer whose absolute value is greater than itself.

Name ____________________ Date ________

1.1 Practice B

Copy and complete the statement using <, >, or =.

1. $|-23|$ __?__ 23 **2.** $|-142|$ __?__ $|-157|$ **3.** $-|-78|$ __?__ 52

4. You and your friend are swimming against the current. You move forward 15 feet. Your friend is not a strong swimmer, so he moves back 6 feet. Write each amount as an integer.

Order the values from least to greatest.

5. $14, |-25|, -|-34|, 28, |0|$ **6.** $|-16|, 10, |25|, -16, |-43|$

Simplify the expression.

7. $|-249|$ **8.** $-|183|$ **9.** $-|-153|$

10. The boiling point of a liquid is the temperature at which the vapor pressure of the liquid equals the environmental pressure surrounding the liquid.

Substance	Hydrogen	Oxygen	Iodine	Phosphorus
Boiling Point (°C)	–253	–183	184	280

a. Which substance in the table has the highest boiling point?

b. Is the boiling point of oxygen or iodine closer to 0°C?

11. You are riding on a rollercoaster.

a. Your velocity is 13 feet per second. Are you moving up or moving down?

b. What is your speed in part (a)? Give the units.

c. Your velocity is -17 feet per second. Are you moving up or moving down?

d. What is your speed in part (c)? Give the units.

12. There is one integer for which there does not exist another integer with the same absolute value. What is that integer?

Determine whether the statement is *true* or *false*. Explain your reasoning.

13. The absolute value of 3 above par is the same as the absolute value of 3 below par.

14. If $x < 0$, then $|x| < x$.

Name____ Date__________

1.1 Enrichment and Extension

Reasoning with Integers

Assume $a > 0$ and $b < 0$. Determine whether the statement is *always, sometimes,* or *never* true.

1. $a > b$ **2.** $a < b$ **3.** $|a| > |b|$ **4.** $|a| < |b|$

5. $-a > -b$ **6.** $-a < -b$ **7.** $-|a| < -|b|$ **8.** $|-a| < |-b|$

Let n be an integer. Determine whether the possible values of n are *all integers, all positive integers, all negative integers, all positive integers and zero, all negative integers and zero,* or *none.*

9. $n < 0$ **10.** $n > 0$ **11.** $|n| < 0$ **12.** $|n| \geq 0$

13. $|-n| < 0$ **14.** $|-n| \geq 0$ **15.** $-|n| \leq 0$ **16.** $-|n| > 0$

17. $-|-n| \leq 0$ **18.** $-|-n| > 0$ **19.** $|n| = -n$ **20.** $-|n| = n$

21. $|n| = n$ **22.** $|-n| = n$ **23.** $-|-n| = -n$ **24.** $-|-n| = n$

25. $|n| < n$ **26.** $|n| > n$ **27.** $|n| > -n$ **28.** $-|n| < n$

29. $|n| > |-n|$ **30.** $|n| \geq -|n|$ **31.** $-|-n| < n$ **32.** $-|-n| \leq |n|$

33. $|n| + |n| < 2n$ **34.** $|n| + |n| = 2|n|$ **35.** $-|n| + |n| = 0$ **36.** $-|-n| + |-n| = 0$

For each exercise number, use your answers and the key below to color the cell. Do not color the cells that have a zero in them.

Always = Orange

Sometimes = Green

Never = Brown

All Integers = Brown

All Positive Integers = Brown

All Negative Integers = Green

All Positive Integers and Zero = Green

All Negative Integers and Zero = Brown

None = Brown

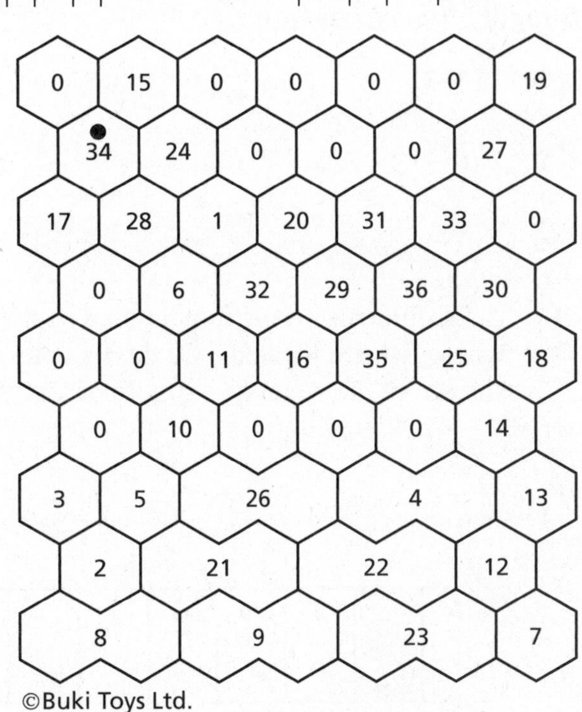

©Buki Toys Ltd.

Name ______________________________ Date ________

1.1 Puzzle Time

What Can You Serve, But Never Eat?

Write the letter of each answer in the box containing the exercise number.

Find the absolute value.

1. $|12|$
2. $|-9|$
3. $|20|$
4. $|-10|$

Complete the statement using $<, >,$ or $=$.

5. $3 \underline{\ ?\ } |-8|$
6. $4 \underline{\ ?\ } |-4|$
7. $|-6| \underline{\ ?\ } -6$

Simplify the expression.

8. $-|-13|$
9. $|-55|$
10. $-|2|$
11. A fishfinder is an instrument on a boat that indicates where fish are located. Are the fish closest to the surface of the water at -20 feet or -30 feet?

Answers

S. 10	**B.** -13
E. -2	**T.** 20
G. -55	**A.** 9
I. 12	**F.** 2
S. -20	**B.** -9
H. -12	**D.** 13
L. 55	**R.** -10
A. $<$	**N.** $>$
L. $=$	**N.** -20 feet
P. -30 feet	

5		3	10	11	7	1	4		8	2	9	6

Activity 1.2 Start Thinking!

For use before Activity 1.2

Explain how playing a round of golf is like solving an integer problem.

Activity 1.2 Warm Up

For use before Activity 1.2

Add.

1. $10 + 12$

2. $14 + 28$

3. $26 + 32 + 19$

4. $47 + 35 + 68$

5. $12 + 33 + 59 + 18$

6. $82 + 13 + 29 + 97$

Lesson 1.2

Start Thinking!

For use before Lesson 1.2

The temperature first rises 10 degrees and then falls 12 degrees. Is the end temperature greater than or less than the starting temperature? How does this compare to adding integers?

Lesson 1.2

Warm Up

For use before Lesson 1.2

Add.

1. $9 + 4$

2. $-5 + (-1)$

3. $-8 + 13$

4. $12 + (-8)$

5. $-10 + 6$

6. $-14 + 14$

Name_______________________________________ Date__________

1.2 Practice A

Add.

1. $8 + 2$ **2.** $-5 + (-3)$ **3.** $-9 + (-3)$ **4.** $6 + (-6)$

5. $4 + (-4)$ **6.** $9 + (-6)$ **7.** $5 + (-2)$ **8.** $7 + (-13)$

9. $-18 + 1$ **10.** $-12 + (-5)$ **11.** $0 + (-7)$ **12.** $12 + (-15)$

13. Your bank account has a balance of $-\$21$. You deposit $\$50$. What is your new balance?

Tell how the Commutative and Associative Properties of Addition can help you find the sum mentally. Then find the sum.

14. $8 + (-5) + (-8)$ **15.** $-4 + 9 + 4$ **16.** $-5 + 12 + (-7)$

Add.

17. $7 + 5 + (-2)$ **18.** $-13 + 7 + (-3)$ **19.** $17 + (-5) + (-1)$

20. $4 + 8 + (-8)$ **21.** $-12 + (-4) + 9$ **22.** $-10 + 10 + (-3)$

23. $(-11) + 5 + (-12)$ **24.** $7 + 15 + (-7)$ **25.** $-12 + (-5) + (-10)$

Use mental math to solve the equation.

26. $n + (-8) = 5$ **27.** $4 + c = 0$ **28.** $-6 + k = -14$

29. In golf, a golfer must have a score of 0 in order to be at par. A golfer scores 2 above par on the first hole, 1 below par on the second hole, and 2 below par on the third hole. Which expression can be used to decide whether the golfer is at par after the first three holes?

$(-2) + 1 + 2$ $2 + (-1) + 2$ $2 + (-1) + (-2)$

30. Copy and complete the magic square so that each row and column has a magic sum of 0. Use each integer from -4 to 4 exactly once.

3		−2
		2

Name ______________________________ Date ________

1.2 Practice B

Add.

1. $15 + 24$
2. $-13 + (-35)$
3. $29 + (-29)$
4. $31 + (-72)$

5. The elevation of your plot of land is 2 feet below sea level. You add 7 feet of dirt to your land. What is the new elevation of your land?

Tell how the Commutative and Associative Properties of Addition can help you find the sum mentally. Then find the sum.

6. $18 + (-25) + (-18)$
7. $-22 + 45 + (-8)$
8. $28 + (-12) + 4$

Add.

9. $17 + (-33) + (-12)$
10. $(-41) + 25 + 19$
11. $(-43) + (-27) + 43$
12. $71 + 27 + (-42)$
13. $(-63) + 81 + 0$
14. $(-39) + (-21) + (-19)$
15. $52 + (-38) + 23$
16. $101 + (-51) + (-36)$
17. $(-117) + 125 + (-67)$

Use mental math to solve the equation.

18. $n + (-20) = 5$
19. $c + (-71) = 0$
20. $-30 + k = -110$

21. Write three integers that do not all have the same sign that have a sum of -20. Write three integers that do not all have the same sign that have a sum of 10.

22. The temperature at 6 A.M. is $-12°$F. During the next twelve hours, the temperature increases $25°$F. During the following 5 hours, the temperature decreases $23°$F. What is the temperature at 11 P.M.?

23. Copy and complete the magic square so that each row and column has a magic sum of 0. Use integers from -9 to 9, without repeating an integer.

9		-3
		2

Name ______________________________ Date __________

1.2 Enrichment and Extension

Magic Squares with Integers

According to a legend, the Chinese Emperor Yu-Huang saw a magic square on the back of a turtle. In a *magic square*, the sum of the numbers in each row, column, and diagonal are the same. This sum is called the magic sum.

This magic square uses integers -6 to 2 exactly once. The magic sum is -6.

1	-6	-1
-4	-2	0
-3	2	-5

Diagonal 1: $-3 + (-2) + (-1) = -6$

Row 1: $1 + (-6) + (-1) = -6$

Row 2: $-4 + (-2) + 0 = -6$

Row 3: $-3 + 2 + (-5) = -6$

Diagonal 2: $1 + (-2) + (-5) = -6$

Column 1: $1 + (-4) + (-3) = -6$

Column 2: $-6 + (-2) + 2 = -6$

Column 3: $-1 + 0 + (-5) = -6$

Complete the magic square using each integer only once. The magic sum is given.

1. Use -9 to -1; Magic Sum $= -15$

-8		-4
	-7	

2. Use -5 to 3; Magic Sum $= -3$

-2		
		1
	-5	

3. Use -7 to 8; Magic Sum $= 2$

	7		-4
		-1	
0		3	-3
	-5		

4. Use -10 to 5; Magic Sum $= -10$

-4	1		
-9		-3	0
	-8		-6
		4	

5. Create your own magic square with integers having the magic sum 6.

Name ______________________________ Date __________

1.2 Puzzle Time

Why Did The Golfer Wear Two Pairs Of Pants?

Write the letter of each answer in the box containing the exercise number.

Add.

1. $12 + 5$

2. $7 + (-7)$

3. $-10 + 2$

4. $9 + (-6)$

5. $-15 + 27$

6. $23 + (-23)$

7. $-17 + 12$

8. $13 + (-15)$

9. $-9 + (-9)$

10. $-14 + (-11)$

11. $12 + (-10) + 16$

12. $15 + (-15) + 12$

13. $-22 + 30 + (-26)$

14. $-8 + (-8) + (-9)$

15. $37 + (-21) + (-16)$

16. $-42 + 8 + 17$

17. $-30 + 34 + (-9)$

18. $14 + (-21) + 7$

19. $-25 + 17 + 6$

20. $-4 + (-8) + (-6)$

21. A roller coaster climbs 84 feet on the first hill then drops 60 feet down. On the second hill the roller coaster climbs another 32 feet then drops 44 feet. What is the height at the end of the second hill?

Answers

S. 18	**N.** -18
O. 12	**C.** -8
L. -17	**H.** -25
I. -5	**E.** 0
T. 17	**G.** 3
A. -2	

7	13		3	8	11	15		14	18		4	21	1		19		10	12	16	2

17	9		5	20	6

Activity 1.3 Start Thinking!

For use before Activity 1.3

You plan a hiking trip. Discuss with a partner the math involved when hiking from one point to the next.

Activity 1.3 Warm Up

For use before Activity 1.3

Subtract.

1. $45 - 11$

2. $87 - 23$

3. $91 - 14$

4. $76 - 69$

5. $87 - 29 - 13$

6. $65 - 52 - 11$

Lesson 1.3 Start Thinking!

For use before Lesson 1.3

How can you tell if the difference of two integers is positive?

How can you tell if the difference of two integers is negative?

How can you tell if the difference of two integers is zero?

Lesson 1.3 Warm Up

For use before Lesson 1.3

Subtract.

1. $5 - 8$

2. $12 - (-3)$

3. $-5 - (-6)$

4. $-4 - 6$

5. $9 - 16$

6. $-7 - 10$

Name________________________________ Date__________

1.3 Practice A

Subtract.

1. $3 - 8$ **2.** $4 - (-5)$ **3.** $-6 - 4$ **4.** $-9 - (-6)$

5. $10 - (-9)$ **6.** $12 - 4$ **7.** $-15 - 7$ **8.** $-6 - (-14)$

9. $-1 - (-3)$ **10.** $15 - (-7)$ **11.** $20 - (-10)$ **12.** $-31 - 14$

13. You are scuba diving at -8 feet. You dive 5 feet deeper. What is your position in the water?

14. Write $7 - 3$ using addition.

15. Write $5 + (-3)$ using subtraction.

Evaluate the expression.

16. $8 - 12 - (-6)$ **17.** $8 - (-8) - 3$ **18.** $0 - (-4) - 8$

19. $9 - (-4) + 1$ **20.** $7 - 12 - (-4)$ **21.** $-11 - (-8) - (-3)$

22. $-14 - 6 - (-2)$ **23.** $8 + 0 - (-11)$ **24.** $8 + 13 - (-5)$

Use mental math to solve the equation.

25. $a - 7 = 3$ **26.** $b - (-8) = -3$ **27.** $6 - c = 10$

28. Write two different pairs of negative integers, x and y, that make the statement $x - y = 2$ true.

29. The table shows the highest and lowest elevations for two cities.

City	Highest elevation (feet)	Lowest elevation (feet)
Long Beach, CA	360	–7
New Orleans, LA	25	–8

a. Find the range of elevations for Long Beach.

b. Find the range of elevations for New Orleans.

c. One of the cities has an average elevation of about 2 feet below sea level. Which city is it?

Name ________________________ Date ________

1.3 Practice B

Subtract.

1. $8 - 13$
2. $18 - (-11)$
3. $-14 - 35$
4. $-51 - (-36)$
5. $100 - (-91)$
6. $-82 - 64$
7. $35 - 47$
8. $-36 - (-54)$
9. A dolphin is at -28 feet. It swims up and jumps out of the water to a height of 8 feet. Write a subtraction expression for the vertical distance the dolphin travels.

Evaluate the expression.

10. $15 - 42 - (-36)$
11. $17 - (-22) - 22$
12. $0 - (-41) - 28$
13. $87 - (-34) + 13$
14. $-35 - 27 - (-14)$
15. $-51 - (-23) + (-16)$
16. $-14 - 63 - (-52)$
17. $-28 + 10 - (-121)$
18. $8 - (-103) - (-95)$

Use mental math to solve the equation.

19. $a - 24 = 47$
20. $26 - b = 9$
21. $c - (-15) = 38$
22. $-5 - k = -12$
23. $t - (-14) = -24$
24. $-25 - m = 28$
25. The table shows the record monthly high and low temperatures in International Falls, Minnesota.

	Jan	Feb	Mar	Apr	May	Jun	Jul	Aug	Sep	Oct	Nov	Dec
High (°F)	48	58	76	93	95	99	98	95	95	88	73	57
Low (°F)	–46	–45	–38	–14	11	23	34	30	20	2	–32	–41

   a. Find the range of temperatures for each month.

   b. What are the all-time high and all-time low temperatures?

   c. What is the range of the temperatures in part (b)?

26. For what values of a and b is the statement $|a - b| = |a| - |b|$ false?

Name ____________________ Date ________

1.3 Enrichment and Extension

Dart Subtraction Strategy

To play the Dart Subtraction Game, you throw two darts and find your score by subtracting the value of the second dart from the value of the first dart. It is possible to land on the same value twice.

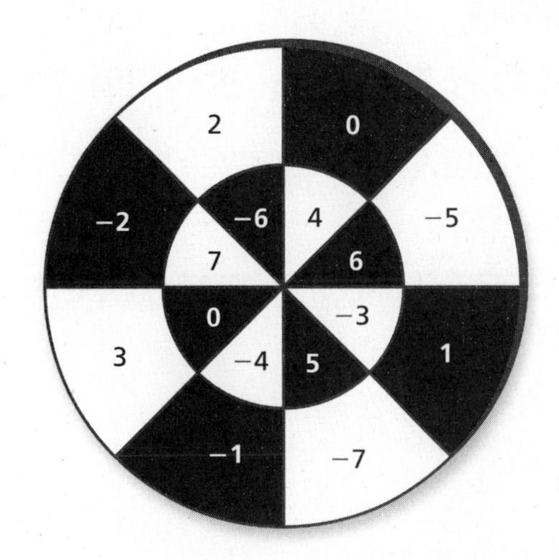

1. What is the highest possible score? What would have to be the value of the first dart? the second dart?

2. What is the lowest possible score? What would have to be the value of the first dart? the second dart?

3. List all the ways to get a score of 4. There are 11 different possibilities.

First Dart											
Second Dart											

4. List all the ways to get a score of −6.

5. How many ways could you get a score of 0? a score of 14? a score of −14?

6. If you are trying to get the *lowest* score, should you try to land on a positive or negative integer with your first dart? with your second dart? Explain your reasoning.

7. Play the game with a partner for ten rounds. Use paper punches as your darts. Hold your hand about 12 inches from the paper and drop the paper punches onto the target. Record your score after each round. Add up the score from each round to get your total score. Did you use any particular strategy? Describe.

Name ______________________________ Date ________

1.3 Puzzle Time

What Did The Sea Say To The Sand?

Write the letter of each answer in the box containing the exercise number.

Subtract.

1. $3 - 11$
2. $-5 - 12$
3. $14 - (-10)$
4. $-9 - (-7)$
5. $25 - (-8)$
6. $-13 - (-13)$

Evaluate the expression.

7. $-6 + 15 - (-4)$
8. $11 - 22 - (-8)$
9. $-14 - 7 - (-25)$
10. $17 + 8 - (-15)$
11. $-9 - (-4) - 2$
12. $-16 + 5 - 12$
13. The high temperature for a day in January was 7 degrees Fahrenheit. The low temperature that day was –5 degrees Fahrenheit. What is the difference in temperatures?
14. The top of a sailboat mast is 22 feet above the water surface. The bottom of the sailboat is 3 feet below the water surface. What is the difference in the elevations?

Answers

J.	24	**H.**	12
W.	-8	**G.**	40
O.	33	**E.**	-7
D.	-2	**I.**	0
S.	25	**N.**	-17
V.	-3	**A.**	-23
U.	13	**T.**	4

2	5	9	13	6	2	10		6	9		3	7	14	9		1	12	8	11	4

Activity 1.4

Start Thinking!

For use before Activity 1.4

How are deposits and withdrawals at a bank similar to adding integers?

Activity 1.4

Warm Up

For use before Activity 1.4

Add.

1. $9 + 9$

2. $-7 + (-7)$

3. $-3 + (-3) + (-3)$

4. $5 + 5 + 5$

5. $6 + 6 + 6 + 6$

6. $-4 + (-4) + (-4) + (-4)$

Lesson 1.4 Start Thinking!

For use before Lesson 1.4

Without using your notes from Activity 1.4, explain to a partner how repeated addition is like multiplication.

Lesson 1.4 Warm Up

For use before Lesson 1.4

Multiply.

1. $7 \bullet 2$

2. $9(-7)$

3. $-6(8)$

4. $-8(-10)$

5. $6 \bullet (-5)$

6. $-5 \bullet (-12)$

Name______________________________ Date__________

1.4 Practice A

Multiply.

1. $4 \bullet (-3)$ **2.** $-6 \bullet 5$ **3.** $-8(-2)$ **4.** $9 \bullet 6$

5. $0 \bullet (-7)$ **6.** $-12(-3)$ **7.** $11 \bullet 7$ **8.** $5(-5)$

9. $-13 \bullet 7$ **10.** $-1 \bullet 9$ **11.** $2(-12)$ **12.** $-9 \bullet (-9)$

13. A water tank leaks 5 gallons of water each day. What integer represents the change in the number of gallons of water in the tank after 7 days?

Multiply.

14. $2 \bullet (-3) \bullet 5$ **15.** $-5(-4)(-1)$ **16.** $7 \bullet 2 \bullet (-3)$

17. $0 \bullet (-8) \bullet 6$ **18.** $-6 \bullet 4 \bullet (-2)$ **19.** $5(-4)(-5)$

Evaluate the expression.

20. $(-3)^2$ **21.** -3^2 **22.** $(-2)^3$

23. -5^2 **24.** $-3 \bullet (-4)^2$ **25.** $(-7)^2 \bullet 2$

26. $|-3| \bullet (-6)$ **27.** $-5(-2) - 3(-4)$ **28.** $2 \bullet (-3)^2 - 5^2$

Find the next two numbers in the pattern.

29. 6, −12, 24, −48, … **30.** 9, −27, 81, −243, …

31. An elevator is 180 feet above the first floor. Each second it descends 12 feet.

a. What integer is the change in the height of the elevator each second?

b. Copy and complete the table.

Time	3 sec	6 sec	9 sec
Height			

c. Estimate how many seconds it takes the elevator to get to the first floor. Explain your reasoning.

d. From the first floor, it takes 4 seconds to reach the basement floor. What is the height of the basement floor with respect to the first floor?

Name ______________________________ Date __________

1.4 Practice B

Multiply.

1. $(-8)(-12)$
2. $10 \bullet (-14)$
3. $-21 \bullet 4$
4. $-15 \bullet (-8)$

5. The water in a pool evaporates at a rate of 16 gallons per week. What integer represents the change in the number of gallons of water in the pool after 24 weeks?

Multiply.

6. $5 \bullet (-11) \bullet (-4)$
7. $-15(-3)(-6)$
8. $-9 \bullet 0 \bullet (-3)$
9. $13 \bullet 2 \bullet (-6)$
10. $-16 \bullet 2 \bullet (-3)$
11. $-9(-9)(-9)$

Evaluate the expression.

12. $(-12)^2$
13. -12^2
14. $(-7)^3$
15. $-(-2)^3$
16. $(-2)^3 \bullet (-3)^2$
17. $(-11)^2 \bullet 7$
18. $-|-3| \bullet (-6)$
19. $11(-3) - (-2)(7)$
20. $-5 \bullet 8 - (-4)^3$

21. The gym offers a discount when more than one member of the family joins. The first member $(n = 0)$ pays \$550 per year. The second member to join $(n = 1)$ gets a discount of \$75 per year. The third member $(n = 2)$ gets an additional \$75 discount. The price for the nth member is given by $550 + (-75n)$.

    a. What is the price for the fourth member to join $(n = 3)$?

    b. For a large family, is it possible that a member would join for free? If so, which member would it be? Explain your reasoning.

    c. Other than \$0, what is the lowest amount that a member would pay to join? Which member would it be? Explain your reasoning.

22. Two integers, a and b, have a product of -48.

    a. What is the greatest possible sum of a and b?

    b. Is it possible for a and b to have a sum of 13? If so, what are the integers?

    c. What is the least possible difference of a and b?

Name________________________ Date__________

1.4 Enrichment and Extension

Multiplying Negative Integers

In Exercises 1–4, find the product.

1. $-3 \bullet (-7) \bullet (-15) \bullet (-5)$

2. $-6 \bullet (-8) \bullet (-1) \bullet (-9)$

3. $-7 \bullet (-18) \bullet (-5) \bullet (-6) \bullet (-3)$

4. $-5 \bullet (-5) \bullet (-4) \bullet (-2) \bullet (-10)$

In Exercises 5–8, use the table.

$(-3)^2$	=	$(-3)(-3)$	=	$(-3)^1(-3)$	=	9
$(-3)^3$	=	$(-3)(-3)(-3)$	=	$(-3)^2(-3)$	=	-27
$(-3)^4$	=				=	
$(-3)^5$	=				=	
$(-3)^6$	=				=	
$(-3)^7$	=				=	
$(-3)^8$	=				=	

5. Complete the table.

6. Describe the pattern in the products.

7. Will $(-3)^{18}$ be a positive or negative integer? Explain.

8. Will $(-3)^{45}$ be a positive or negative integer? Explain.

Name ______________________ Date __________

1.4 Puzzle Time

When Do Kangaroos Celebrate Their Birthdays?

A	B	C	D	E	F		G	H	I	J		K	L	M	N

Complete each exercise. Find the answer in the answer column. Write the letter under the answer in the box containing the exercise letter.

Answer	Letter
36	G
12	O
−105	E
25	P
50	C
−60	D
72	S
45	N
−25	T
−42	U
49	P
−36	H
52	W
110	R

Multiply.

A. $5 \bullet (-12)$

B. $-14 \bullet 3$

C. $-10(-11)$

D. $8 \bullet (-7)$

E. $-9 \bullet (-5)$

F. $6(-2)(-3)$

G. $-4 \bullet 5 \bullet (-4)$

H. $(-7)(-3)(-5)$

I. $-15 \bullet 0 \bullet (-12)$

J. $(-5)^2$

K. -7^2

L. $-3^2 \bullet 8$

M. $(-4)^3$

N. You are making a necklace that is 9 inches long. You use 6 beads for each inch. What integer is the change in your supply of beads after making the necklace?

Answer	Letter
0	A
−25	M
−49	Y
−64	A
80	L
−50	U
100	B
64	F
−56	I
−110	J
66	S
−54	R
−72	E
54	K

Activity 1.5 Start Thinking!
For use before Activity 1.5

What are the different ways you can multiply two integers?

Activity 1.5 Warm Up
For use before Activity 1.5

Multiply.

1. $-10 \bullet 5$
2. $-5 \bullet (-6)$
3. $8 \bullet (-9)$
4. $-15 \bullet 7$
5. $-22 \bullet (-8)$
6. $32 \bullet (-4)$

Lesson 1.5 Start Thinking!

For use before Lesson 1.5

Find the next 2 numbers in the pattern: 12, 6, 24, 12, 48, …. Explain your reasoning.

Lesson 1.5 Warm Up

For use before Lesson 1.5

Divide.

1. $12 \div (-3)$

2. $-32 \div 4$

3. $-56 \div (-8)$

4. $\dfrac{-45}{9}$

5. $\dfrac{81}{-9}$

6. $\dfrac{-144}{-12}$

Name____________________ Date__________

1.5 Practice A

Divide, if possible.

1. $8 \div (-4)$

2. $-15 \div (-3)$

3. $\dfrac{-10}{5}$

4. $0 \div (-7)$

5. $-35 \div 7$

6. $\dfrac{18}{-6}$

7. $-72 \div 9$

8. $-5 \div 5$

9. $\dfrac{15}{0}$

10. $12 \div (-2)$

11. $\dfrac{-56}{-8}$

12. $21 \div (-3)$

13. Your team dives for 28 lobsters over 7 days. What is the average daily lobster catch?

Find the mean of the integers.

14. $5, -7, 12, -10, 15$

15. $-16, -27, 21, -19, 14, -3$

Evaluate the expression.

16. $6 - 12 \div (-3)$

17. $|-16| \div (-2)^2 - 4^2$

18. $\dfrac{-10 + (-2)^3}{-3}$

Find the next two numbers in the pattern.

19. $-96, 48, -24, 12, \ldots$

20. $12{,}500, -2500, 500, -100, \ldots$

21. A skateboarder descends on a ramp from 172 feet to 67 feet in 15 seconds. What is the average change in height per second?

22. The velocity (in feet per second) of a bouncing ball was recorded every second. The table shows the velocity for each second.

Time (sec)	1	2	3	4	5
Velocity (ft/sec)	−15	−6	2	10	−11

a. What is the average velocity of the bouncing ball over the 5 seconds?

b. What is the highest recorded speed of the bouncing ball? Is the ball going up or down at this speed?

c. During the 5 second period, did the ball spend more time going up or going down? Explain your reasoning.

d. Between which two seconds did the ball change from going up to going down? Explain your reasoning.

Name ______________________ Date __________

1.5 Practice B

Divide, if possible.

1. $51 \div (-3)$ **2.** $\frac{-63}{21}$ **3.** $\frac{0}{25}$ **4.** $\frac{-144}{-9}$

5. $105 \div (-5)$ **6.** $-82 \div 0$ **7.** $-96 \div 8$ **8.** $-15 \div (-15)$

9. $\frac{99}{-9}$ **10.** $225 \div (-25)$ **11.** $\frac{-156}{3}$ **12.** $-48 \div (-3)$

13. Your team catches 42 Mahi Mahi over 2 weeks. What is the average daily Mahi Mahi catch?

Evaluate the expression.

14. $-10 + 16 \div (-2) + 7$ **15.** $(-68) \div (-4) + 5 \bullet (-3)$

16. $10 - 12^2 \div (-2)^3$ **17.** $-|-16| \div (-8) \bullet 5^2$ **18.** $\frac{3 + 7 \bullet (-3^2)}{-5}$

19. PI-Squared and Euler Circles are in a math competition consisting of 10 two-part questions. Both parts correct earns 5 points, one part correct earns 2 points, and no parts correct earns –1 point.

a. What is the mean points per question for PI-Squared?

b. What is the mean points per question for Euler Circles?

Team	Both	One	None
PI-Squared	4	2	4
Euler Circles	2	6	2

c. Which team should win the competition? Explain your reasoning.

20. A 155-pound person burns about 500 calories per hour playing racquetball.

a. One pound is equal to 3500 calories. How long will it take to burn 1 pound playing racquetball?

b. How long will it take to burn 5 pounds playing racquetball? Explain your reasoning.

c. If the person were to rest 5 minutes for every 30 minutes of playing, how long would it take to burn 1 pound?

Name____ Date__________

1.5 Enrichment and Extension

Absolute Value and Integers

Assume $a > 0$ and $b < 0$. Determine whether the statement is *always*, *sometimes*, or *never* true.

1. $a + b > 0$ **2.** $a + b < 0$ **3** $|a| + |b| < 0$ **4.** $|a| + |b| > 0$

5. $a - b > 0$ **6.** $a - b < 0$ **7.** $b - a > 0$ **8.** $b - a < 0$

9. $|a| - |b| > 0$ **10.** $|a| - |b| < 0$ **11.** $-|a| - |b| < 0$ **12.** $|a| - \left(-|b|\right) < 0$

13. $ab > 0$ **14.** $ab < 0$ **15.** $\frac{a}{b} < 0$ **16.** $\frac{a}{b} > 0$

Assume *a*, *b*, and *n* are integers, with $a > 0$ and $b < 0$. Determine whether the possible values of *n* are *all integers, all positive integers and zero, all negative integers and zero,* or *none.*

17. $-|-n| - |n| > 0$ **18.** $-|n| + |-n| = 0$ **19.** $|n| + (-n) = 0$ **20.** $-n + n = 0$

21. $an \geq 0$ **22.** $an \leq 0$ **23.** $\frac{n}{b} \geq 0$ **24.** $\frac{n}{b} \leq 0$

25. $|a| + |n| = a + n$ **26.** $|a| + |n| = a - n$ **27.** $|an| = |a| \bullet |n|$ **28.** $\left|\frac{n}{b}\right| = \frac{|n|}{|b|}$

29. $\left|\frac{n}{a}\right| = \frac{n}{a}$ **30.** $|an| = an$ **31.** $|bn| = bn$ **32.** $\left|\frac{n}{b}\right| = \frac{n}{b}$

For each exercise number, use your answers and the key below to color the cell. Do not color the cells that have a zero in them.

Always = Purple

Sometimes = Blue

Never = Red

All Integers = Blue

All Positive Integers and Zero = Orange

All Negative Integers and Zero = Red

None = Purple

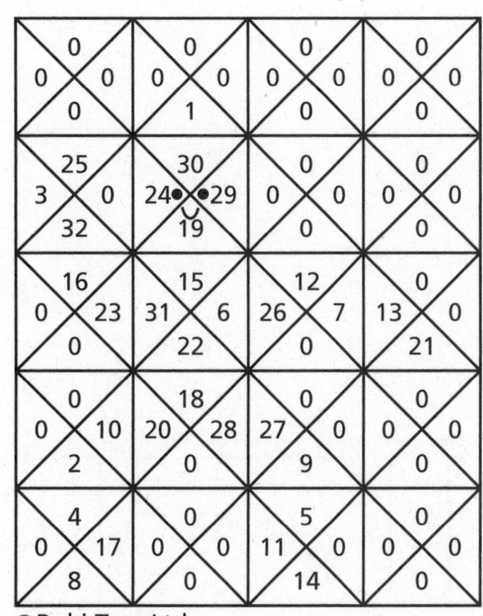

©Buki Toys Ltd.

Name ______________________________ Date __________

1.5 Puzzle Time

What Did The Baseball Mitt Say To The Ball?

Circle the letter of each correct answer in the boxes below. The circled letters will spell out the answer to the riddle.

Divide.

1. $6 \div (-3)$

2. $-52 \div (-4)$

3. $-27 \div 3$

4. $-36 \div 2$

5. $56 \div (-8)$

6. $-24 \div (-3)$

7. $\dfrac{-18}{6}$

8. $\dfrac{25}{-5}$

9. $\dfrac{-16}{-4}$

10. $\dfrac{-66}{11}$

Evaluate the expression.

11. $32 \div (-2) + (-25) \div 5$

12. $4 \bullet (-3) + 12 \div (-4)$

13. You improve your time running a course by 5 seconds in week one, by 3 seconds in week two, and by 4 seconds in week three. What is the average weekly change in your running time?

B	C	A	R	T	S	C	H	E	Y	D	O	U	N	L	A	S	T	E	O	R
14	−21	4	12	13	2	8	−9	20	−7	15	−15	−3	9	−6	−18	−8	−2	−5	−13	−4

Name______________________________ Date__________

Chapter 1

Technology Connection

For use after Section 1.2

Entering Negative Numbers on a Calculator

You can use the [+ | −] key on a calculator to enter negative numbers.

EXAMPLE 1 How does your calculator display negative numbers? Find the value of 0 − 3.

SOLUTION

Press 0 [−] 3 [=].

The integer that is 3 less than 0 is −3. Study the calculator display and determine where the negative sign is displayed. It may be immediately to the left of the 3, or it may be at the far left of the display.

Displays

[−3] or [− 3]

Now clear the display and try entering −2 using the [+ | −] key, also called the change-sign key. Press 2 [+ | −]. The display should be similar to the one you found in the first example. A negative sign displayed to the far left can be easy to miss if you are not looking for it; this will make a big difference in your calculations.

EXAMPLE 2 The temperature is −2°. It rises 6°. What is the new temperature?

SOLUTION

You need to find −2 + 6.

Press 2 [+ | −] + 6 [=]. The answer is 4.

Use the [+ | −] key on a calculator to evaluate the expression.

1. 0 − 8 **2.** −5 + 5 **3.** −1 + 0

4. 1 − 2 **5.** 0 − 7 **6.** −4 + 2

7. What do you think happens when you subtract 1 from a negative number? Explain your reasoning. Check your answer with a calculator.

Chapter 2

Name______________________________ Date__________

Chapter 2 Rational Numbers

Dear Family,

Every family has a favorite pizza. Pizzas are unusual in that the ingredients are normally customized. The variety of meats, vegetables, sauces, crusts, herbs and spices produces a dizzying number of possibilities.

You and your student might enjoy making homemade pizza. First, you have to decide how many pizzas you need and how big each will be. If there are just two people eating, a small pizza might be just right. But for a large group, several pizzas may be needed.

You might estimate that each person will eat about one fourth of a large pizza. Multiply that by the number of people eating and you know how many fourths are needed. From that, work with your student to figure out how many whole pizzas are needed.

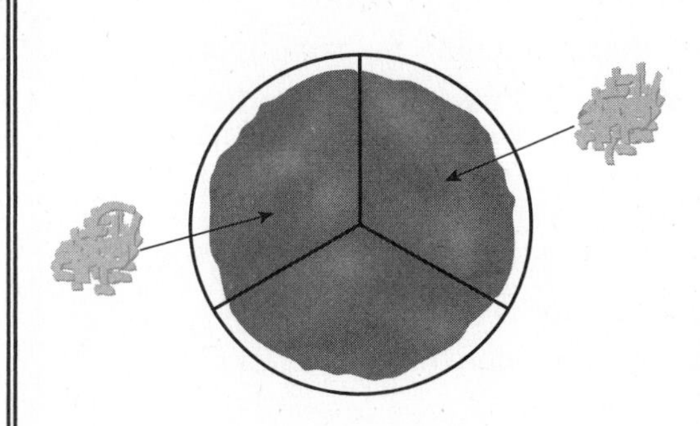

Next you have to choose and prepare the toppings—maybe even customize the toppings for each person. As you choose the toppings, divide them so they will be evenly distributed. If two thirds of your group wants green peppers on the pizza, divide the diced green peppers into two equal piles. Divide the pizza into three equal parts and put the piles on two of the parts. Talk with your student about a strategy for doing this when you have more than one pizza.

Finally, decide how you will cut the pizza. If there are two people, you probably will choose an even number of slices. If there are five people, you might choose to cut the pizza into ten slices, so that each person can have two slices. Count the number in your group and think of a good number of slices to use. What if you have more than one pizza?

Enjoy your pizza—and don't hold the math!

Nombre ______________________________ Fecha ________

Capítulo 2 Números Racionales

Estimada Familia:

Cada familia tiene una pizza favorita. Las pizzas no son comunes en el sentido en que los ingredientes generalmente son personalizados. La variedad de carnes, verduras, salsas, masas, hierbas y especies produce un abrumador número de posibilidades.

Usted y su estudiante pueden disfrutar preparando pizza en casa. Primero, tienen que decidir cuántas pizzas necesitan y qué tan grande será cada una. Si sólo dos personas van a comer una pizza pequeña será suficiente. Pero para un grupo grande, se necesitarán varias pizzas.

Pueden estimar que cada persona comerá alrededor de un cuarto de una pizza grande. Multipliquen eso por el número de personas que van a comer y entonces sabrán cuántos cuartos necesitarán. A partir de ese dato, trabaje con su estudiante para averiguar cuántas pizzas enteras se necesitarán.

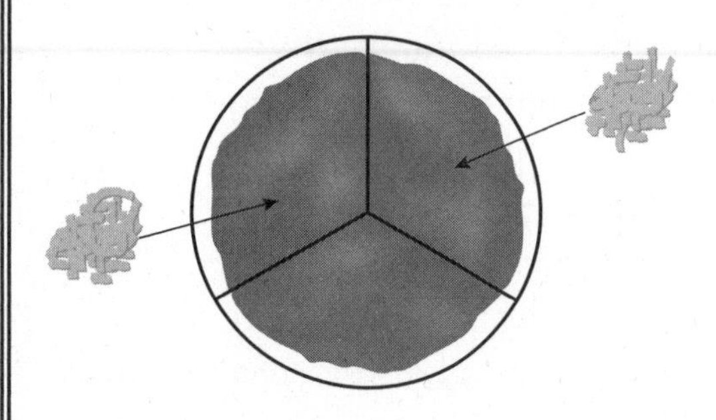

Luego, tendrán que elegir y preparar las cubiertas—quizás incluso personalizar las cubiertas para cada persona. A medida que eligen las cubiertas, divídanlas para que queden distribuidas de manera uniforme. Si dos tercios de su grupo desean pimientos verdes en la pizza, dividan los pimientos verdes picados en dos montones iguales. Dividan la pizza en tres partes iguales y coloquen los montones en dos de las partes. Converse con su estudiante acerca de una estrategia para hacer esto cuando tienen más de una pizza.

Finalmente, decidan cómo cortarán la pizza. Si hay dos personas, probablemente elegirán un igual número de tajadas. Si son cinco personas, querrán cortar la pizza en diez tajadas, para que cada persona reciba dos. Cuenten el número de personas en su grupo y piensen acerca de un buen número de tajadas para usar. ¿Qué pasa si tienen más de una pizza?

¡Disfruten su pizza—y no dejen de usar las matemáticas!

Activity 2.1 Start Thinking!

For use before Activity 2.1

You eat $\frac{3}{8}$ of a pie and your sister eats $\frac{1}{4}$ of the pie. Explain who ate more of the pie.

Activity 2.1 Warm Up

For use before Activity 2.1

Copy and complete the statement using $<$, $>$, or $=$.

1. $\frac{2}{3}$ __?__ $\frac{4}{6}$

2. 0.7 __?__ $\frac{3}{4}$

3. 1.5 __?__ $\frac{2}{3}$

4. -0.6 __?__ -1

5. $3\frac{7}{10}$ __?__ $3\frac{4}{5}$

6. -2.4 __?__ $-2\frac{2}{5}$

Lesson 2.1

Start Thinking!

For use before Lesson 2.1

You and two friends are playing basketball. You make 7 out of 15 shots. Your first friend makes 6 out of 10 shots and your second friend makes 5 out of 12 shots. Who is the better shooter?

How would you solve this problem using what you know about rational numbers?

Lesson 2.1

Warm Up

For use before Lesson 2.1

Order the numbers from least to greatest.

1. $-\frac{2}{3}, 0.6, \frac{3}{4}, -\frac{7}{4}, -0.3$

2. $1.5, -1.3, \frac{7}{5}, -\frac{6}{5}, 1.65$

3. $-2.75, \frac{11}{4}, \frac{5}{4}, -0.37, 2.65$

4. $\frac{4}{10}, -0.8, \frac{1}{8}, 4.5, -\frac{4}{2}$

5. $3.8, -\frac{9}{3}, -0.3, \frac{6}{4}, \frac{8}{5}$

6. $-1.5, \frac{7}{3}, -\frac{3}{4}, 0.6, \frac{9}{6}$

Name ______________________________ Date __________

2.1 Practice A

Write the rational number as a decimal.

1. $\frac{5}{9}$ **2.** $-\frac{3}{8}$ **3.** $-\frac{3}{11}$ **4.** $\frac{7}{30}$

5. $1\frac{5}{12}$ **6.** $-2\frac{1}{3}$ **7.** $-\frac{13}{22}$ **8.** $5\frac{1}{6}$

Write the decimal as a fraction or mixed number in simplest form.

9. 0.7 **10.** –0.3 **11.** –0.43 **12.** 0.52

13. 1.25 **14.** –2.07 **15.** 4.18 **16.** 3.125

Order the numbers from least to greatest.

17. $1.6, -\frac{2}{3}, -0.5, \frac{3}{2}, -\frac{5}{2}$

18. $\frac{3}{4}, -1.7, 0.6, -\frac{7}{4}, 1.1$

19. $0, -\frac{2}{5}, 0.67, \frac{7}{9}, -0.5$

20. $-\frac{1}{3}, -0.3, \frac{4}{3}, 1.2, \frac{3}{2}$

21. You receive two quarters, one dime, and four pennies back in change.

a. Write the amount as a decimal.

b. Write the amount as a fraction in simplest form.

22. In football, a completion percentage is the number of completions divided by the number of passes. Does Tom or Ian have a higher completion percentage?

Player	Passes	Completions
Tom	22	10
Ian	18	9

23. You get 17 out of 22 questions correct on a math test.

a. What is your percent of correct answers?

b. The lowest score to pass is 70%. Did you pass the test?

c. What is the minimum number of correct answers needed in order to pass the test?

Name ______________________ Date __________

2.1 Practice B

Write the rational number as a decimal.

1. $\frac{5}{8}$
2. $-\frac{3}{22}$
3. $1\frac{2}{9}$
4. $-5\frac{3}{40}$
5. $-7\frac{5}{11}$
6. $4\frac{1}{15}$
7. $-9\frac{1}{9}$
8. $-7\frac{5}{6}$

Write the decimal as a fraction or mixed number in simplest form.

9. 0.68
10. –0.01
11. –3.99
12. 8.745
13. 3.005
14. –13.012
15. –9.98
16. –10.452

17. You caught a red snapper that is $8\frac{5}{12}$ inches long. Your friend caught a red snapper that is $8\frac{6}{13}$ inches long. Who caught the larger red snapper?

Copy and complete the statement using $<$, $>$, or $=$.

18. $0.13 \ \underline{\ ?\ }\ \frac{1}{8}$
19. $-1\frac{2}{9}\ \underline{\ ?\ }\ -\frac{5}{4}$
20. $-5.175\ \underline{\ ?\ }\ -5\frac{1}{6}$

21. Find one terminating decimal and one repeating decimal between $-1\frac{1}{2}$ and $-1\frac{7}{9}$.

22. The table gives the tidal changes in the water level of a lagoon for every six hours of a given day.

Time	4:00 A.M.	10:00 A.M.	4:00 P.M.	10:00 P.M.
Change (feet)	2.25	$-2\frac{6}{7}$	$-\frac{3}{2}$	$2\frac{1}{3}$

a. Order the numbers from least to greatest.

b. At what time(s) did the water level decrease?

c. What was the largest change in water level?

d. Did the tidal change in part (c) involve an increase or a decrease in water level?

e. Will the next tidal change be an increase or decrease in water level? Explain.

Name______________________________ Date__________

2.1 Enrichment and Extension

Where do I Belong?

An **irrational number** is a decimal that goes on forever and does not repeat. Irrational numbers cannot be written as fractions.

Examples: 3.14159…, −0.010110111…, 2.2360679…

A **real number** is any number that is either rational or irrational.

The Venn diagram shows how real numbers, rational numbers, irrational numbers, integers, and whole numbers are related.

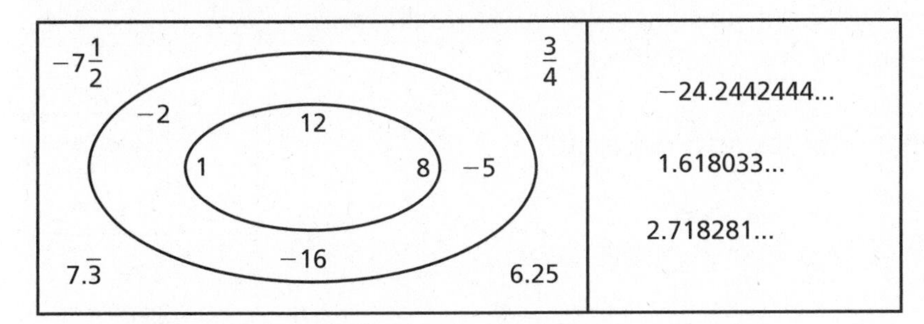

1. Describe any patterns you notice with the numbers.

2. Place the name of each number type in its appropriate spot.

   Real number   Rational number   Irrational number   Integer   Whole number

For each real number, tell whether it is *rational*, *irrational*, an *integer*, and/or a *whole number*. You may have more than one answer.

3. $4\frac{2}{3}$ **4.** $-32.\overline{3}$ **5.** 0.919911…   **6.** 15

7. Can you write 2 as a fraction? Are all whole numbers also rational numbers?

8. Can you write −14 as a fraction? Are all integers also rational numbers?

Complete the statement with *always, sometimes,* or *never*.

9. A **real number** chosen at random is __?__ an **integer**.

10. An **integer** chosen at random is __?__ a **real number**.

11. An **irrational number** chosen at random is __?__ a **rational number**.

12. A **rational number** chosen at random is __?__ an **integer**.

13. A **whole number** chosen at random is __?__ a **rational number**.

Name ______________________________ Date __________

Puzzle Time

Did You Hear About...

A	B	C	D	E	F
G	H	I	J	K	L
M	N	O	P	Q	R

Complete each exercise. Find the answer in the answer column. Write the word under the answer in the box containing the exercise letter.

$4.1\overline{6}$ WRITE
-0.375 STUDENT
-3.875 COULDN'T
$-0.41\overline{6}$ WHO
0.125 WATERPROOF
$5\frac{11}{200}$ HAVE
$\frac{27}{40}$ GOLDFISH
$-1\frac{13}{50}$ HE
$-\frac{7}{10}$ ESSAY

Write the rational number as a decimal.

A. $\frac{8}{9}$

B. $-\frac{3}{8}$

C. $-\frac{5}{12}$

D. $\frac{23}{30}$

E. $1\frac{3}{4}$

F. $-3\frac{7}{8}$

G. $4\frac{1}{6}$

H. $4\frac{4}{25}$

Write the decimal as a fraction or mixed number in simplest form.

I. -0.7

J. 0.84

K. 0.675

L. -0.252

M. -1.26

N. -2.78

O. 5.055

P. -11.688

Q. You eat one slice of a pizza that is cut into 8 even slices. What is the amount you ate written as a decimal?

R. At basketball practice, Charlie makes 52 baskets out of 80 shots. What percentage of baskets did he make?

1.75 HE
$0.7\overline{6}$ SAID
65% INK
$0.\overline{8}$ THE
4.16 HIS
$-2\frac{39}{50}$ DIDN'T
$-11\frac{86}{125}$ ANY
$-\frac{63}{250}$ BECAUSE
$\frac{21}{25}$ ON

Activity 2.2 Start Thinking!

For use before Activity 2.2

The temperature on a given day is 55°. Explain to a partner how to use addition to find how the temperature would change if the temperature in one day increased 15 degrees, decreased 20 degrees, and then increased 8 degrees.

Is the temperature at the end of the day greater or less than the temperature at the beginning of the day? Explain.

Activity 2.2 Warm Up

For use before Activity 2.2

Add.

1. $-54 + (-23)$

2. $78 + (-24)$

3. $-23 + 65$

4. $-45 + 25$

5. $62 + (-29)$

6. $-87 + (-12)$

Lesson 2.2 Start Thinking!

For use before Lesson 2.2

Explain how to find $\frac{4}{5} + \left(-\frac{2}{5}\right)$.

Explain how to find $-2.6 + 5.8$.

Lesson 2.2 Warm Up

For use before Lesson 2.2

Add. Write fractions in simplest form.

1. $3\frac{4}{9} + \left(-\frac{2}{3}\right)$

2. $-7 + 4\frac{5}{9}$

3. $-2\frac{2}{5} + 1\frac{2}{3}$

4. $-12.3 + 5.4$

5. $1.6 + (-19.8)$

6. $5.3 + (-7.8)$

Name____ Date__________

2.2 Practice A

Add. Write fractions in simplest form.

1. $\frac{5}{16} + \left(-\frac{7}{16}\right)$

2. $\frac{3}{5} + \left(-\frac{4}{15}\right)$

3. $-\frac{7}{2} + 3\frac{2}{3}$

4. $5.6 + (-1.3)$

5. $-8.2 + 5.4$

6. $7.15 + (-12.76)$

7. Describe and correct the error in finding the sum.

$$\times \quad \frac{3}{10} + \left(-\frac{1}{10}\right) = \frac{3+1}{10} = \frac{4}{10} = \frac{2}{5}$$

Evaluate the expression when $x = \frac{1}{2}$ and $y = -\frac{2}{5}$.

8. $-x + y$

9. $x + 2y$

10. $|x + y|$

11. The temperature is -12.6 degrees Celsius. The temperature goes up 7.9 degrees. What is the new temperature?

12. You finish $\frac{3}{8}$ of the project. Your friend finishes $\frac{1}{4}$ of the project. What fraction of the project is finished?

Add. Write fractions in simplest form.

13. $5 + \left(-2\frac{1}{3}\right) + \left(-3\frac{1}{6}\right)$

14. $-4\frac{1}{5} + 3\frac{2}{3} + \left(-1\frac{2}{5}\right)$

15. $-12.4 + 19.1 + (-4.3)$

16. Determine if the following statements are *always*, *sometimes*, or *never* true.

a. When adding two negative rational numbers, the sum will be negative.

b. When adding two rational numbers with different signs, the sum will be zero.

c. When adding two positive rational numbers, the sum will be zero.

d. When adding two rational numbers with different signs, the sum will be negative.

Name ______________________________ Date __________

2.2 Practice B

Add. Write fractions in simplest form.

1. $\frac{2}{5} + \left(-\frac{3}{15}\right)$
2. $\frac{7}{12} + \left(-1\frac{2}{3}\right)$
3. $\frac{2}{7} + \left(-3\frac{5}{14}\right)$
4. $7.26 + (-13.43)$
5. $-18.02 + 15.68$
6. $-15.75 + (-12.76)$
7. Describe and correct the error in finding the sum.

$$\text{X} \quad 2\frac{5}{6} + \left(-\frac{8}{15}\right) = \frac{13}{6} + \left(-\frac{8}{15}\right) = \frac{65 + (-16)}{30} = \frac{49}{30} = 1\frac{19}{30}$$

Evaluate the expression when $x = -\frac{1}{5}$ and $y = \frac{3}{4}$.

8. $x + (-y)$
9. $4x + y$
10. $-|x| + y$
11. Your banking account balance is −$1.56. You deposit $10. What is your new balance?
12. You mow $\frac{1}{3}$ of the lawn. Your sister mows $\frac{2}{7}$ of the lawn. What fraction of the lawn is mowed?

Add. Write fractions in simplest form.

13. $1\frac{1}{4} + \left(-4\frac{1}{5}\right) + \left(-2\frac{3}{5}\right)$
14. $-\frac{1}{3} + 2\frac{2}{9} + \left(-5\frac{2}{3}\right)$
15. $-1.5 + (14.2) + 7.3$
16. When is the sum of two rational numbers with different signs positive?
17. The table at the right shows the amount of snowfall (in inches) for three months compared to the yearly average. Is the snowfall for the three-month period greater than or less than the yearly average? Explain.

December	January	February
$1\frac{2}{3}$	$-2\frac{1}{6}$	$2\frac{5}{6}$

18. The table below shows the weekly profits of a concession stand. What must the Week 5 profit be to break even over the 5-month period?

Week 1	Week 2	Week 3	Week 4	Week 5
2.4	−1.7	5.4	−3.75	?

Name________________________________ Date__________

2.2 Enrichment and Extension

Fun with Puzzles

3, 3, 4, 9	1, 2, 3, 5	4, 6, 7, 8	2, 3, 7, 8
3, 4, 5, 6	2, 3, 8, 9	1, 2, 6, 9	1, 2, 3, 4

Choose a set of numbers from above, and fill in the boxes to make each equation true. Each set of numbers will be used once.

1. $\square\frac{\square}{\square} - 4\frac{\square}{2} = -1\frac{1}{10}$

2. $-\square.5\square + \square 7.\square = 35.32$

3. $\square\square.\square - 4\square 8 = -410.2$

4. $-\frac{\square}{\square} - \square\frac{\square}{2} = -3\frac{1}{4}$

5. $-\square.8\square - \square 2.\square = -46.19$

6. $\square\frac{\square}{8} - \square\frac{3}{\square} = -1\frac{7}{8}$

7. $-\square\frac{\square}{3} + 2\frac{\square}{\square} = -4\frac{5}{9}$

8. $-\square 8 + \square\square.\square = 18.4$

9. Use the numbers 1 through 8 to fill in the blanks and make the equation true. There is more than one way to do it.

$\square.\square\square - \square.\square\square = -1.\square\square$

Name ______________________________ Date __________

2.2 Puzzle Time

Where Do Polar Bears Vote?

Write the letter of each answer in the box containing the exercise number.

Add. Write fractions in simplest form.

1. $\frac{5}{6} + \frac{8}{6}$

2. $\frac{7}{10} + \left(-\frac{3}{5}\right)$

3. $-\frac{9}{2} + \frac{5}{12}$

4. $5\frac{1}{3} + \left(-\frac{5}{9}\right)$

5. $\frac{3}{5} + \frac{8}{5}$

6. $-4 + \frac{3}{2}$

7. $3.6 + (-2.4)$

8. $-8.2 + 9.1$

9. $6.8 + (-3.2)$

10. $-4.5 + (-4.7)$

11. $5.327 + (-2.25)$

12. $14.62 + (-11.302)$

13. Sara has $4\frac{3}{4}$ yards of red fleece and $2\frac{2}{3}$ yards of blue fleece fabric. How many yards of red and blue fleece fabric does she have altogether?

14. On Saturday, you biked 7.5 miles. On Sunday, you biked 8.9 miles. How many miles did you bike altogether?

Answers

O. $2\frac{1}{6}$

T. 3.6

E. $-2\frac{1}{2}$

O. $2\frac{1}{5}$

P. 3.077

L. 16.4

T. $\frac{1}{10}$

H. -9.2

R. $7\frac{5}{12}$

A. $4\frac{7}{9}$

E. $-4\frac{1}{12}$

T. 1.2

N. 3.318

H. 0.9

4	9		2	8	3		12	1	13	7	10		11	5	14	6

Activity 2.3 Start Thinking!

For use before Activity 2.3

Think of some sports where you add and subtract rational numbers. Give examples of why you would add rational numbers. Give examples that use negative rational numbers.

Activity 2.3 Warm Up

For use before Activity 2.3

Subtract.

1. $-45 - 25$

2. $62 - (-29)$

3. $-87 - (-12)$

4. $-32 - 43$

5. $-75 - (-87)$

6. $-12 - (-54)$

Lesson 2.3 Start Thinking!

For use before Lesson 2.3

Explain how to find $-\frac{4}{5} - \frac{2}{5}$.

Explain how to find $-2.6 - 5.8$.

Lesson 2.3 Warm Up

For use before Lesson 2.3

Subtract. Write fractions in simplest form.

1. $-7 - \frac{5}{7}$

2. $-5\frac{9}{10} - 7\frac{3}{5}$

3. $-4\frac{1}{4} - 6\frac{3}{8}$

4. $-13 - 5.9$

5. $14.6 - (-9.2)$

6. $-7.4 - 10.6$

Name______________________________ Date__________

2.3 Practice A

Subtract. Write fractions in simplest form.

1. $\frac{3}{7} - \frac{10}{7}$

2. $\frac{7}{12} - \left(-\frac{13}{12}\right)$

3. $-\frac{1}{3} - \left(-\frac{9}{4}\right)$

4. $-3\frac{1}{2} - 1\frac{5}{6}$

5. $-12.41 - (-9.95)$

6. $2 - 8.25$

Find the distance between the two numbers on a number line.

7. $6, -4\frac{1}{4}$

8. $-3.1, -5.7$

9. $-1\frac{1}{3}, -4\frac{2}{5}$

10. Your dog's water bowl is $\frac{3}{4}$ full. After taking a drink, the water bowl is $\frac{1}{3}$ full. What fraction of the bowl did your dog drink?

Evaluate.

11. $\frac{7}{8} - \left(-2\frac{3}{4}\right) + \left(-4\frac{1}{2}\right)$

12. $5.76 - (-2.31) - 10.64$

13. Mary filled a water cooler with $6\frac{1}{2}$ gallons of water. She forgot to close the plug and $2\frac{5}{6}$ gallons leaked out.

a. How many gallons of water remain in the cooler?

b. She adds $1\frac{1}{4}$ gallons. How many gallons of water are now in the cooler?

c. How many gallons of water must she add to the cooler to get the required $6\frac{1}{2}$ gallons?

14. Is the difference of two positive rational number always positive? Explain.

Name ______________________________ Date ________

2.3 Practice B

Subtract. Write fractions in simplest form.

1. $\frac{7}{3} - \frac{8}{15}$

2. $-\frac{7}{24} - \left(-\frac{5}{8}\right)$

3. $1\frac{5}{6} - \left(-2\frac{1}{4}\right)$

4. $-3\frac{7}{8} - 9\frac{5}{6}$

5. $-102.431 - (-59.95)$

6. $12.001 - 8.215$

Find the distance between the two numbers on a number line.

7. $-7\frac{1}{5}, -4\frac{2}{3}$

8. $-9.2, 4.5$

9. $-2, -3.7$

10. The largest orange in a bag has a circumference of $9\frac{5}{8}$ inches. The smallest orange has a circumference of $7\frac{13}{16}$ inches. Write the difference of the circumferences of the smallest orange and the largest orange.

Evaluate.

11. $\frac{5}{12} - \left(-3\frac{1}{4}\right) + \left(-6\frac{1}{2}\right) - 3$

12. $23.706 - (-82.31) - 130.641$

13. $-\frac{3}{8} - (-4.35)$

14. $-\frac{5}{18} - \left|-\frac{1}{6}\right| + \left(-\frac{7}{9}\right)$

15. Your bank account balance is \$32.00. You make the following withdrawals, in the following order: \$15.00, \$7.41, \$35.79, and \$0.53. After each withdrawal that leaves a negative balance, the bank adds a –\$32.00 bank fee to your account. What is your new balance?

16. Fill in the blanks to make the solution correct.

$$3\frac{3}{4} - \square\frac{\square}{8} = 2$$

Name____ Date__________

2.3 Enrichment and Extension

Adding and Subtracting Numbers

Use the information to label the total change.

1. In golf, each hole has a par, or the amount of strokes it should take to complete the hole. Find the total number of strokes above or below par the golfer shot after the first six holes golfing.

Par	5	3	5	5	5	4
# of Strokes	3	7	3	6	6	6

2. Find the amount of money left in the account.

Transaction	Credit	Debit	Balance
Beginning Balance			37.07
Deposit	38.10		
Withdrawal		8.79	
Withdrawal		63.12	

3. Find the total number of yards the football team gained or lost.

   First down: Loss of 17 yards

   Second down: Gain of 5 yards

   Third down: Gain of 9 yards

   Fourth down: Gain of 1 yard

Determine if the situation would result in a *positive number*, a *negative number* or *zero*. Explain.

4. Your bank account has a balance of \$111.13. You write a check for \$113.10.

5. You walk to your friends' house then back home. Your friend lives $\frac{7}{10}$ of a mile away.

6. The temperature at the beginning of the day was 9° Fahrenheit. The temperature dropped 14° Fahrenheit.

7. You cut a piece of wood into a $3\frac{1}{3}$-foot piece and a $2\frac{2}{3}$-foot piece. The original length of the board was 6 feet.

Name ______________________ Date __________

2.3 Puzzle Time

Where Does A Salad Dressing Get A Good Night's Sleep?

Write the letter of each answer in the box containing the exercise number.

Subtract. Write the fractions in simplest form.

1. $\frac{3}{4} - \frac{9}{4}$

2. $-3 - \frac{7}{2}$

3. $-\frac{1}{5} - \left(-\frac{5}{11}\right)$

4. $-\frac{5}{8} - \frac{2}{7}$

5. $-2\frac{2}{3} - 4\frac{1}{6}$

6. $-3\frac{1}{9} - \left(-2\frac{1}{3}\right)$

7. $-7 - 3.2$

8. $6.1 - 5.8$

9. $-4.125 - (-2.8)$

10. $-12.33 - 7.21$

11. $5.67 - (-3.142)$

12. $2.567 - 6.814$

Find the distance between the two numbers on a number line.

13. $-3\frac{1}{4}, 4\frac{1}{2}$

14. $-6.1, 8.4$

15. Your project requires a board that has a length of $5\frac{3}{16}$ inches. You found a board that has a length of $9\frac{1}{8}$ inches. How much of the board needs to be cut to use it for your project?

Answers

O. $-\frac{7}{9}$

A. $-6\frac{1}{2}$

T. 8.812

E. $-1\frac{1}{2}$

O. -10.2

B. -4.247

T. 0.3

E. $\frac{14}{55}$

C. $7\frac{3}{4}$

F. -1.325

L. $3\frac{15}{16}$

E. $-\frac{51}{56}$

D. 14.5

N. -19.54

U. $-6\frac{5}{6}$

6	10		2		12	4	14		7	9		15	3	11	8	5	13	1

Activity 2.4 Start Thinking!
For use before Activity 2.4

Draw a diagram to represent the following football plays: gain of 2 yards, loss of 8 yards, gain of 20 yards, loss of 5 yards, and gain of 3 yards.

Activity 2.4 Warm Up
For use before Activity 2.4

Multiply.

1. $-12 \bullet 9$ **2.** $11(-10)$ **3.** $14 \bullet 12$

Divide.

4. $\frac{-48}{-6}$ **5.** $\frac{140}{-10}$ **6.** $\frac{-81}{3}$

Lesson 2.4 Start Thinking!

For use before Lesson 2.4

A company's profits for a week are as follows: Monday: +$32.65, Tuesday: −$75.32, Wednesday: −$125.75, Thursday: +$100.89, and Friday: +$65.30. Does the company show a gain or loss at the end of the week?

Lesson 2.4 Warm Up

For use before Lesson 2.4

Multiply. Write fractions in simplest form.

1. $-2\frac{1}{4} \bullet \frac{4}{5}$

2. $\left(-\frac{3}{5}\right)^3$

3. $-1.2(-3.05)$

Divide. Write fractions in simplest form.

4. $-2\frac{4}{7} \div (-3)$

5. $-9.6 \div 8$

6. $6.45 \div (-30)$

Name______________________________ Date__________

2.4 Practice A

Tell whether the expression is *positive* or *negative* without evaluating.

1. $\dfrac{-7.5}{4.25}$
2. $\dfrac{4}{9} \times \left(-\dfrac{6}{7}\right)$
3. $-\dfrac{1}{5} \div \left(-\dfrac{2}{3}\right)$
4. $-3.2 \times (-1.7)$

Divide. Write fractions in simplest form.

5. $-\dfrac{2}{7} \div \dfrac{10}{7}$
6. $-\dfrac{1}{2} \div \left(-\dfrac{3}{4}\right)$
7. $\dfrac{2}{3} \div (-14)$
8. $-1\dfrac{1}{6} \div \dfrac{5}{3}$
9. $-0.72 \div (-0.9)$
10. $5.4 \div (-3.6)$

Multiply. Write fractions in simplest form.

11. $\dfrac{2}{5} \times \left(-\dfrac{10}{7}\right)$
12. $-\dfrac{3}{4} \bullet \left(-\dfrac{10}{9}\right)$
13. $\dfrac{3}{2}\left(-2\dfrac{2}{9}\right)$
14. $\left(-1\dfrac{3}{8}\right)^2$
15. -3.7×2.1
16. $-5.7 \bullet (-2.06)$

17. There are 15 people in a room. Each person ate $\dfrac{2}{3}$ of a pizza. There was no pizza remaining. How many pizzas were in the room?

18. During a drought, a river's height decreases by 0.35 inch every day. What is the change in the river's height after 7 days?

Evaluate.

19. $-3^2 + 4.6 \times (-0.1)$
20. $-2\dfrac{2}{3} \div 1\dfrac{5}{6} + 2$
21. $-4.31 \bullet 3.09 + (-0.98)$
22. $-3 \times \left(-1\dfrac{7}{12}\right) - \left(-\dfrac{3}{2}\right)^2$

23. Write two fractions, both not positive, whose product is $\dfrac{5}{8}$.

24. Fill in the blank to make the solution correct.

$$5.6 \times \underline{\ ?\ } = -19.04$$

Name ______________________________ Date __________

2.4 Practice B

Divide. Write fractions in simplest form.

1. $-\frac{3}{7} \div \frac{11}{35}$

2. $-\frac{1}{9} \div \left(-\frac{13}{30}\right)$

3. $1\frac{5}{6} \div (-30)$

4. $-2\frac{4}{5} \div 10\frac{2}{3}$

5. $-0.801 \div (-0.09)$

6. $14.616 \div (-2.32)$

Multiply. Write fractions in simplest form.

7. $-\frac{2}{15} \times \left(-\frac{25}{6}\right)$

8. $-\frac{3}{14} \bullet \frac{21}{12}$

9. $1\frac{2}{3}\left(-2\frac{9}{10}\right)$

10. $-\left(3\frac{2}{5}\right)^2$

11. -2.75×3.1

12. $-1.27 \bullet (-2.02)$

13. How many three-quarter pound burgers can be made with twelve pounds of hamburger?

14. The table shows the changes in your times (in seconds) at the new skateboard ramp. What is your mean change?

Trial	1	2	3	4	5
Change	2.2	−1.4	0.6	−2.3	−1.7

Evaluate.

15. $-0.2^3 - 4.15(-0.06)$

16. $5 - 3\frac{9}{10} \div 2\frac{3}{5}$

17. $-14.01 \bullet 2.39 + |-4.89|$

18. $2\frac{1}{3} \times \left(-4\frac{5}{7}\right) - \left(-\frac{3}{5}\right)^2$

19. A gallon of gasoline costs \$2.96. Your car has a 25-gallon gas tank and can travel 28.8 miles on each gallon of gasoline.

a. Find the cost of filling your gas tank if it is already $\frac{3}{8}$ full.

b. You take a trip of length $705\frac{3}{5}$ miles. How much money do you spend on gasoline?

Name____ Date__________

2.4 Enrichment and Extension

The Zweezam Factory

The Zweezam Factory manufactures Zweenubs, Zweedulls, and Zweebuds.

The table shows the costs and income for each type of item.

	Zweenubs	Zweedulls	Zweebuds
Cost per item manufactured	$2.53	$6.58	$8.72
Income per item sold	$4.50	$8.89	$9.99

Answer the following questions about this week at the Zweezam Factory.

1. This week the Zweezam Factory manufactured 46 items altogether including 24 Zweenubs. They manufactured two-thirds as many Zweedulls as Zweenubs and one-fourth as many Zweebuds as Zweenubs. How many of each type did the factory manufacture?

2. What is the factory's total manufacturing cost?

3. Also this week, the Zweezam Factory sold 32 parts altogether including 10 Zweedulls. They sold half as many Zweedulls as Zweenubs and one-fifth as many Zweebuds as Zweedulls. How many of each type did the factory sell?

4. What is the factory's total income?

5. What was the Zweezam Factory's profit this week? (*Hint:* profit = total income − total cost)

6. If the Zweezam Factory continues to manufacture and sell the same number of items for four more weeks, what will be the total profit for five weeks?

7. The Zweezam Factory plans to manufacture fewer items next week. They will manufacture 40 items altogether: two-fifths as many Zweedulls as Zweenubs and one-half as many Zweebuds as Zweedulls. If the Zweezam Factory manufactured 5 Zweebuds, how many of each type will the Zweezam Factory manufacture?

8. They already have pre-orders for next week's batch: 12 Zweenubs, 5 Zweedulls, and 1 Zweebud. Each week the parts must be thrown out, and only the new batch can be sold. So, how much more income do they have to make in order to break even (i.e. have a profit of zero) next week? What is the most profit they could earn next week?

9. If they sold Zweenubs only next week, could they make enough income to break even? Explain your reasoning.

Name ______________________________ Date __________

2.4 Puzzle Time

When Is A Baby Like A Basketball Player?

Write the letter of each answer in the box containing the exercise number.

Multiply. Write fractions in simplest form.

1. $-\frac{4}{5} \bullet \left(-\frac{5}{7}\right)$

2. $2\frac{2}{3} \bullet \left(-4\frac{1}{4}\right)$

3. $\left(-\frac{3}{4}\right)^3$

4. $0.8 \times (-2.1)$

5. $-7.5 \times (-0.3)$

6. $(-0.8)^3$

Divide. Write fractions in simplest form.

7. $\frac{5}{8} \div \left(-\frac{1}{4}\right)$

8. $-1\frac{1}{6} \div \frac{2}{9}$

9. $-6\frac{2}{5} \div \left(-2\frac{2}{7}\right)$

10. $0.3 \div (-1.5)$

11. $-5.415 \div (-2.85)$

12. $-16.29 \div 3.62$

13. What is the square foot area of a room with a length of $10\frac{3}{4}$ feet and a width of $8\frac{1}{2}$ feet?

14. For a fundraiser, the seventh grade class sells 45 submarine sandwiches. They collect a total of $150.75. What is the cost per sub?

Answers

R. 2.25	**E.** $-\frac{27}{64}$
S. $-2\frac{1}{2}$	**D.** $91\frac{3}{8}$
H. -0.512	**E.** $-5\frac{1}{4}$
B. 3.35	**I.** -0.2
L. -4.5	**W.** 1.9
E. $\frac{4}{7}$	**B.** -1.68
N. $2\frac{4}{5}$	**H.** $-11\frac{1}{3}$

11	6	3	9		2	8		13	5	10	14	4	12	1	7

Name____ Date__________

Technology Connection

For use after Section 2.3

Adding and Subtracting Rational Numbers

In this activity, you will use a scientific calculator to add and subtract rational numbers. To enter a fraction on a scientific calculator, use the [a b/c] key.

To convert a mixed number to an improper fraction, use the [d/c] key.

EXAMPLE Use a scientific calculator to find the sum or difference.

a. $\frac{1}{4} + \frac{2}{3}$ **b.** $-\frac{21}{8} - \frac{17}{12}$ **c.** $6\frac{3}{10} + \left(-2\frac{5}{8}\right)$

SOLUTION

Enter the following keystrokes:

a. 1 [a b/c] 4 [+] 2 [a b/c] 3 [=]

The answer is $\frac{11}{12}$.

b. 21 [+/−] [a b/c] 8 [−] 17 [a b/c] 12 [=]

The answer is $-4\frac{1}{24}$.

c. 6 [a b/c] 3 [a b/c] 10 [+] 2 [+/−] [a b/c] 5 [a b/c] 8 [=]

The answer is $3\frac{27}{40}$.

Notice in parts (b) and (c) that the final display shows a mixed number. To convert the mixed number in part (b) to an improper fraction, press [2nd] or [Shift] then [d/c]. The answer as an improper fraction is $-\frac{97}{24}$.

Use a scientific calculator to find the sum or difference. Write your answer as a mixed number and an improper fraction, if possible.

1. $\frac{1}{5} + \frac{7}{10}$ **2.** $\frac{11}{4} - \frac{3}{8}$ **3.** $-\frac{4}{9} - \left(-\frac{1}{12}\right)$

4. $\frac{14}{25} + \frac{38}{15}$ **5.** $3\frac{5}{16} + \left(-2\frac{1}{2}\right)$ **6.** $-5\frac{1}{4} + 6\frac{5}{9}$

Chapter 3

Name _______________________________________ Date __________

Expressions and Equations

Dear Family,

Algebra is used to describe relationships in general terms. Consider the following statements.

• Game tickets are $7 each.	The cost of n tickets is $7n$ dollars.
• It takes 5 minutes to get shoes and car keys and walk to the car.	For a drive of m minutes, allow $m + 5$ minutes.
• Each question on a 20-question test is worth 1 point.	If you miss x questions, your score on the test will be $20 - x$.

On the left, the rule is stated in words, the way you might remember it. On the right, the rule is stated as a mathematical expression with a variable. The number of tickets, the length of the drive, and the number of questions missed are all variables—that is, they might have many different values. The cost of a ticket, the time to get to the car, and the total number of questions on the test are constants—that is, they remain the same. Ask your student to answer each question, using the information above.

- What is the cost of 3 game tickets?
- You want to arrive at baseball practice at 4:30. The drive is 15 minutes. What time should you get ready to leave?
- You miss 2 questions on the test. What is your score?

(Answers: $21, 4:10, 18 points)

Rather than remember all possible ticket costs, driving times, or test scores, you remember the rule for finding them. These examples are uses of algebra in daily life.

With your student, find another algebraic rule you could use in daily life. What are the variables? What are the constants? Have your student evaluate your rule for two different values of the variable(s).

Have fun exploring expressions together!

Nombre ____ Fecha __________

Expresiones y ecuaciones

Estimada Familia:

El álgebra se utiliza para describir relaciones en términos generales. Consideren los siguientes enunciados.

- Los boletos para el juego cuestan \$7 cada uno. — n boletos cuestan $7n$ dólares.
- Se necesitan 5 minutos para ponerse los zapatos, tomar las llaves del auto y caminar hacia él. — Para un trayecto de m minutos, considere $m + 5$ minutos.
- Cada pregunta en una prueba de 20 preguntas vale 1 punto. — Si falla x respuestas, su puntaje en la prueba será $20 - x$.

A la izquierda, la regla se expresa en palabras, la manera en que podrían recordarla. A la derecha, la regla se indica en una expresión matemática con una variable. El número de boletos, la duración del trayecto y el número de preguntas erróneas son todas variables—es decir, que pueden tener muchos valores diferentes. El costo de un boleto, el tiempo para llegar al auto y el número total de preguntas en la prueba son constantes—es decir, siguen siendo los mismos. Pida a su estudiante que responda cada pregunta utilizando la información anterior.

- ¿Cuánto cuestan 3 boletos para el juego?
- Desean llegar al entrenamiento de béisbol a las 4:30. El trayecto dura 15 minutos. ¿A qué hora deben alistarse para salir?
- Se equivoca en 2 preguntas en la prueba. ¿Cuál es su puntaje?

(Respuestas: \$21, 4:10, 18 puntos). En lugar de recordar todos los gastos posibles de boletos, los tiempos de manejo o los resultados de la prueba, debe recordar la regla para hallarlos. Estos ejemplos son usos del álgebra en la vida diaria.

Con su estudiante, busquen otra regla algebraica que se pueda utilizar en la vida diaria. ¿Cuáles son las variables? ¿Cuáles son las constantes? Haga que su estudiante evalúe la regla para dos valores diferentes de la(s) variable(s).

Activity 3.1

Start Thinking!

For use before Activity 3.1

Explain to a partner how an expression and an equation are different.

Give an algebraic and numerical example for each.

Write the phrase as an expression.

1. 7 increased by a number x

2. negative 14 minus y

3. negative 19 increased by n

4. the product of 14 and y

5. 10 divided by the sum of a number n and 6

6. 6 times the quotient of a number x and 3

Lesson 3.1 Start Thinking!

For use before Lesson 3.1

Given the problem $6x + 4 - 2x$, your brother says the answer is $8x + 4$. Explain to your brother why his answer is incorrect. Give the correct answer.

Lesson 3.1 Warm Up

For use before Lesson 3.1

Simplify the expression.

1. $10x + 4x$

2. $7y + 12 - 15$

3. $6x - 4x + 16$

4. $3.8y - 4 + 7.2y$

5. $7 + 11x + 8.4 - x$

6. $\frac{4}{7}y + 8 - 2\frac{1}{2} + \frac{3}{7}y$

Name____ Date__________

3.1 Practice A

Identify the terms and like terms in the expression.

1. $-4y + 7 + 9y - 3$

2. $3n^2 - 1.4n + 5n^2 - 6.4$

3. $\frac{1}{2}b^3 - b^3 + 2b$

Simplify the expression.

4. $-15m + 9m$

5. $8k - 2(4 - 3k)$

6. $3.2 - 9x + 7.1 - 3x$

7. $25 - 6x - 12 - 2x$

8. $19a - 7 - 3a + 12a$

9. $\frac{5}{2}(6x - 7) + \frac{4}{3}(2 + 9x)$

10. $\frac{1}{8}h + 7 - \frac{3}{4}h$

11. $\frac{2}{3}y + 5 - 3 - \frac{11}{12}y$

12. Write an expression in simplest form that represents the perimeter of the polygon.

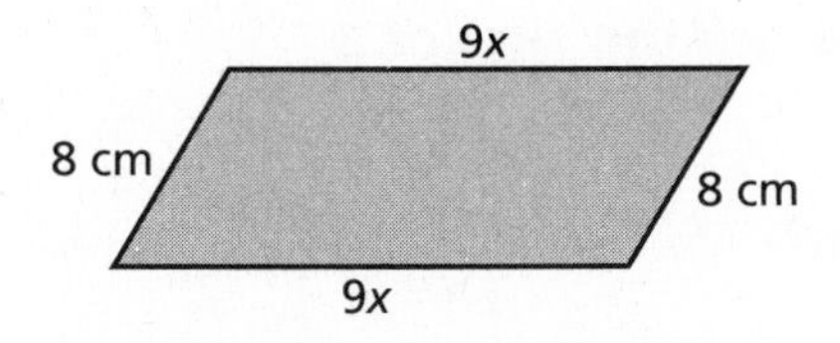

13. Each runner is carrying an 8 ounce bottle of water, a 2.1 ounce energy bar, and a 3 ounce energy drink. Write an expression in simplest form that represents the weight carried by y runners. Interpret the expression.

14. John weighs 65 kilograms, Sam weighs $22x$ kilograms, and Mark weighs $13x$ kilograms. Write an expression in simplest form for their combined weight.

15. Are the expressions $8a^2 - 4b + 7a^2$ and $5(3a^2 - 2b) + 6b$ equivalent? Explain your reasoning.

Name ______________________________ Date __________

3.1 Practice B

Identify the terms and like terms in the expression.

1. $1.3x - 2.7x^2 - 5.4x + 3$

2. $10 - \frac{3}{10}m + 6m^2 + \frac{2}{5}m$

Simplify the expression.

3. $-\frac{15}{4}b + \frac{5}{6}b$

4. $60m - 15(4 - 8m) + 20$

5. $4(5.8 - 9x) + 8.2 + 22x$

6. $9y - 15y + 12 - 6y$

7. $v + 13 - 8(v + 2)$

8. $\frac{5}{3}(5x + 9) + \frac{4}{5}(1 - 9x)$

9. Write an expression in simplest form that represents the perimeter of the polygon.

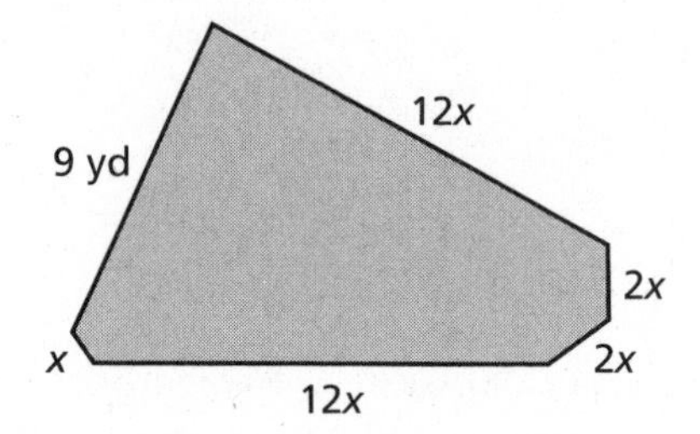

Draw a diagram that shows how the expression can represent the area of a figure. Then simplify the expression.

10. $8(3x - 1)$

11. $(5 + 2)(x + 3x)$

12. Danielle is x years old. Her sister is 5 years older and her brother is half Danielle's age. Write an expression in simplest form for the sum of their ages.

13. The length of a rectangular field is 30 more than twice its width. Write an expression in simplest form for the perimeter of the field in terms of its width w.

14. You buy x packs of pencils, twice as many packs of erasers, and three times as many rolls of tape. Write an expression in simplest form for the total amount of money you spent.

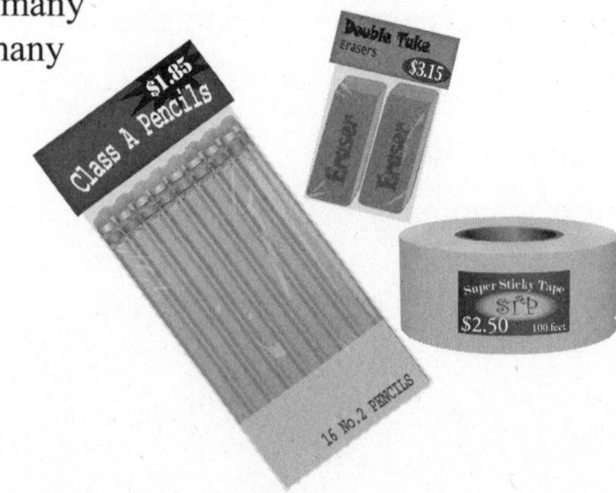

Name______________________________ Date__________

3.1 Enrichment and Extension

Matching

Simplify the expressions on the left by using the Distributive Property and combining like terms. Then, match it to an equal expression on the right by connecting the two with a line.

1. $6x + 2x$

2. $14x - 12 - x - 3$

3. $-5x + 14 - x - 2$

4. $-3 - 5x - 3x + 11x + 3$

5. $-2(-5 - x) + x - x + 1$

6. $\frac{1}{2}(12) + 4x - (x - 1)$

7. $6(x^2 - 2) + 1 - 16 + x$

8. $4\left(\frac{1}{2}x + 4\right) + 1 - 16 + x$

9. $5(x^2 + x)$

10. $x + \left(1 - \frac{1}{2}x\right)$

11. $x^3 + x^2 + x + x - x^2 - x^3$

a. $8x$

b. $\frac{1}{2}x + 1$

c. $13x - 15$

d. $2x + 11$

e. $2x$

f. $6x^2 + x - 27$

g. $3x$

h. $3x + 1$

i. $3x + 7$

j. $-6x + 12$

k. $5x^2 + 5x$

12. Write an expression containing x-terms and constants. The x-terms should combine to $7x$ and the constants should sum to 13.

13. Write an expression containing x^2-terms, x-terms, and constants. The x^2-terms should combine to $-2x^2$, the x-terms should subtract to $3x$, and the constants should sum to 3.

Name ______________________________ Date __________

3.1 Puzzle Time

How Can You Turn A Pumpkin Into A Squash?

A	B	C	D	E	F
G	H	I	J	K	L

Complete each exercise. Find the answer in the answer column. Write the word under the answer in the box containing the exercise letter.

$2x + 4$ SMASH
$13x - 2$ THE
$-2x + 6.2$ COME
$2.4x + 2.9$ AND
$21x$ THROW
$-1.5x - 7$ WILL
$7x + 14$ SQUASH

Simplify the expression.

A. $8x + 13x$

B. $15x + 10 - 6$

C. $7x - 4x + 3$

D. $5.3x - 9 + 7.6x$

E. $6x - 4x - 2 + 11x$

F. $\frac{3}{4}x + 11 - 5\frac{1}{2} + \frac{1}{4}x$

G. $5(x + 8) + 3$

H. $3.6x - 7 - 5.1x$

I. $4 + 8x + 2.2 - 10x$

J. $\frac{5}{6}x - 9 + 3 - \frac{2}{3}x$

K. $2.4(x + 3) - 4.3$

L. The length of a rectangle is 7 inches and the width is $(x + 2)$ inches. Write an expression in simplest form that represents the area of a rectangle.

$x + 5\frac{1}{2}$ AIR
$3x + 3$ UP
$5x + 43$ IT
$x - 4\frac{1}{2}$ TOSS
$12.9x - 9$ IN
$\frac{1}{6}x - 6$ DOWN
$15x + 4$ IT

Activity 3.2 Start Thinking!

For use before Activity 3.2

Give two examples of *like terms*. Give two examples of *unlike terms*. Explain what the difference is between like terms and unlike terms.

Activity 3.2 Warm Up

For use before Activity 3.2

Simplify the expression.

1. $4x + 2 - 3x$

2. $8y - 3 - 10y - 6$

3. $-2x + 3 - 8x$

4. $\frac{1}{2}y + 3y - \frac{2}{5}$

5. $4.8x - 4.6 + 3.9x$

6. $3y - 4.6 + 1.3 + 2.1y$

Lesson 3.2

Start Thinking!

For use before Lesson 3.2

Explain how to simplify $(5x - 4) + (3x - 6)$ using algebra tiles.

Lesson 3.2

Warm Up

For use before Lesson 3.2

Write the sum or difference of two algebraic expressions modeled by the algebra tiles. Then use the algebra tiles to simplify the expression.

1.  (+ − − − −) + (+ + +)

2.  (+ + − − −) + (+ − − −)

3. 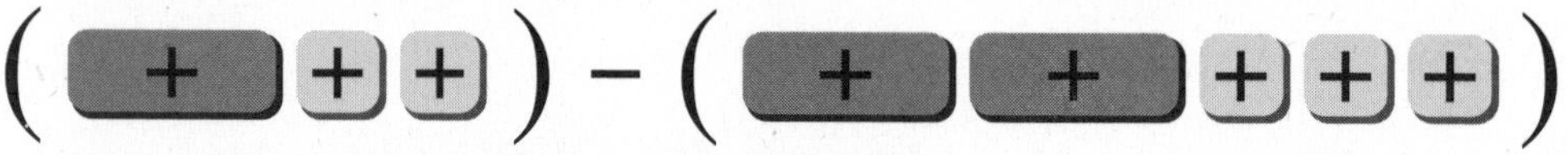 (+ + +) − (+ + + + +)

4.  (+ + − − − −) − (+ + +)

Name ______________________ Date __________

3.2 Practice A

Find the sum.

1. $(p - 3) + (p - 7)$

2. $(3n - 1) + (4 - n)$

3. $(-3r + 8) + (5r - 1)$

4. $6(x - 3) + (2x - 9)$

5. $(3c + 2) + 4(1.3c - 5)$

6. $10(2.1q - 2) + (7.5q + 18)$

7. $(-6y - 2) + 5(3 + 2.5y)$

8. $\frac{1}{2}(6x - 10) + \frac{1}{3}(6 + 9x)$

9. After a week of rain, tadpoles appeared in your pond. After t minutes, you have $(7t + 5)$ tadpoles and your friend has $(8t - 3)$ tadpoles.

a. Write an expression that represents the number of tadpoles you and your friend caught together.

b. Who has more tadpoles after 9 minutes?

Find the difference.

10. $(k + 3) - (3k - 5)$

11. $(-6d + 2) - (7 + 2d)$

12. $(10j - 7) - (-9j + 2)$

13. $(3x + 8) - 6(2.5x - 3)$

14. $(7 - 3t) - 5(-1.6t + 5)$

15. $\frac{1}{2}(12w + 8) - \frac{1}{5}(10w - 5)$

16. The admission to a local fair is \$10 for each adult and \$6 for each child. Each rides costs \$1.50 for an adult and \$1 for a child.

a. Write an expression that represents how much more an adult will spend at the fair.

b. An adult and a child each go on 7 rides. How much more did the adult spend?

17. Write an expression that represents the perimeter of the triangle.

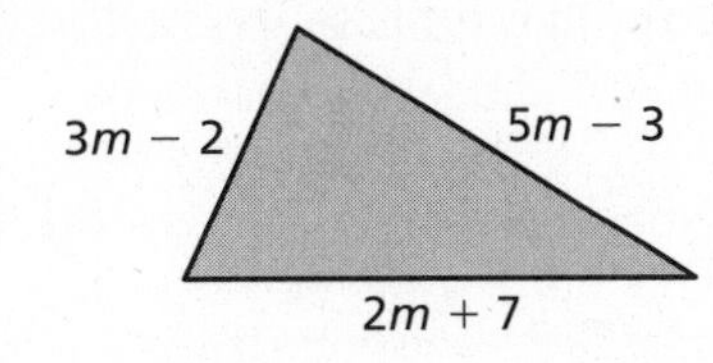

Name ______________________________ Date __________

3.2 Practice B

Find the sum.

1. $(5 - t) + (3t + 2)$

2. $(7k + 9) + (4k - 3)$

3. $2(-5y + 6) + (2y - 8)$

4. $4(2.5g - 4) + 3(1.2g - 2)$

5. $5(-0.3s - 2) + 2(5 - 3.4s)$

6. $\frac{1}{3}(6p - 3) + \frac{1}{7}(7p + 14)$

7. $\frac{1}{5}(-15w - 20) + \frac{1}{2}(3 - 4w)$

8. $\frac{1}{8}(16k - 24) + \frac{1}{5}(2 + 10k)$

9. You are selling tickets to a play. You have sold $(3t + 2)$ tickets for \$5 each and $(2t + 5)$ tickets for \$7 each.

a. Write an expression that represents the total number of tickets sold so far.

b. Write an expression that represents the total amount of money received for the tickets that have been sold.

c. When $t = 3$, what is the total amount of money received?

Find the difference.

10. $(8 - u) - (5u + 1)$

11. $(2x + 7) - 6(5x - 8)$

12. $(3h + 4) - 6(5 - 1.4h)$

13. $(12 + 7.2b) - 3(0.9b - 4)$

14. $\frac{1}{4}(16j - 12) - \frac{1}{9}(18j + 45)$

15. $\frac{2}{3}(12n + 6) - \frac{1}{5}(10n - 2)$

16. You are collecting pairs of socks and toothbrushes for a local charity. After d days, you have collected $(4d + 5)$ pairs of socks and $(3d + 7)$ toothbrushes.

a. Write an expression that represents the total number of items that have been collected.

b. How many more pairs of socks than toothbrushes have been collected on day 7?

Name________________________________ Date__________

3.2 Enrichment and Extension

Using the Distributive Property

When working with algebraic expressions and the Distributive Property, the exponents of the variables are added.

Example: Simplify $x(x + 6)$.

Distribute x to each term inside the parentheses. (Remember that x can be rewritten as $1 \bullet x^1$.) Then, multiply the coefficients.

$x(x + 6) = (1x \bullet 1x) + (1x \bullet 6)$	Distibute x to each term.
$= (1x^1 \bullet 1x^1) + (1x^1 \bullet 6)$	Rewrite to show exponents.
$= 1x^{1+1} + 6$	Multiply coefficients and add exponents.
$= x^2 + 6x$	Simplify.

Simplify the expression.

1. $x^2(x + 1)$

2. $-x(2x - 8)$

3. $x(x^4 - 4)$

4. $x(3x - 1)$

5. $3x(x - 1)$

6. $2x(x - 1)$

7. $4x(-4x - 3)$

8. $n(n - 4)$

9. $-b(3b + 9)$

10. $2w(-4w - 14)$

11. $2x(4x - 9) - 3x(4x - 2)$

12. $3k(-5k + 21) + 2(2.5k + 9)$

13. $4(1 + 1.8h) + h(2.2h + 5)$

14. $\frac{1}{2}m(10 + 6m) - \frac{1}{5}m(10m + 10)$

15. $\frac{1}{3}(6z - 6) - \frac{1}{4}z(4z + 16)$

16. $3d^6(24d^3 + 6) + 24d^9$

Name ____ Date __________

Puzzle Time

What Did The Candle Say To The Match?

Write the letter of each answer in the box containing the exercise number.

Find the sum.

1. $(x + 10) + (x - 14)$

2. $(9 - 2x) + (6x + 4)$

3. $(3x - 7) + (-4x - 8)$

4. $(2x - 7) + 5(x - 3)$

5. $6(-2.3x - 5) + (4x + 11)$

6. $(8 - 2x) + 3(4.5x + 9)$

7. $\frac{1}{2}(8 - 4x) + \frac{1}{3}(9x - 6)$

8. $-\frac{3}{4}(3x + 7) + \frac{1}{4}(12x + 20)$

Find the difference.

9. $(-3x + 8) - (x + 10)$

10. $(5x + 4) - (1 - 2x)$

11. $(3 - 4x) - 3(2.4x - 7)$

12. $(4x - 8) - 4(-6.5x + 5)$

13. $\frac{1}{9}(-9x + 18) - \frac{1}{5}(10 + 15x)$

14. $\frac{4}{7}(4x + 3) - \frac{1}{7}(9x + 5)$

15. $\frac{1}{2}(-4x + 8) - \frac{1}{4}(8x - 12)$

16. Your class project involves recycling aluminum cans. After x weeks, your class has $(13x + 50)$ aluminum cans. The class goal is to collect $(80x + 120)$ aluminum cans. How many more aluminum cans does your class need to collect?

Answers

U. $-4x - 2$

P. $30x - 28$

T. $-9.8x - 19$

E. $x + 2$

I. $2x - 4$

L. $67x + 70$

H. $-11.2x + 24$

Y. $7x - 22$

I. $4x + 13$

U. $\frac{3}{4}x - \frac{1}{4}$

G. $x + 1$

L. $-4x + 7$

Y. $11.5x + 35$

F. $-4x$

M. $7x + 3$

O. $-x - 15$

6	3	9		16	2	14	11	5		8	12		10	4		15	1	13	7

Extension 3.2 Start Thinking!

For use before Extension 3.2

Your friend says the greatest common factor of 15 and 30 is 5. Is your friend correct? Explain your reasoning

Extension 3.2 Warm Up

For use before Extension 3.2

Find the GCF.

1. 2, 10 **2.** 3, 24 **3.** 9, 27

4. 14, 49 **5.** 24, 42 **6.** 50, 90

Name ______________________ Date ________

Extension 3.2 Practice

Factor the expression using the GCF.

1. $8 - 22$
2. $25 + 30$
3. $6y + 3$
4. $2t - 10$
5. $16p - 8$
6. $21s + 15$
7. $32v + 24w$
8. $9b + 24c$
9. $12y - 42z$

Factor out the coefficient of the variable.

10. $\frac{1}{2}m + \frac{1}{2}$
11. $\frac{2}{3}j - \frac{2}{9}$
12. $1.2k + 2.4$
13. $1.5a - 4.5$
14. $3f + 5$
15. $\frac{3}{10}x - \frac{3}{5}$
16. Factor $-\frac{1}{3}$ out of $-\frac{1}{3}x - 12$.
17. Factor $-\frac{1}{6}$ out of $-\frac{1}{3}x + \frac{5}{6}y$.
18. The area of the rectangle is $(18x - 12)$ square inches. Write an expression that represents the length of the rectangle (in inches).

6 in.

19. A concession stand sells hamburgers. The revenue from the hamburgers is $(30x + 45)$ dollars.

    a. The price of a hamburger is \$5. Write an expression that represents the number of hamburgers sold.

    b. The revenue from drinks is $(63x + 84)$ dollars. The price of a drink is \$3. Write an expression that represents the number of drinks sold.

    c. Write and simplify an expression that represents how many more drinks were sold.

Activity 3.3

Start Thinking!

For use before Activity 3.3

You have 7 less points than your cousin. Your brother has 8 more points than your sister.

Write an expression to model each situation. Use p as your variable. Can each expression be written in more than one way? Explain.

Warm Up

For use before Activity 3.3

Add.

1. $65 + (-23)$ **2.** $-12 + (-34)$ **3.** $-35 + 42$

Subtract.

4. $-15 - 24$ **5.** $29 - 35$ **6.** $52 - (-13)$

Lesson 3.3 Start Thinking!

For use before Lesson 3.3

Discuss with a partner, using an example, how inverse operations are used to solve equations.

Lesson 3.3 Warm Up

For use before Lesson 3.3

Solve the equation. Check your solution.

1. $x + 5 = 10$

2. $y - 2 = 16$

3. $n - 13 = 65$

4. $18 = p + 3$

5. $34 = t - 23$

6. $z + 14 = 21$

Name______________________________ Date__________

3.3 Practice A

Solve the equation. Check your solution.

1. $x + 3 = 10$

2. $b - 6 = -14$

3. $5 = n + 9$

4. $y - 2.1 = 7.5$

5. $-6.4 = x + 4.3$

6. $k - \frac{1}{3} = \frac{5}{6}$

7. $10.5 + p = -8.32$

8. $3\frac{3}{4} = r + \frac{1}{8}$

9. $m + 1.06 = 5$

10. $-\frac{7}{12} = 1\frac{5}{6} + d$

11. $t - \frac{2}{7} = \frac{1}{2}$

12. $-10.2 + c = -8.14$

Write the word sentence as an equation. Then solve.

13. 5 more than a number y is -2.

14. The sum of 8 and a number h is 12.

15. -13 is 4 less than a number n.

In Exercises 16–20, write an equation. Then solve.

16. You earn \$9 per hour babysitting. This is \$2 more than what you earned per hour last year. What did you earn per hour last year?

17. Your mother asked you to turn the oven down to 325°F. This is 75°F less than it was. What was the original temperature?

18. The difference between the heights of your chair and your desk is $-10\frac{1}{4}$ inches. The height of your desk is $29\frac{3}{4}$ inches. What is the height of your chair?

19. Your Two-Day-Pass to a theme park is \$76.50. This is \$31.41 less than your uncle's Two-Day-Pass. What is the price of your uncle's pass?

20. The perimeter of a triangle is 25 feet. What is the length of the unknown side?

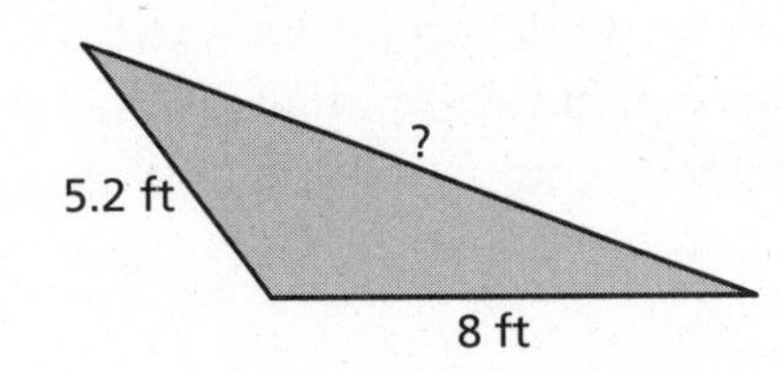

21. Find the value of $3x + 2$ when $7 + x = 5$.

Name ______________________ Date ________

3.3 Practice B

Solve the equation. Check your solution.

1. $x + 12 = 100$
2. $g - 16 = -52$
3. $-4.5 = m + 1.9$
4. $j - 12.1 = 7.53$
5. $y + 4.003 = -3.14$
6. $z - 4\frac{1}{3} = 2\frac{2}{3}$
7. $90.8 + q = -18.24$
8. $10\frac{2}{3} = r + 12\frac{1}{6}$
9. $b + 4.006 = 9$
10. $-7\frac{5}{8} = 1\frac{5}{6} + d$
11. $f - \frac{2}{15} = 6\frac{3}{5}$
12. $-10.216 + c = -12.014$

Write the word sentence as an equation. Then solve.

13. 27 is 12 more than a number x.
14. The difference of a number p and -9 is 12.
15. 35 less than a number m is -72.

In Exercises 16–18, write an equation. Then solve.

16. You swim the 50-meter freestyle in 28.12 seconds. This is 0.14 second less than your previous fastest time. What was your previous fastest time?
17. The perimeter of a rectangular backyard is $32\frac{1}{2}$ meters. The two shorter sides are each $7\frac{3}{8}$ meters long. What is the length of the two longer sides? (*Hint*: The sum of the shorter side and the longer side is equal to half of the perimeter.)
18. The temperature of dry ice is $-109.3°F$, which is $183.6°F$ less than the outside temperature. What is the outside temperature?
19. Your cell phone bill in August was \$61.43, which was \$21.75 more than your bill in July. Your cell phone bill in July was \$13.62 less than your bill in June. What was your cell phone bill in June?

Find the values of *x*.

20. $|x| - 10.5 = 4.3$
21. $|x + 2| - 7 = 5$

Name____ Date__________

3.3 Enrichment and Extension

You Be the Teacher

In Exercises 1 and 2, use the student solutions below.

Courtney	Karen
$\lvert x\rvert = 7$	$\lvert x\rvert = 7$
$x = 7$	$x = 7$ or $x = -7$

1. Did both students get a correct solution?
2. Is one student's answer more complete? If so, which one? Explain.
3. A student asks you how to solve $\lvert x + 8\rvert = 12$. Describe your explanation and any math steps you would show.
4. Describe Mario's solution. Did he get the correct answers? Explain.

Mario

$\lvert x - 5\rvert = 9$

$x - 5 = 9$ or $x - 5 = -9$

$x = 14$ or $x = -4$

5. Kelly looks at Mario's solutions and does not understand why -4 is a solution. She thought absolute value could not be negative. Explain Kelly's error.
6. Pat says the solutions of $\lvert x\rvert + 7 = 2$ are $x = 5$ and $x = -5$. What mistake did Pat make? Explain.
7. A student asks you if every absolute value equation has two solutions. How would you respond? Explain.
8. Give an example of an absolute value equation with (a) one solution, (b) two solutions, and (c) no solutions.

Name ______________________________ Date __________

Puzzle Time

What Did The Digital Clock Say To Its Mother?

Circle the letter of each correct answer in the boxes below. The circled letters will spell out the answer to the riddle.

Solve the equation.

1. $x + 8 = 21$

2. $3 = a - 12$

3. $y - 7 = -4$

4. $g + 11 = -13$

5. $z - 1.75 = 3.82$

6. $4.9 = h - 2.6$

7. $8.7 + b = 14.5$

8. $-10.3 = w - 5.8$

9. $\frac{3}{5} = c + \frac{1}{4}$

10. $r + 3\frac{1}{2} = -4\frac{2}{3}$

11. $5\frac{3}{4} = d - 2\frac{1}{8}$

12. $-7\frac{1}{3} = p - \frac{4}{9}$

13. The second book in your favorite series has 9 more chapters than the first book in the series. The second book has 38 chapters. How many chapters does the first book have?

14. Emily has a Springer Spaniel that weighs 48.5 pounds. She also has a Cocker Spaniel that weighs 24.8 pounds less than the Springer Spaniel. How many pounds does the Cocker Spaniel weigh?

L	M	O	O	I	K	E	T	M	S	O	R	D	M	A	E
23.7	4.7	3	$-8\frac{1}{6}$	$2\frac{3}{7}$	5.8	$-5\frac{2}{5}$	7.2	13	−42	−4.5	$1\frac{1}{9}$	−8.9	$-6\frac{8}{9}$	12	8
H	**N**	**E**	**R**	**O**	**S**	**H**	**T**	**U**	**A**	**N**	**Y**	**D**	**M**	**E**	**S**
52	7.5	$4\frac{3}{8}$	$12\frac{1}{3}$	−24	$-1\frac{2}{5}$	29	−3.9	17	5.57	$\frac{7}{20}$	$-\frac{1}{6}$	15	33	65.5	$7\frac{7}{8}$

Activity 3.4 Start Thinking!
For use before Activity 3.4

Explain how buying a number of items at a store is like solving a math problem.

Activity 3.4 Warm Up
For use before Activity 3.4

Multiply.

1. $(-21)(8)$

2. $(-18) \bullet (-12)$

3. $5 \bullet (-13)$

Divide.

4. $\dfrac{-96}{3}$

5. $\dfrac{108}{-12}$

6. $\dfrac{-128}{-8}$

Lesson 3.4 Start Thinking!

For use before Lesson 3.4

With a partner, write and solve a real-life word problem using the equation $13x = 39$.

Then rewrite the word problem using division.

Lesson 3.4 Warm Up

For use before Lesson 3.4

Solve the equation. Check your solution.

1. $4x = 24$

2. $2x = 56$

3. $-7x = 77$

4. $\frac{x}{5} = 12$

5. $\frac{x}{7} = 9$

6. $\frac{x}{-8} = 6$

Name____ Date__________

3.4 Practice A

Solve the equation. Check your solution.

1. $4b = 24$

2. $-7n = 35$

3. $\frac{y}{-3} = 33$

4. $\frac{p}{5} = -32$

5. $-3t = -4.2$

6. $1.5q = -8.4$

7. $\frac{1}{5}d = -3$

8. $14 = 3y$

9. $\frac{g}{2.1} = -6.8$

10. $-\frac{3}{5}a = 2$

11. $\frac{k}{-9} = -\frac{1}{3}$

12. $\frac{5}{8}j = -10$

Write the word sentence as an equation. Then solve.

13. A number multiplied by $\frac{1}{2}$ is $-\frac{5}{12}$.

14. The quotient of a number and 0.2 is -2.6.

In Exercises 15–19, write an equation. Then solve.

15. You earn $7.50 per hour at a fast food restaurant. You earned $123.75 last week. How many hours did you work last week?

16. Your family took a road trip on Saturday. You were in the car for 4.5 hours and averaged 70 miles per hour. How many miles did you travel?

17. The area of a rectangle is $\frac{1}{2}$ square inch. The length of the rectangle is $\frac{3}{8}$ inch. What is the width of the rectangle?

18. You are in a room with other students and are asked to get in groups of 3. When finished, there are 21 groups of 3. How many students are in the room?

19. The perimeter of a square is 26.46 inches. What is the side length of the square?

20. Write a multiplication equation that has a solution of $\frac{2}{7}$.

21. Write a division equation that has a solution of -20.

Name ______________________________ Date __________

3.4 Practice B

Solve the equation. Check your solution.

1. $16t = 60$

2. $-14p = -21$

3. $\frac{q}{5} = -7.35$

4. $\frac{d}{1.2} = -3.3$

5. $-\frac{8}{15}k = -4$

6. $-7.24q = 17.014$

7. $\frac{1}{8}d = -\frac{3}{5}$

8. $1.5 = 3.3y$

9. $\frac{g}{0.003} = -2.8$

10. $-\frac{10}{21}c = -\frac{15}{28}$

11. $\frac{k}{-9} = -1$

12. $18 = -\frac{6}{11}h$

In Exercises 13 and 14, write an equation. Then solve.

13. You order an entree for \$12.00. You pay \$0.78 in taxes. What is the tax rate?

14. If a project is handed in late, you receive $\frac{8}{9}$ of your earned points. You received 72 points on your late project. How many points did you lose?

15. Write a multiplication equation that has a solution of -14.8.

16. Write a division equation that has a solution of $-\frac{9}{14}$.

17. There are 92 students in a room. They are separated into 18 groups. How many students are in each group? How many students are not in a group?

18. A bus token costs \$1.75.

a. You spend \$15.75 on tokens. Write and solve an equation to find how many tokens you purchase.

b. If you purchase 10 tokens, you get 2 free tokens. Write and solve an equation to find the approximate reduced price of each token.

c. You also receive free tokens if you purchase 20 tokens. The reduced price for each token is \$1.40. Write and solve an equation to find how many free tokens you receive.

19. Solve $\frac{1}{3}|z| = 2$.

Name____ Date__________

3.4 Enrichment and Extension

Equations with No Solution

For some equations, there is no value that could be substituted for the variable to make the equation true. In this case, the equation has no solution.

Example: Solve $4|x| = -12$.

$4|x| = -12$ Rewrite the equation.

$|x| = -3$ Divide both sides by 4.

Because absolute value is always nonnegative, no number has an absolute value of -3. So, this equation has *no solution.*

Without solving the equation, tell whether it has *one solution, two solutions,* or *no solution.* If the equation has one solution, tell whether the solution is *positive* or *negative.* Explain your reasoning.

1. $-5x = -16$

2. $\dfrac{n}{-5} = -12$

3. $\dfrac{g}{7} = -8$

4. $-12t = 100$

5. $-8|v| = -16$

6. $\dfrac{|k|}{-9} = 6$

7. $\left|\dfrac{x}{-5}\right| = -15$

8. $|-6p| = 42$

9. $|2.7u| = 10.8$

10. $\left|1\frac{1}{2}b\right| = -13\frac{4}{5}$

11. $3h = |8|$

12. $|-9|y = -12$

13. $\dfrac{|b|}{-3} = -7$

14. $\left|\dfrac{a}{-9}\right| = -2.5$

For each exercise number, use your answers and the key below to color the cell. Do not color the cells that have a zero in them.

One Positive Solution = Orange **One Negative Solution = Blue**

Two Solutions = Green **No Solution = Green**

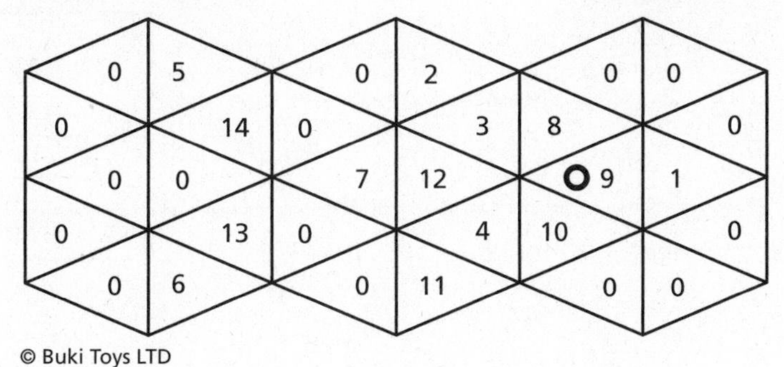

© Buki Toys LTD

Name ____________________ Date ________

3.4 Puzzle Time

Did You Hear About...

A	B	C	D	E	F
G	H	I	J	K	L
M					

Complete each exercise. Find the answer in the answer column. Write the word under the answer in the box containing the exercise letter.

−12 PLANET
−60 MOON
−14 GOOD
4 THE
−1.7 EARTH
−56 REALLY
9 ATMOSPHERE
$-13\frac{1}{2}$ THAT
$\frac{3}{11}$ ORBIT

Solve the equation.

A. $6x = 24$

B. $-7a = 35$

C. $-3g = -33$

D. $\frac{c}{4} = -8$

E. $\frac{z}{-12} = 5$

F. $\frac{2}{3}h = -9$

G. $-\frac{4}{5} = 2b$

H. $32 = -\frac{4}{7}y$

I. $-1.8m = 25.2$

J. $\frac{p}{3.7} = 5.1$

K. $20.3 = -2.9c$

L. $-12.6w = -16.38$

M. Tyler has $11.25. How many ride tickets can he buy for himself and his friends if the ride tickets cost $1.25 each?

8.3 EAT
11 ON
−5 RESTAURANT
−7 BUT
−32 THE
1.3 NO
18.87 FOOD
$-1\frac{3}{5}$ BAD
$-\frac{2}{5}$ HAS

Activity 3.5 Start Thinking!

For use before Activity 3.5

In Sections 3.3 and 3.4, you learned how to solve one-step problems. Without using your notes, give an example and then describe to a partner how to solve each of the four types of one-step problems.

Activity 3.5 Warm Up

For use before Activity 3.5

Solve the equation. Check your solution.

1. $-9x = -108$

2. $x - 3 = -12$

3. $\frac{x}{5} = -8$

4. $12x = 144$

5. $x - 3.6 = 5.44$

6. $x + \frac{1}{2} = 4\frac{2}{3}$

Lesson 3.5 Start Thinking!

For use before Lesson 3.5

Explain to a partner how to solve the equation $3x - 4 = 11$.

Lesson 3.5 Warm Up

For use before Lesson 3.5

Solve the equation. Check your solution.

1. $3x - 10 = 14$

2. $5x + 24 = -36$

3. $17 = 3x + 2$

4. $-9x + 17 = 71$

5. $4x + 13.2 = 2.6$

6. $-3x - 3.4 = 12.2$

Name____ Date__________

3.5 Practice A

Solve the equation. Check your solution.

1. $3k - 2 = 10$
2. $5p + 2 = -10$
3. $-4x + 3 = -11$
4. $12 = 2d + 3.2$
5. $-1 - 5h = 14$
6. $1.25r - 7 = 2.5$
7. $-4k + 3.6 = 7.8$
8. $6 + 2n = 3$
9. $4y - 16.3 = 53.1$
10. $\frac{1}{2}b + \frac{9}{4} = \frac{7}{4}$
11. $\frac{5}{6} + 3j = -\frac{2}{3}$
12. $-\frac{9}{10}p - 3 = \frac{3}{5}$

In Exercises 13–15, write an equation. Then solve.

13. It costs $4 to enter the fair. Each ride costs $2.50. You have $21.50. How many rides can you go on?

14. The cable company charges a monthly fee of $45. Each movie rental is $1.99. You owe $68.88. How many movies did you rent?

15. The perimeter of the rectangle is 24 feet. What is the width of the rectangle?

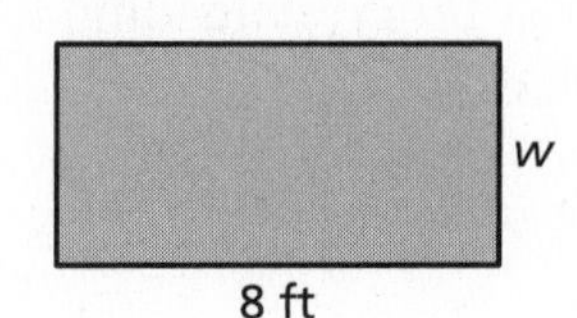

Solve the equation. Check your solution.

16. $7c - 2c = 45$
17. $3(k - 5) = -16$
18. $-2(m + 1) = 10$

19. The senior class has 412 students. They are assigned to different homerooms. There are 28 students in the smallest homeroom and the remaining 12 homerooms have the same number of students. How many students are in each of the remaining 12 homerooms?

20. You purchased paint for the rooms in your house. You have 1.5 cans of paint left. You painted 4 rooms and each room required 2 cans of paint. You spilled $\frac{1}{2}$ of a can of paint. How many cans of paint did you purchase?

    a. Solve the problem by working backwards.

    b. Solve the equation $\frac{x - 2}{4} = 2$. How does the answer compare to part (a)?

Name ______________________ Date __________

3.5 Practice B

Solve the equation. Check your solution.

1. $5k - 8 = 7$
2. $6b + 9 = -15$
3. $-3.2w - 2 = -4.5$
4. $13 - 2n = 27$
5. $25 = 4.5z + 12$
6. $5.25s - 2.01 = -8.94$
7. $84 = 51 - 14p$
8. $81 + 7t = 81$
9. $16 + 2.4c = 22.5$
10. $4h + \frac{1}{3} = \frac{3}{4}$
11. $\frac{1}{7}f - 5\frac{1}{2} = \frac{9}{14}$
12. $-\frac{1}{2}u + \frac{3}{5} = \frac{1}{6}$

In Exercises 13 and 14, write an equation. Then solve.

13. You purchased \$132.49 worth of wheels and bearings for your skateboards. The shop charges \$15 per board to install them. The total cost is \$192.49. How many skateboards will be repaired?

14. A music download service charges a flat fee each month and \$0.99 per download. The total cost for downloading 27 songs this month is \$42.72. How much is the flat fee?

Solve the equation. Check your solution.

15. $-5x - 2x + 3x = 9$
16. $-5(m + 4) = 27$
17. $-12(a - 2) = -50$

18. The perimeter of a triangle is 60 feet. One leg is 12 feet long. Of the two unknown sides, one of them is twice as long as the other. Find the lengths of the two unknown sides.

19. Sally picks seashells by the seashore. She lost 17 of them on her way home. She planned to fill 5 jars with the same amount of seashells in each. How many seashells did Sally pick?

    a. You do not have enough information to solve this problem. The number of seashells in each jar is the same as the number portion of her street address, which is a 2-digit number. The first digit is 5. The last digit is 9 less than 3 times the first digit. How many seashells did Sally plan to put in each jar?

    b. By working backwards, determine how many seashells Sally picked.

    c. The 5 jars that Sally chose would not each hold that many seashells. In her search for a 6th jar, she discovered a few seashells in her pocket. What are possible values for the number of seashells in each of the 6 jars and the number of seashells discovered in her pocket, such that there are no seashells left over?

Name______________________________ Date__________

3.5 Enrichment and Extension

Solving Equations with Fractions

1. If you multiply each term by this number, the equation $\frac{3x}{2} - \frac{4}{5} = 5\frac{1}{5}$ will contain no fractions. What number could this be?

2. Are there other numbers you can multiply by to rewrite the equation in Exercise 1 without fractions? Explain.

3. What number do you think is best to use at the multiplier? Explain.

4. Why can you multiply each term and not change the solution of the equation?

5. Describe to someone how to rewrite an equation with fractions so that there are no fractions left in it.

6. Solve each equation by rewriting it without fractions first.

   a. $\frac{x}{8} - 5 = \frac{3}{4}$

   b. $\frac{x}{4} - \frac{1}{2} = 3\frac{1}{4}$

   c. $2\frac{1}{3} - \frac{x}{4} = \frac{5}{6}$

   d. $-\frac{2}{7} + \frac{x}{2} = \frac{9}{14}$

   e. $6\frac{7}{9} = \frac{2}{3} - 5x$

   f. $\frac{x + 10}{6} = \frac{2}{3}$

Name ______________________ Date __________

3.5 Puzzle Time

What Did One Bowling Ball Say To The Other Bowling Ball?

Write the letter of each answer in the box containing the exercise number.

Solve the equation.

1. $2c - 5 = 9$

2. $3m + 7 = -8$

3. $-7x - 3 = 12$

4. $15 = 4a + 3$

5. $5y - 6 = -20$

6. $9f + 3.6 = 10.8$

7. $-4p - 5.7 = 11.1$

8. $-20.3 = 6w + 3.1$

9. $2 + 5.3k = 18.43$

10. $7.8b - 2.14 = -42.7$

11. $\frac{1}{4}z - \frac{2}{7} = \frac{5}{7}$

12. $3 - \frac{r}{8} = -\frac{9}{2}$

13. $-\frac{1}{3} + 5e = -\frac{3}{4}$

14. $14d - 2d = -84$

15. $-5g - 13g = 54$

16. $-3(t - 8) = 32$

17. Kayla's age is 3 less than twice her brother's age. Kayla is 13 years old. How old is her brother?

18. Mario spent $23.85 at the bookstore on one book and some magazines. The book cost $12.60 and the magazines cost $2.25 each. How many magazines did Mario buy?

19. Ethan planted a tree that is 37.5 inches tall. If the tree grows 3 inches each year, how long will it take for the tree to reach a height of 54 inches?

Answers

T. -5.2	**N.** 3
S. 5	**M.** 8
E. $-2\frac{4}{5}$	**O.** 3.1
L. $-2\frac{1}{7}$	**T.** 7
N. -7	**O.** $-\frac{1}{12}$
M. -5	**I.** 4
P. $-2\frac{2}{3}$	**A.** 0.8
O. -3.9	**D.** -3
L. 60	**R.** 5.5
O. -4.2	

15	7	4	10		18	1	13	16		17	5		11	2		8	14		6		19	9	12	3

Name______________________________ Date__________

Chapter 3

Technology Connection

For use after Section 3.5

Solving Multi-Step Equations

You may find it helpful to use a calculator to solve multi-step equations with large numbers or decimals.

EXAMPLE 1 Solve $19.39 + 4.8x = 34.51$.

SOLUTION

Step 1 Enter 34.51 [−] 19.39 to undo addition.

Step 2 Most calculators will store the answer, 15.12, so that you don't need to write it down. Just enter [÷] 4.8 to undo the multiplication.

ANSWER 3.15

You can use this method to solve equations that require more steps.

EXAMPLE 2 Solve $\frac{x}{1.2} + 1.8 = 9 - 1.45$.

SOLUTION

Step 1 Enter 9 [−] 1.45 to simplify the right side of the equation.

Step 2 Enter [−] 1.8 to undo addition.

Step 3 Enter [×] 1.2 to undo division.

ANSWER 6.9

Use a calculator to solve the equation.

1. $106y + 317 = 5299$
2. $0.93x - 0.904 = 0.2864$
3. $325.85 = 0.14m - 40.67$
4. $\frac{x}{52} - 73 = 36$
5. $\frac{w}{27} + 372 = 431$
6. $203 = \frac{m}{67} + 189$
7. $212x - 5216 = 9310 + 2646$
8. $4.838 - 1.202 = 1.85z - 3.209$
9. $41 + \frac{a}{109} = 22.3 + 21.5$
10. $0.65 + 9.85 = \frac{c}{0.65} + 2.6$

Chapter 4

Name ____ Date __________

Chapter 4 Inequalities

Dear Family,

Gardeners are familiar with uncertainty. Will there be enough sun? Will there be enough rain? Did I use too much fertilizer? Planning a garden can be a challenge, whether in a small container or over several acres.

You might work with your student to plan and create a small potted garden. Make sure you plant more seeds than you need—some will not germinate and some will produce weak plants. Make sure the plants get enough sun but not too much heat. Have your student write an *inequality* to represent each of these situations.

As your garden grows, ask your student to keep track of the growing conditions. Track your garden's basic needs and have your student write an inequality to represent these situations:

- How tall are the plants likely to get?
- Research how much water the plants need. Check the soil's moisture content every day—plants need water to survive. However, too much water can be just as bad as too little.
- Make sure the recommended amount of sunshine is available. The seed packet will usually tell you the minimum amount required.
- In a potted garden, your plants will probably need some fertilizer to stay healthy. Keep an eye out for signs of overfeeding, however.

Not all problems in mathematics involve a single answer. Many problems have answers that fall into a range. Your plants need at least enough fertilizer to grow, but you must limit the amount of fertilizer to what the plant can safely use. You must make sure the water stays in the right range.

It's hard to beat the satisfaction of growing a successful garden—and the fruits of your labor are beautiful to behold!

Nombre ____ Fecha __________

Desigualdades

Estimada Familia:

Los jardineros están familiarizados con la incertidumbre. ¿Habrá suficiente sol? ¿Habrá suficiente lluvia? ¿Usé demasiado fertilizante? Planear un jardín puede ser un desafío, ya sea en un contenedor pequeño o a lo largo de muchos acres.

Puede trabajar con su estudiante para planear y crear un pequeño jardín en maceta. Asegúrese de plantar más semillas de las que necesita—algunas no germinarán y otras producirán plantas débiles. Asegúrese que las plantas tengan suficiente sol, pero no demasiado calor. Haga que su estudiante escriba una *desigualdad* para representar cada una de estas situaciones.

A medida que su jardín crece, pida a su estudiante que registre las condiciones de crecimiento. Registre las necesidades básicas de su jardín y haga que su estudiante escriba una desigualdad para representar estas situaciones:

- ¿Qué tan altas van a crecer las plantas?
- Investiguen la cantidad de agua que necesitan las plantas. Revisen la humedad del suelo todos los días—las plantas necesitan agua para sobrevivir. sin embargo, demasiada agua puede ser tan malo como demasiado poco.
- Asegúrese que la cantidad de luz solar recomendada esté disponible. Normalmente, el paquete de semillas indicará la cantidad mínima requerida.
- En un jardín en maceta, probablemente sus plantas necesitarán algo de fertilizante para estar sanas. No obstante, revise que no esté sobrealimentándolas.

No todos los problemas en matemáticas implican una respuesta única. Muchos problemas tienen respuestas que caen dentro de un rango. Sus plantas necesitan al menos fertilizante para crecer, pero debe limitar la cantidad de fertilizante que la planta puede usar con seguridad. Debe asegurarse que el agua se mantenga dentro del rango correcto.

Es difícil tener una mayor satisfacción que la de hacer crecer un jardín exitoso—¡y el fruto de su labor es algo bello para contemplar!

Activity 4.1 Start Thinking!

For use before Activity 4.1

How are inequalities related to height or age restrictions on amusement park rides?

Research the ride restrictions at an amusement park and write inequalities to describe them.

Activity 4.1 Warm Up

For use before Activity 4.1

Plot and label each number on the same number line.

1. 8

2. -2

3. $2\frac{1}{2}$

4. $-3\frac{1}{2}$

5. 0

6. -4

Lesson 4.1 Start Thinking!
For use before Lesson 4.1

Write a sentence involving a real-life situation that can be modeled using an inequality.

Which inequality symbol applies: $<$, $\leq$, $>$, or $\geq$?

Lesson 4.1 Warm Up
For use before Lesson 4.1

Write an inequality for the graph. Then, in words, describe all the values of *x* that make the inequality true.

1. 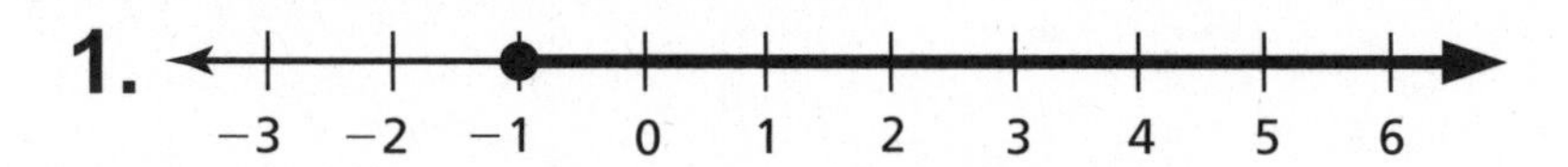

2. 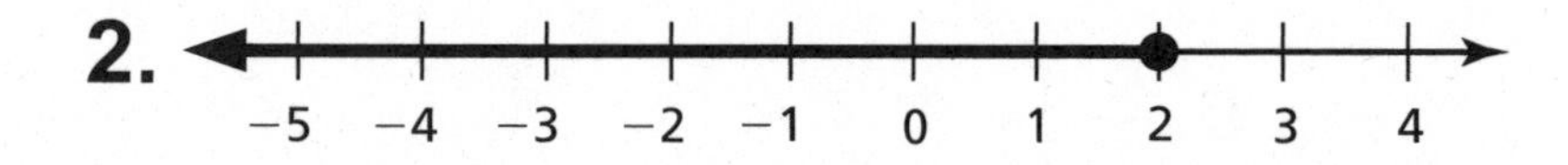

3. 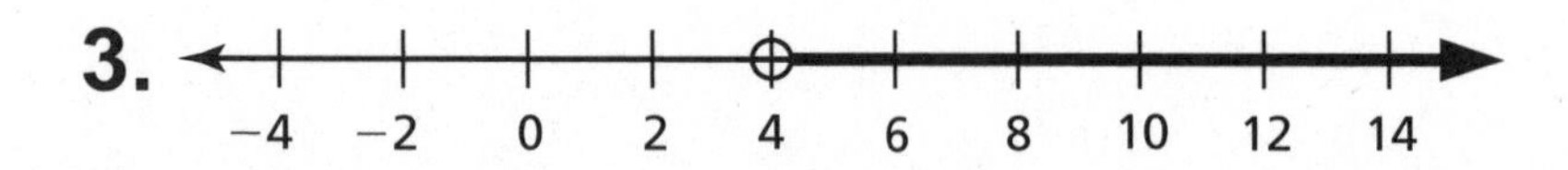

4. 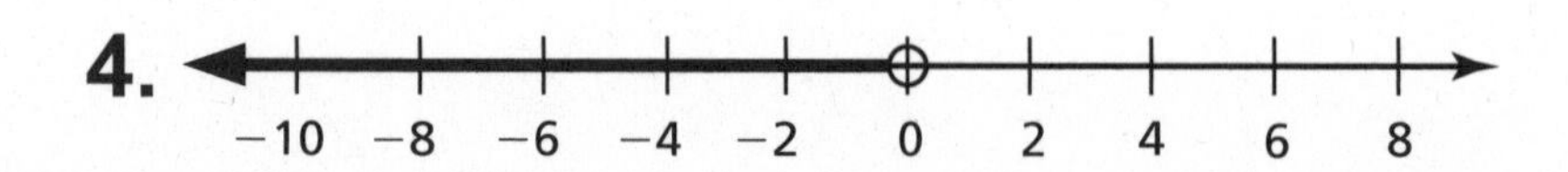

Name ______________________________ Date __________

4.1 Practice A

Write an inequality for the graph. Then, in words, describe all the values of *x* that make the inequality true.

1. 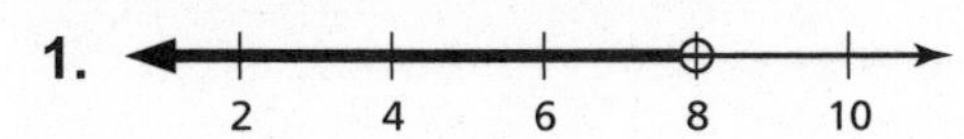

2. −6 −5 −4 −3 −2

Write the word sentence as an inequality.

3. A number x is at most 3.

4. A number y added to 2 is greater than 7.

5. A number c multiplied by 3 is less than −12.

6. A number m minus 1.5 is no less than 2.

Tell whether the given value is a solution of the inequality.

7. $t - 3 \geq 2; t = 10$

8. $6w < -2; w = 1$

9. $p + 1.6 \leq 4; p = 5$

10. $\frac{1}{2}d > -3; d = 0$

Graph the inequality on a number line.

11. $k > 1$

12. $n \leq -2.5$

13. In order to try out for one of the parts in a play at the local theater, you must be at most 12 years old. Write an inequality that represents this situation.

Tell whether the given value is a solution of the inequality.

14. $3h - 7 < h; h = 2$

15. $q + 8 \geq \frac{q}{4}; q = -12$

16. Consider the inequalities $-2x < 10$ and $-6 < -2x$.

a. Is $x = 0$ a solution to both inequalities?

b. Is $x = 4$ a solution to both inequalities?

c. Find another value of x that is a solution to both inequalities.

Name ______________________________ Date __________

4.1 Practice B

Write an inequality for the graph. Then, in words, describe all the values of *x* that make the inequality true.

1. 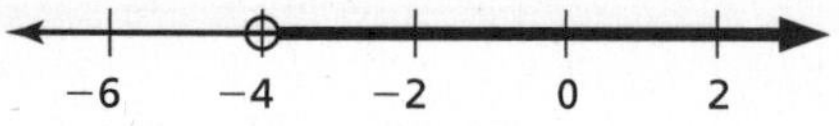

2. 

Write the word sentence as an inequality.

3. A number x is at least 15.

4. A number r added to 3.7 is less than 1.2.

5. A number h divided by 2 is more than -5.

6. A number a minus 8.2 is no greater than 12.

Tell whether the given value is a solution of the inequality.

7. $p + 1.7 \geq -4;\ p = -9$

8. $-3y < -5;\ y = 1$

9. $1.5g \leq 6;\ g = 0$

10. $\frac{3}{4} - d > \frac{1}{3};\ d = \frac{1}{2}$

Graph the inequality on a number line.

11. $\ell \leq 3.5$

12. $m > -15$

13. To get a job at the local restaurant, you must be at least 16 years old. Write an inequality that represents this situation.

Tell whether the given value is a solution of the inequality.

14. $5t < 4 - t;\ t = -3$

15. $\frac{q}{5} < q - 20;\ q = 15$

16. In order to qualify for a college scholarship, you must have acceptable scores in either the SAT or the ACT along with the following requirements: a minimum GPA of 3.5; at least 12 credits of college preparatory academic courses; and at least 75 hours of community service.

   a. Write and graph three inequalities that represent the requirements.

   b. Your cousin has a GPA of 3.6, 15 credits of college preparatory class, and 65 hours of community service. Other than the test scores, does your cousin satisfy the requirements? Explain.

Name____ Date__________

4.1 Enrichment and Extension

Compound Inequalities

Little League is a commercially sponsored baseball league for boys and girls.

A *compound inequality* is a special type of inequality that places both an upper and lower boundary on a variable. Write a compound inequality that describes the Little League rule.

Example: The maximum number of innings in a Little League game is 6. Each player must play at least 2 innings. Write a compound inequality that represents the number of innings a player plays.

Let n represent the number of innings a player plays. Because 2 is the minimum number of innings and 6 is the maximum number of innings, the compound inequality that represents the number of innings a player plays is $2 \leq n \leq 6$.

1. To be eligible to play Little League, a player must be at least 9 years old and at most 12 years old. Let a represent the player's age.

a. Write an inequality that represents the minimum age a player must be to participate in Little League.

b. Write an inequality that represents the maximum age a player can be to participate in Little League.

c. Use the inequalities from parts (a) and (b) to write a compound inequality that represents the age restrictions of Little League players.

2. For health and safety reasons, the number of pitches p a player can make per game is limited based on his or her age. A 12-year-old may pitch a maximum of 85 pitches in a game day.

a. Write an inequality that represents the minimum number of pitches a player could make during a game.

b. Write an inequality that represents the maximum number of pitches a player could make during a game.

c. Use the inequalities from parts (a) and (b) to write a compound inequality that represents the number of pitches that a player can throw per game.

3. A Little League game lasts for at least 3.5 innings and at most 6 innings. Write a compound inequality that represents the number of innings n that a Little League game lasts.

Name ______________________________ Date __________

4.1 Puzzle Time

What Do You Call A Bull That's Sleeping?

Write the letter of each answer in the box containing the exercise number.

Write the word sentence as an inequality.

1. A number x is greater than 25.8.

2. Twice a number x is at most $-\frac{3}{5}$.

3. A number x minus 9.3 is more than 4.6.

4. A number x added to 11.7 is less than 14.

Tell whether the given value is a solution of the inequality.

5. $x - 3.6 \leq 2.8;\ x = 6.7$

6. $\frac{5}{6}x > -10;\ x = -6$

Match each inequality with its graph.

7. $x \leq -7$

8. $x > 3.2$

9. $x < 3\frac{1}{4}$

10. $x \geq -11$

Answers

U. $11.7 + x < 14$

L. $x > 25.8$

A.

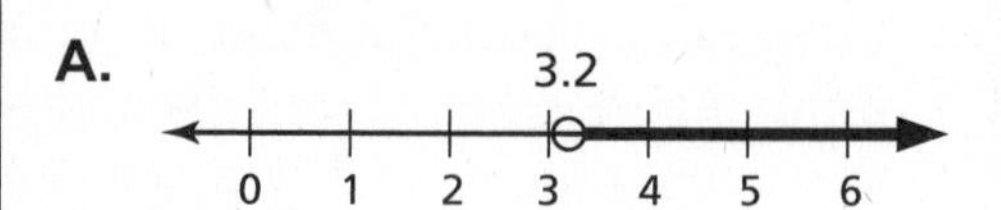

D. yes

E. $2x \leq -\frac{3}{5}$

R.

−10 −9 −8 −7 −6 −5 −4

L. $x - 9.3 > 4.6$

Z.

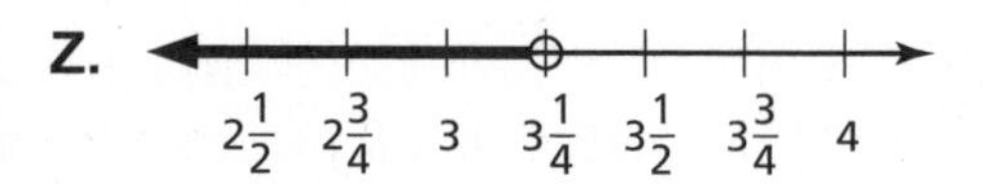

B. no

O.

−14 −13 −12 −11 −10 −9 −8

8		5	4	1	3	6	10	9	2	7

Activity 4.2 Start Thinking!

For use before Activity 4.2

An elevator can carry at most 15 people. Write an inequality that models this statement.

Explain to a friend what the inequality means.

Activity 4.2 Warm Up

For use before Activity 4.2

Graph the inequality on a number line.

1. $x \leq 13$

2. $x \geq -4$

3. $x < -15$

4. $x > 6$

5. $x \leq -1$

6. $x \geq 0$

Lesson 4.2

Start Thinking!

For use before Lesson 4.2

Describe a real-life situation that can be represented by the inequality $x + 5 \leq 20$.

Lesson 4.2

Warm Up

For use before Lesson 4.2

Solve the inequality. Graph the solution.

1. $x - 9 < 8$

2. $6 + h > 9$

3. $10 \geq y - 3$

4. $y - 2 \geq 14$

5. $t - 4 > -2$

6. $x + 7 \leq 10$

Name____ Date__________

4.2 Practice A

Solve the inequality. Graph the solution.

1. $p - 4 < 2$

2. $s + 1 \geq -5$

3. $k - 14 \leq -10$

4. $2 < n + \frac{3}{2}$

5. $z - \frac{2}{3} \geq \frac{1}{3}$

6. $-\frac{1}{2} > -\frac{1}{6} + t$

7. $d - 2.4 \leq -5.1$

8. $-4.5 + q > 2.5$

9. To stay within your budget, the area of the house and the garage combined is at most 3000 square feet. The area of the garage is 528 square feet. Write and solve an inequality that represents the area of the house.

10. You have \$137.26 in a bank account. The bank requires you to have at least \$50 in your account or else you are charged a fee. Write and solve an inequality that represents the amount you can write your next check for without being charged a fee.

Write and solve an inequality that represents *x*.

11. The perimeter is less than 20 meters.

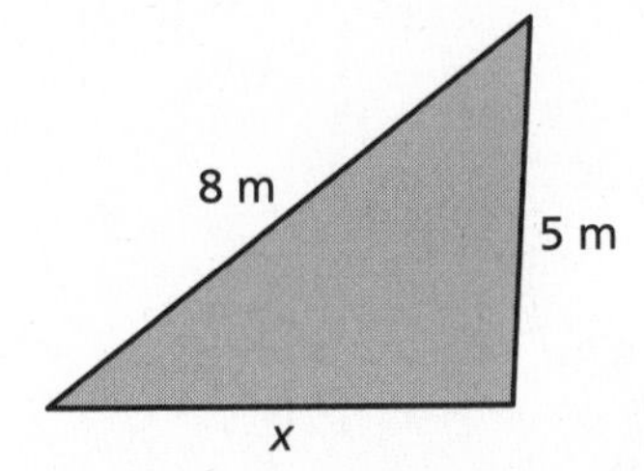

12. The perimeter is at least 18 feet.

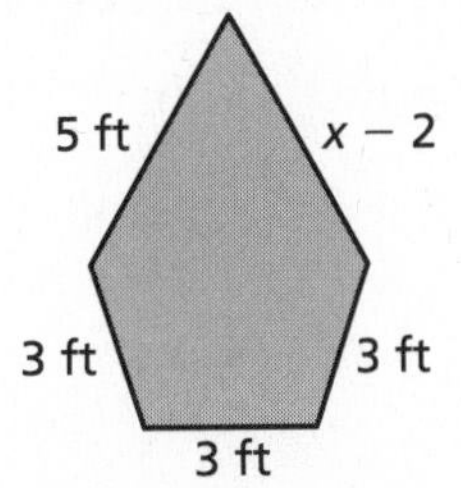

13. You need at least 5000 points to earn a gift card from your bank. You currently have 2700 points.

a. Write and solve an inequality that represents the number of points you need to earn a gift card.

b. You deposit money in your savings account and earn an additional 400 points. How does this change the inequality?

Name ______________________ Date ________

4.2 Practice B

Solve the inequality. Graph the solution.

1. $-12 \le y - 17$

2. $w - 1.8 < 2.5$

3. $v + \frac{1}{3} > 8$

4. $\frac{2}{5} < \frac{4}{5} + k$

5. $q + \frac{3}{4} \ge -\frac{1}{4}$

6. $-\frac{3}{2} + r < \frac{1}{2}$

7. $7.4 > c + 3.9$

8. $p - 10.2 > 3.5$

9. You and two friends are diving for lobster. The maximum number of lobsters you may have on your boat is 18. You currently have 7 lobsters.

a. Write and solve an inequality that represents the additional lobsters that you may catch.

b. Another friend comes on your boat and he has 3 lobsters. You may now have 24 lobsters on your boat. Write and solve an inequality that represents the additional lobsters that you may catch.

c. How many lobsters is each person allowed to catch?

Write and solve an inequality that represents *x*.

10. The length is greater than the width.

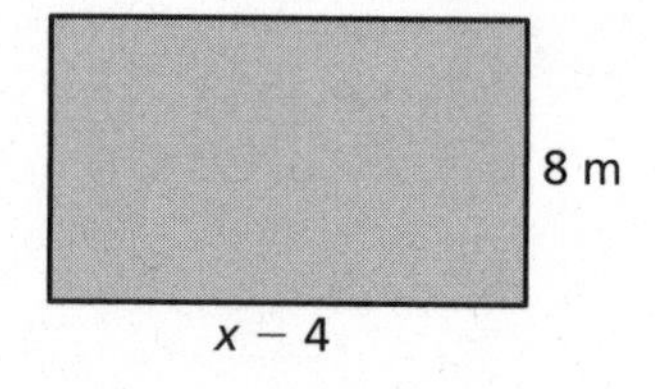

11. The perimeter is less than or equal to 50 inches.

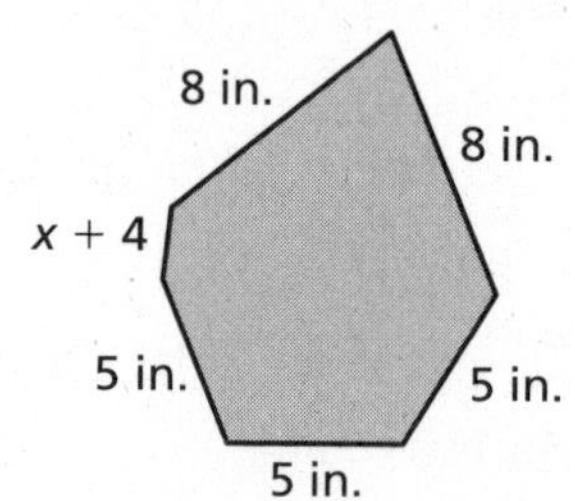

12. The solution of $w - c > -3.4$ is $w > -1.4$. What is the value of c?

13. Describe all numbers that are solutions to $|x| < 5$.

14. The *triangle inequality theorem* states that the sum of the lengths of any two sides of a triangle is greater than the length of the third side. A triangle has side lengths of 6 inches and 17 inches. What are the possible values for the length of the third side? Explain how you found your answer.

Name____ Date__________

4.2 Enrichment and Extension

Airplanes

The cabin (interior) of an airplane is partitioned into 3 distinct sections, or classes. The cost of a seat in each of the classes is different and the amenities in each class vary.

Seats	Width of Seat	Pitch of Seat
First class	51 cm	p cm
Business class	x cm	140 cm
Economy class	43 cm	81 cm

1. The width of a seat in business class is the average of the widths of the seats in first class and economy class. Find the width of a seat in business class.

2. The cabin has a minimum width requirement so that each passenger on the plane has comfortable accommodations.

   a. The width of four business class seats and an aisle y must be at least 304.75 centimeters. Write an inequality that represents this situation.

   b. What is the width of the aisle?

   c. An airplane row contains 8 business class seats and 2 aisles. What is the minimum width w of the cabin?

3. The pitch of an airplane seat refers to the distance between the backs of two consecutive seats.

   a. Thirty-eight less than the pitch p between first class seats is at least as big as 2 times the pitch between economy class seats. Write an inequality that models the pitch between seats in first class.

   b. What is the pitch between seats in first class?

4. Which class contains the greatest number of seats? the least number of seats? Explain your reasoning.

Name ____________________ Date __________

4.2 Puzzle Time

Did You Hear About The...

A	B	C	D	E	F
G	H	I	J	K	L
M	N				

Complete each exercise. Find the answer in the answer column. Write the word under the answer in the box containing the exercise letter.

$x \geq -3$ GAME
$x \leq \frac{2}{3}$ IN
$x \leq -6$ AND
$x \leq 16$ TIE
$x \geq 15$ THE
$x < -5$ CATCHER
$x < 3$ COLLARS
$x > 2.8$ THAT
$x \leq \frac{2}{5}$ SHIRTS

Solve the inequality.

A. $x + 5 \geq 20$

B. $x - 4 > 6$

C. $6 \leq 9 + x$

D. $3 + x \leq -2$

E. $-17 \leq x - 8$

F. $x - 1 < 2$

G. $x - 10 \leq -16$

H. $x + \frac{1}{3} \geq 3$

I. $\frac{3}{5} \geq x + \frac{1}{5}$

J. $-4.4 < x - 7.2$

K. $\frac{11}{4} > x + \frac{9}{4}$

L. $-\frac{5}{12} \geq x - \frac{13}{12}$

M. $x + 0.4 < -0.8$

O. To play on the football team, a seventh grader must weigh no more than 110 pounds. Your neighbor is in seventh grade and weighs 94 pounds. Write and solve an inequality that represents how much weight your neighbor can gain and still meet the requirement.

$x \geq 2\frac{2}{3}$ THE
$x > 10$ BASEBALL
$x < \frac{1}{2}$ ENDED
$x \geq 1.1$ WHICH
$x > 1$ MITT
$x \geq -9$ THE
$x \leq 2$ SOCKS
$x < -1.2$ A
$x \leq -5$ BETWEEN

Activity 4.3 Start Thinking!

For use before Activity 4.3

Explain to a partner how to solve the following inequalities. Then graph the inequalities.

$x + 7 \leq -4$ $\qquad$ $x - 6 > 8$

Activity 4.3 Warm Up

For use before Activity 4.3

Complete the statement with < or >.

1. -7 __?__ -5

2. 2 __?__ -2

3. 7 __?__ -10

4. -13 __?__ -11

5. -1 __?__ -2

6. -8 __?__ 8

Lesson 4.3 Start Thinking!

For use before Lesson 4.3

Are the solutions to the following inequalities the same? Explain why or why not.

$2x < -12$ $\qquad$ $-2x < 12$

Lesson 4.3 Warm Up

For use before Lesson 4.3

Use a table to solve the inequality.

1. $-2x > 4$

2. $-4x < -8$

3. $-2x \leq -12$

4. $-3x \geq 18$

5. $\dfrac{x}{-3} < 21$

6. $\dfrac{x}{-5} > \dfrac{1}{2}$

Name____ Date__________

4.3 Practice A

Solve the inequality. Graph the solution.

1. $8x > 8$

2. $\frac{r}{5} \le 2$

3. $-32 > 1.6h$

4. $\frac{u}{8} \ge 2.1$

5. $1.5j < -6.6$

6. $-\frac{3}{2} < 3x$

Write the word sentence as an inequality. Then solve the inequality.

7. Five times a number is not less than 15.

8. The quotient of a number and 4 is less than -1.

9. An SUV averages 16.5 miles per gallon. The maximum average number of miles that can be driven on a full tank of gas is 363 miles. Write and solve an inequality that represents the number of gallons in a tank.

Solve the inequality. Graph the solution.

10. $-2p \ge 10$

11. $-2 > \frac{v}{-3}$

12. $\frac{g}{-3.2} > 4$

13. $-\frac{y}{3} \le 1.4$

14. $-12 > -9h$

15. $\frac{a}{-3.5} \le -1.7$

16. You are creating a decorative rope that is at least 20 feet long.

a. To create the rope you are using beads that are 6 inches long. Write and solve an inequality that represents the number of beads that you can use.

b. You do not have enough 6-inch beads to make the rope, so you will use 10-inch beads instead. Write and solve an inequality that represents the number of 10-inch beads that you can use.

Name ______________________________ Date __________

4.3 Practice B

Solve the inequality. Graph the solution.

1. $3y \leq \frac{3}{4}$
2. $-3.2 < \frac{p}{10}$
3. $1.6g \geq 0.48$
4. $2.5k < -100$
5. $\frac{s}{3.1} \geq 4.5$
6. $-\frac{4}{5} < 2x$

Write the word sentence as an inequality. Then solve the inequality.

7. A number divided by 5 is at least 4.
8. The product of 2 and a number is at most –6.
9. The solution of $cx \geq -4$ is $x \geq -8$. What is the value of c?

Solve the inequality. Graph the solution.

10. $-6t < 24$
11. $-\frac{2}{5} \leq \frac{u}{-1}$
12. $\frac{q}{-0.4} \leq 1.9$
13. $-\frac{d}{2} > \frac{3}{8}$
14. $-1.2 \leq -0.8r$
15. $\frac{j}{-5.2} \leq -1.5$
16. The height of a room is 10 feet. You are building shelving from the floor to the ceiling.

    a. Each shelf requires 8 inches. Write and solve an inequality that represents the number of shelves that can be made.

    b. You forgot to include the thickness of each shelf in your measurements. The amount of space needed for each shelf is actually 10 inches. Write and solve an inequality that represents the number of shelves that can be made.

Describe all numbers that satisfy *both* inequalities. Include a graph with your description.

17. $3x < 12$ and $-3x < -3$
18. $\frac{y}{5} \leq -2$ and $-\frac{y}{4} \geq 1$

Name____ Date__________

4.3 Enrichment and Extension

Margin of Error

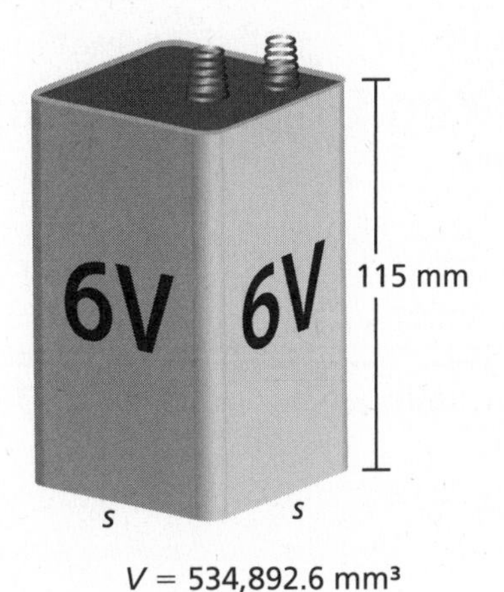

A 6-volt lantern battery has the given dimensions. Any company that manufactures the batteries must make sure their product meets the specifications. The batteries are made using machines. No machine is perfect and so there will always be a slight variation of the size of the batteries.

The allowable difference between the required dimensions of the battery and its actual dimensions is called the *margin of error*.

1. What is the ideal side length of the base of a 6-volt lantern battery?

2. The side length of the base of the battery has a margin of error of 0.002 millimeter. Write an inequality that models the margin of error e of the base's side length.

3. A margin of error in the side length will produce a margin of error in the volume of the battery.

a. What is the smallest side length allowed by the margin of error? What is the volume of a battery with this side length? Round your answer to the nearest thousandth.

b. What is the greatest side length allowed by the margin of error? What is the volume of a battery with this side length? Round your answer to the nearest thousandth.

c. Write an inequality that models the range of acceptable side lengths s of a battery.

d. Write an inequality that models the range of acceptable volumes V of a battery.

4. The side length of a battery is the value you calculated in Exercise 1 and the height is 115.002 millimeters.

a. Find the volume of the battery. Round your answer to the nearest thousandth.

b. Which margin of error has a greater impact on the volume, the side length or the height? Why? Explain your reasoning.

Name ______________________________ Date __________

Puzzle Time

What Do You Do When Your Smoke Alarm Goes Off?

Write the letter of each answer in the box containing the exercise number.

Solve the inequality.

1. $4x < 24$

2. $\frac{x}{6} \geq -3$

3. $-2.3x > 23$

4. $-15 \geq \frac{x}{3}$

5. $\frac{x}{4} > -4.1$

6. $9 \leq -1.5x$

7. $-6x > -\frac{1}{4}$

8. $4.2x \geq -12.6$

9. Three times a number x is at least -18.

10. The quotient of -7 and a number x is less than 8.

Answers

N. $x < -10$

U. $x \leq -6$

R. $x < \frac{1}{24}$

A. $x > -16.4$

I. $x \geq -6$

T. $x < 6$

F. $x \geq -3$

E. $x > -56$

T. $x \leq -45$

R. $x \geq -18$

7	6	3		5	8	1	10	2		9	4

Activity 4.4

Start Thinking!

For use before Activity 4.4

Write a word problem involving using a gift card to make an online purchase. The situation must be able to be solved using a one-step inequality. Exchange problems with a classmate and solve your classmate's problem.

Warm Up

For use before Activity 4.4

Solve the inequality.

1. $x + 9 < 12$

2. $x - 3 \geq 1$

3. $x - 2 > -5$

4. $-3x > 5$

5. $4x > -16$

6. $-5x \leq -21$

Lesson 4.4 Start Thinking!

For use before Lesson 4.4

Explain to a partner how to find the value of x so that the area of the rectangle is more than 36 square units. Justify your answer.

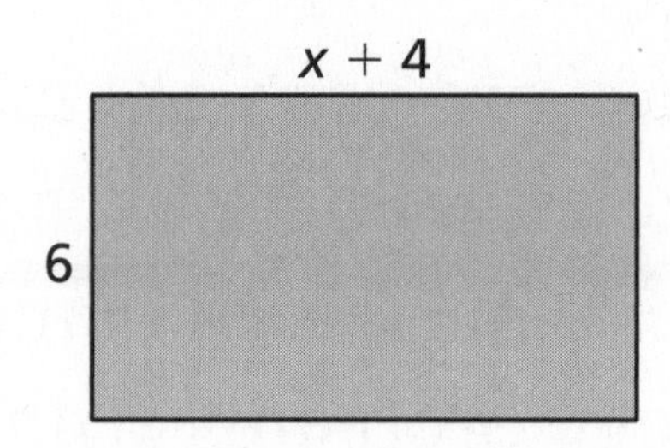

List two values of the variable that satisfy the inequality you wrote above.

Lesson 4.4 Warm Up

For use before Lesson 4.4

Match the inequality with its graph.

1. $\frac{x}{4} + 2 < 1$

A.

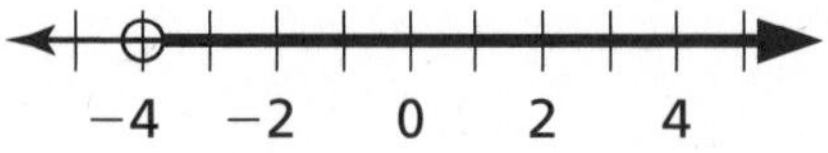

B.

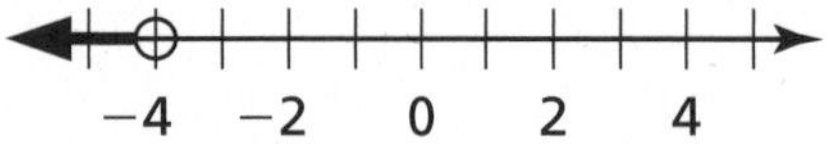

C.

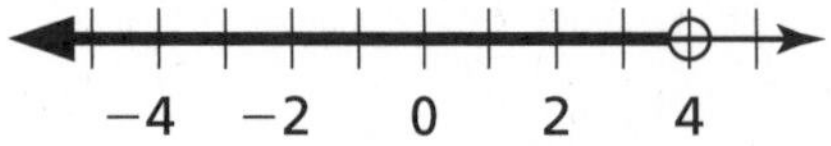

2. $2x - 3 \geq 1$

A.

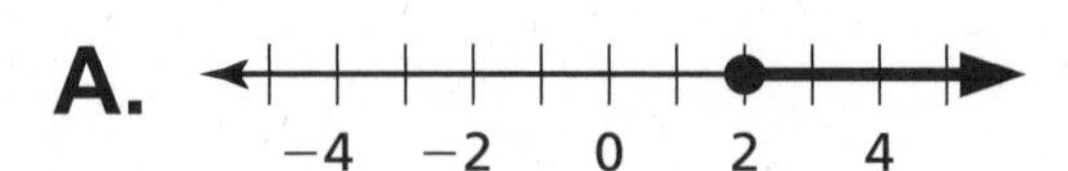

B.

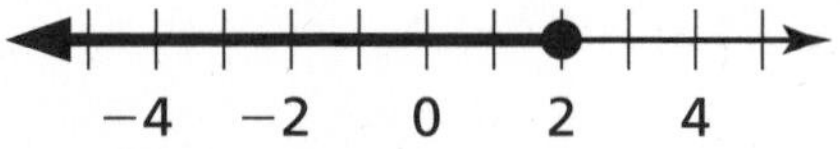

C.

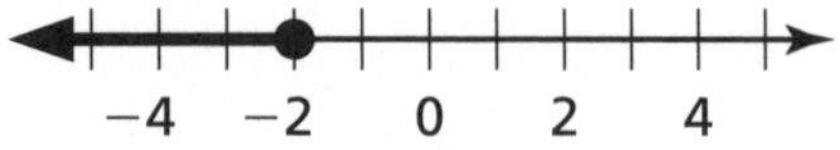

Name________________________________ Date__________

4.4 Practice A

Solve the inequality. Graph the solution.

1. $3m - 7 < 2$

2. $-13 \leq -5r + 2$

3. $2k + \frac{1}{3} > 1$

4. $4.3 - 1.5c \leq 10$

5. You are renting a moving truck for a day. There is a daily fee of \$20 and a charge of \$0.75 per mile. Your budget allows a maximum total cost of \$65. Write and solve an inequality that represents the number of miles you can drive the truck.

Solve the inequality. Graph the solution.

6. $2(b - 4) > -6$

7. $-8(p + 3) \leq 16$

8. $15 \geq \frac{5}{3}(d - 6)$

9. $3.4 < 0.4(a + 12)$

10. Write and solve an inequality that represents the values of x for which the area of the rectangle will be at least 35 square feet.

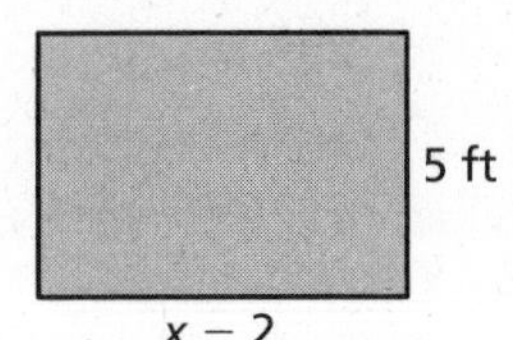

Solve the inequality. Graph the solution.

11. $3x - 7x + 2 < 10 - 12$

12. $14w - 8w - 5.4 \geq 7.3 - 10$

13. Your weekly base salary is \$150. You earn \$20 for each cell phone that you sell.

a. What is the minimum amount you can earn in a week?

b. Write and solve an inequality that represents the number of cell phones you must sell to make at least \$630 a week.

c. Write and solve an inequality that represents the number of cell phones you must sell to make at least \$750 a week.

d. The company policy is that as a part-time employee, the maximum you can earn each week is \$950. Write and solve an inequality that represents the number of cell phones you can sell each week.

Name ______________________________ Date __________

4.4 Practice B

Solve the inequality. Graph the solution.

1. $2 - \frac{q}{3} > 6$

2. $7 \leq 0.5v + 10$

3. $-\frac{1}{4} \leq 4k + \frac{7}{4}$

4. $3.6 - 0.24n < 1.2$

5. An RV park receives $300 per month from each residential site that is occupied as well as $2000 per month from their overnight sites. Write and solve an inequality to find the number of residential sites that must be occupied to make at least $14,000 in revenue each month.

Solve the inequality. Graph the solution.

6. $-5(m - 2) > 30$

7. $-\frac{3}{2}(f + 6) \leq -6$

8. $10.5 < 1.5(p - 3)$

9. $30 \geq -7.5(w - 4.2)$

10. Write and solve an inequality that represents the values of x for which the area of the rectangle will be at most 45 square meters.

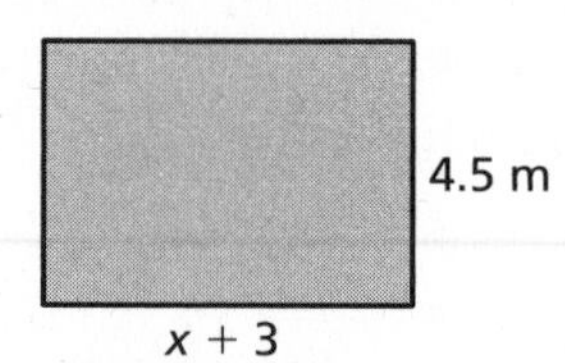

Solve the inequality. Graph the solution.

11. $12x - 5x - 4 \geq 60 - 8$

12. $4v + 6v + 3.2 < 6.8 - 9.2$

13. An animal shelter has fixed weekly expenses of $750. Each animal in the shelter costs an additional $6 a week.

   a. During the summer months, the weekly expenses are at least $1170. Write and solve an inequality that represents the number of animals at the shelter for expenses to be at least $1170 a week.

   b. During the winter months, the weekly expenses are at most $900. Write and solve an inequality that represents the number of animals at the shelter for expenses to be at most $900 a week.

   c. The cost for each animal has increased by $2. What will be the maximum weekly expenses during the winter months?

Name ______________________________ Date __________

4.4 Enrichment and Extension

Solving Multi-Step Inequalities

Solve the inequality. Graph the solution.

1. $4(x + 1) < -6$

2. $2(x - 3) \geq 10$

3. $\frac{1}{2}(x + 28) \leq 11$

4. $\frac{x + 5}{6} > 2$

5. $\frac{2x - 1}{4} \geq 1$

6. $\frac{3x + 2.3}{5} \leq 7$

7. $8.5 < \frac{4 + 10 + x}{2}$

8. $-5(2x + 6) \geq 40$

9. $3(7 - 10x) < 21$

10. You get scores of 85 and 91 on two history tests. Write and solve an inequality to find the scores you can get on your next history test to have an average of at least 90.

11. You and a group of friends wait $\frac{1}{2}$ hour to ride an amusement park ride. You go on the ride a second time and wait $\frac{1}{3}$ hour. You want to go on a third time. Write and solve an inequality to find how many minutes you can wait for your average waiting time to be at most $\frac{1}{3}$ hour. (*Hint:* Convert the waiting times to minutes.)

12. Write and solve an inequality to find the possible values of x so that the rectangle has an area of more than 130 square units.

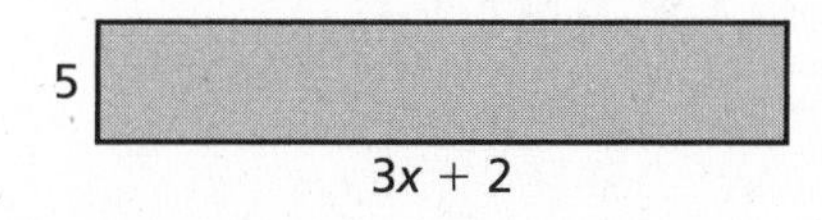

Name ______________________ Date __________

Puzzle Time

What Did Ernie Say When Bert Asked Him If He Wanted Ice Cream?

Write the letter of each answer in the box containing the exercise number.

Solve the inequality.

1. $8x - 11 < 13$
2. $3x - 5 \geq 16$
3. $2 - \dfrac{x}{4} \geq 4$
4. $\dfrac{6}{7} > -2x - \dfrac{8}{7}$
5. $4.6 > 1.2 + 1.7x$
6. $8(x - 4) \geq 40$
7. $-30 \leq -\dfrac{3}{4}(x + 4)$
8. $-6.8 \geq 0.8(x + 1)$

Answers

R. $x < 2$

T. $x \leq 36$

E. $x \geq 7$

S. $x \leq -9.5$

B. $x > -1$

U. $x < 3$

R. $x \leq -8$

E. $x \geq 9$

8	1	3	6		4	2	5	7

Name____ Date__________

Technology Connection

For use after Section 4.4

Using Logical Tests in a Spreadsheet

Expressions involving "greater than," "less than," and "equal to" are called **logical tests** because they can be either True or False. For example, the logical test "10 > 15" returns an answer of "False." The logical test functions of a spreadsheet can be used to compare and categorize numbers.

ACTIVITY Use logical tests to solve the following problem:

A soccer team is in a tournament where each win earns 3 points, a tie earns 1 point, and a loss earns 0 points. Finish the spreadsheet to tally the points for the home team after five games.

	Home Team	Away Team	Win	Loss	Tie
	4	2			
	7	5			
	2	3			
	1	1			
	5	0			
			Total Points	________	

SOLUTION

Step 1 Copy the chart above into a spreadsheet.

Step 2 In cell C2, enter: **=IF(A2>B2, 3, "")**. The "IF" function performs the logical test that asks if the number in cell A2 is greater than the number in cell B2. If it is true, then the cell will display a "3." If it is false, the double quotation marks will keep that cell blank. Next, copy and paste the formula from cell C2 to cells C3, C4, C5, and C6. Your spreadsheet should change the formulas to correspond with the correct row.

Step 3 In cell D2, enter: **=IF(A2<B2, 0, "")**. This formula puts a zero in the loss column if the score in cell A2 is less than the score in cell B2. Copy and paste the formula down the column.

Step 4 In cell E2, enter: **=IF(A2=B2, 1, "")**. This formula will find and award a single point for a tie. Copy and paste the formula down the column.

Step 5 Lastly, to find the sum of all your points, in cell D7 enter: **=SUM(C2:E6)**.

So, the home team earned ______ points.

Chapter 5

Name ______________________ Date __________

Chapter 5 Ratios and Proportions

Dear Family,

An emergency evacuation plan is required in most commercial buildings. A good plan shows the locations of exits in the building as well as the locations of fire extinguishers and other emergency equipment. The plan is usually shown on a scale drawing of the building's floor plan.

Creating an emergency evacuation plan for your home is a good idea as well. You and your student can work together to make a scale drawing of your home.

Choose a scale that will make measurements relatively easy and allow the plan to fit on a single piece of paper. A common scale is $\frac{1}{4}$ inch for every foot. This scale will allow a building as large as 34 x 44 feet to fit on a letter-sized piece of paper. If your home won't fit within those dimensions, you can choose larger paper or a smaller scale—such as $\frac{1}{8}$ inch for every foot.

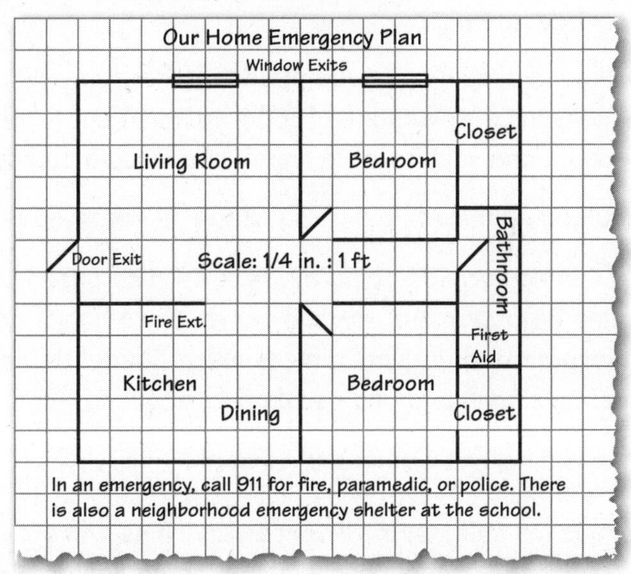

Mark the exits, fire extinguishers, and any alarms in red. If you have emergency medical equipment available, such as a first aid kit, mark those in blue.

Ask your student to help you with the following.

- Make measurements of each room in the home. Include measurements of doors and windows that will work as exits.
- Convert your measurements to the scale you have chosen.
- Draw the plan on $\frac{1}{4}$-inch graph paper. The sides of each square have a length of $\frac{1}{4}$-inch which makes it easier to use for $\frac{1}{4}$-inch and $\frac{1}{8}$-inch scales.

It's a good idea to include other information on the plan as well, such as the numbers for fire, paramedic, and police services (provided by 911 in many towns). Work with your student to decide what information should be included.

Being prepared will give you great peace of mind!

Nombre ____ Fecha __________

Capítulo 5 Razones y Proporciones

Estimada Familia:

En la mayoría de edificios comerciales se requiere un plano de evacuación de emergencia. Un buen plano muestra las ubicaciones de las salidas del edificio, así como las ubicaciones de los extintores de incendios y otros equipos de emergencia. El plano generalmente se muestra en un dibujo a escala del plano del piso del edificio.

Crear un plano de evacuación de emergencia para su hogar también puede ser una buena idea. Usted y su estudiante pueden trabajar juntos para hacer un dibujo a escala de su hogar.

Elijan una escala en la que puedan hacerse medidas de manera relativamente fácil y permita que el plano quepa en una sola hoja de papel. Una escala común es $\frac{1}{4}$ pulgada por cada pie. Esta escala permite que un edificio tan grande como de 34 x 44 pies quepa dentro de una hoja de papel tamaño carta. Si su hogar no cupiera dentro de esas dimensiones, puede elegir un papel más grande o una escala más pequeña—como por ejemplo $\frac{1}{8}$ pulgada por cada pie.

Marque las salidas, extintores y alarmas en rojo. Si tiene disponible equipo médico de emergencia, como por ejemplo un equipo de primeros auxilios, márquelo en azul.

Pida a su estudiante que lo ayude con lo siguiente:

- Hagan medidas de cada habitación de la casa. Incluyan medidas de las puertas y ventanas, ya que funcionarán como salidas.
- Conviertan sus medidas en la escala elegida.
- Dibujen el plano en un papel para gráficos de $\frac{1}{4}$ pulgada. Los lados de cada cuadrado miden $\frac{1}{4}$ pulgada de largo, lo que los hace más fácil de usar con las escalas de $\frac{1}{4}$ pulgada y $\frac{1}{8}$ pulgada.

Es una buena idea incluir también otra información en el plano, como los teléfonos de los servicios de bomberos, paramédicos y policías (proporcionados por 911 en muchas ciudades). Trabaje con su estudiante para decidir qué información debe incluirse.

¡Estar preparado le dará una gran tranquilidad!

Activity 5.1 Start Thinking!

For use before Activity 5.1

You are planning a trip to Atlanta, Georgia. Discuss with a partner the factors you need to consider to determine if it would be better to fly or to drive. Determine which method of transportation you would use and explain why.

Activity 5.1 Warm Up

For use before Activity 5.1

Convert the measurement.

1. 30 min = __?__ h

2. 4 h = __?__ min

3. 15 sec = __?__ min

4. 60 h = __?__ days

5. 3 days = __?__ h

6. 1 wk = __?__ h

Lesson 5.1 Start Thinking!

For use before Lesson 5.1

You and your sister go to a store. You buy 3 erasers for \$1.00. Your sister buys 4 erasers for \$1.25. Explain who gets the better deal.

Lesson 5.1 Warm Up

For use before Lesson 5.1

Find the product. List the units.

1. $6 \text{ h} \times \frac{\$7}{\text{h}}$

2. $2 \text{ gal} \times \frac{\$24}{\text{gal}}$

3. $8 \text{ h} \times \frac{25 \text{ mi}}{\text{h}}$

4. $9 \text{ mo} \times \frac{\$650}{\text{mo}}$

5. $12 \text{ lb} \times \frac{\$2.50}{\text{lb}}$

6. $6 \text{ yr} \times \frac{35 \text{ in.}}{\text{yr}}$

Name________________________________ Date__________

5.1 Practice A

Find the product. List the units.

1. $12\text{ h} \times \frac{\$5}{\text{h}}$ **2.** $6\text{ oz} \times \frac{\$0.59}{\text{oz}}$ **3.** $9\text{ h} \times \frac{70\text{ mi}}{\text{h}}$

Write the ratio as a fraction in simplest form.

4. 12 to 15 **5.** 24 : 9 **6.** 14 tetras : 6 angelfish

Find the unit rate.

7. 360 miles in 6 hours **8.** 18 bowlers on 6 lanes **9.** $28 for 7 people

Use the ratio table to find the unit rate with respect to the specified units.

10. Laps per minute

Minutes	0	2	4	6
Laps	0	1	2	3

11. Grams of protein per serving

Servings	0	1	2	3
Grams of Protein	0	15	30	45

12. At 9 A.M. you have run 2 miles. At 9:24 A.M. you have run 5 miles. What is your running rate in minutes per mile?

13. Are the two statements equivalent? Explain your reasoning.

- The ratio of orange to blue is 3 to 4.
- The ratio of blue to orange is 12 to 9.

14. There are 234 students in 9 different classrooms. What is the ratio of students to classrooms?

15. Dishwasher detergent is sold in individual packs. It is sold in 20-, 60-, and 90-pack containers.

a. Which container do you think has the lowest unit rate of dollars per pack? Why?

b. The 20-pack container sells for $5.49. What is the unit rate in dollars per pack? Round your answer to the nearest cent.

c. The 60-pack container sells for $10.97. What is the unit rate in dollars per pack? Round your answer to the nearest cent.

d. The 90-pack container sells for $18.95. What is the unit rate in dollars per pack? Round your answer to the nearest cent.

e. Which container has the lowest unit rate? How does this compare with your answer in part (a)?

Name ______________________ Date ________

5.1 Practice B

Write the ratio as a fraction in simplest form.

1. 35 to 63
2. 10.8 seconds : 36 feet
3. 198 women to 110 men
4. 1000 songs : 2 megabytes
5. 26.1 miles : 3.6 hours
6. 12 completions to 28 attempts

Find the unit rate.

7. $5.40 for 24 cans
8. $1.29 for 20 ounces
9. 50 meters in 27.5 seconds

10. There are 16 bacteria in a beaker. Four hours later there are 228 bacteria in the beaker. What is the rate of change per hour in the number of bacteria?

11. The table shows nutritional information for three energy bars.

Energy Bar	Calories	Protein	Fiber	Sugar
A	220	20 g	12 g	14 g
B	130	12 g	8 g	10 g
C	140	4 g	9 g	9 g

   a. Which has the most protein per calorie?

   b. Which has the least sugar per calorie?

   c. Which has the highest rate of sugar to fiber?

   d. Compare bar A with bar B. Which nutritional item do you think has the highest ratio: calories, protein, fiber, or sugar?

   e. Calculate the ratios in part (d). Which one has the highest ratio?

12. The graph shows the cost of buying scoops of gelato.

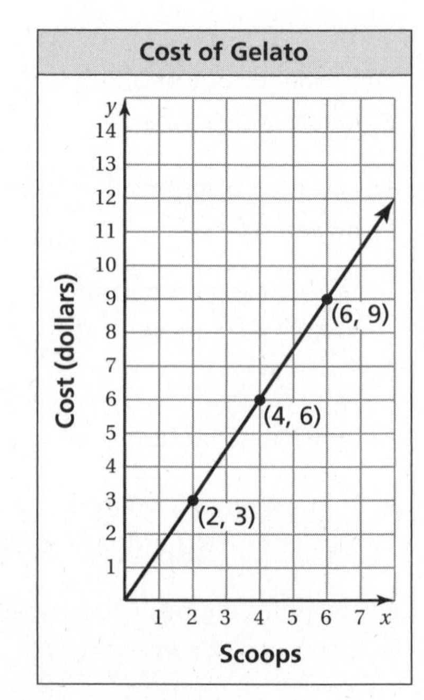

   a. What does the point $(4, 6)$ represent?

   b. What is the unit cost?

   c. What is the cost of 12 scoops?

   d. Explain how the graph would change if the unit rate was $1.75 per scoop.

   e. How would the coordinates of the point in part (a) change if the unit rate was $1.75 per scoop?

Write a situation for the ratio.

13. $\frac{9}{5}$

14. 2 : 3

Name______________________________ Date__________

5.1 Enrichment and Extension

Ratios, Rates, and On-the-Job Decisions

Sally Smith has been offered two employment opportunities. Help her decide which job is better by answering the following questions.

1. Job A will pay $32,448 per year. How much money would Sally be making per hour? Assume pay is based on fifty-two 40-hour weeks.

2. Job B pays $14.80 per hour. How much money would she make in a year? Assume the pay is based on fifty-two 40-hour weeks.

3. Which job pays better? Explain your reasoning.

4. Sally lives 18 miles from Job A. A work week is five days. How many miles would she have to drive each week just to get to and from work?

5. For Job A, Sally would have to put 7.5 gallons of gas in her tank every 3 days that she drives to and from work only. How many gallons of gas would she use in a five-day work week?

6. Based on your answers to Exercises 4 and 5, what is Sally's gas mileage? (*Hint:* Gas mileage is a unit rate that is calculated as miles per gallon.)

7. If gas costs $4 per gallon, how much will gas cost Sally per day just to get to and from Job A per day? per week? per year?

8. The ratio of the length of Sally's drive to Job A to the length of Sally's drive to Job B is 2 : 3. How much money would she save on gas in a week with Job A as opposed to Job B?

9. Other than distance and cost of gas, what other factors should Sally consider when comparing commutes to work?

10. Macaroni and cheese is one of Sally's favorite lunch foods. In the cafeteria at Job A, $1\frac{1}{4}$ cups of macaroni and cheese contains 6.5 grams of fat. The macaroni and cheese at Job B contains 10 grams of fat per pint. Which one has a lower fat content? Explain your reasoning.

11. Both places sell macaroni and cheese by weight. Job A's cafeteria charges $7.25 per pound and Job B's cafeteria charges $0.58 per ounce. Which cafeteria has the cheaper macaroni and cheese? Explain your reasoning.

12. Based on the information on this page, would you recommend that Sally take Job A or Job B? Explain your reasoning.

13. Other than pay, transportation costs, and cafeteria selections, what other factors should Sally consider when choosing a job?

Name ______________________ Date ________

5.1 Puzzle Time

What Do You Get If You Cross A Duck With A Firework?

Write the letter of each answer in the box containing the exercise number.

Find the product.

1. $4 \text{ tbsp} \times \frac{20 \text{ cal}}{\text{tbsp}}$

P. 5 cal **Q.** 80 cal **R.** 5 tbsp

2. $3 \text{ lb} \times \frac{\$1.29}{\text{lb}}$

E. \$0.43 **F.** \$3.87 **G.** 3.87 lb

3. $4 \text{ gal} \times \frac{17.5 \text{ mi}}{\text{gal}}$

C. 70 mi **D.** 35 mi **E.** 70 gal

4. $40 \text{ h} \times \frac{\$8.50}{\text{h}}$

T. 340 h **U.** \$340 **V.** \$300

Write the ratio as a fraction.

5. 12 to 36

I. $\frac{1}{6}$ **J.** $\frac{1}{4}$ **K.** $\frac{1}{3}$

6. 15 : 75

E. $\frac{1}{5}$ **F.** $\frac{1}{4}$ **G.** $\frac{1}{3}$

7. 10 out of 25

E. $\frac{2}{5}$ **F.** $\frac{1}{5}$ **G.** $\frac{1}{25}$

8. 52 males to 28 females

P. $\frac{27}{14}$ **Q.** $\frac{7}{13}$ **R.** $\frac{13}{7}$

Find the unit rate.

9. 48 cups in 12 quarts

R. $\frac{4 \text{ c}}{\text{qt}}$ **S.** $\frac{3 \text{ c}}{\text{qt}}$ **T.** $\frac{2 \text{ c}}{\text{qt}}$

10. \$17.85 for 3 pounds

A. $\frac{\$5.95}{\text{lb}}$ **B.** $\frac{\$5.56}{\text{lb}}$ **C.** $\frac{\$3.65}{\text{lb}}$

11. 26.2 miles in 4 hours

A. $\frac{6.55 \text{ mi}}{\text{h}}$ **B.** $\frac{7.5 \text{ mi}}{\text{h}}$ **C.** $\frac{6.5 \text{ mi}}{\text{h}}$

12. \$12.60 for 3 boxes

G. $\frac{\$3.15}{\text{box}}$ **H.** $\frac{\$4.15}{\text{box}}$ **I.** $\frac{\$4.20}{\text{box}}$

10		2	12	8	6		1	4	11	3	5	7	9
						-							

Activity 5.2 Start Thinking!

For use before Activity 5.2

Give an example of two fractions that are equivalent. Show that they are equivalent. Give an example of two fractions that are not equivalent. Explain why they are not equivalent.

Activity 5.2 Warm Up

For use before Activity 5.2

Write the fraction in simplest form. Do not change improper fractions to mixed numbers.

1. $\frac{24}{48}$

2. $\frac{64}{48}$

3. $\frac{30}{24}$

4. $\frac{9}{21}$

5. $\frac{45}{63}$

6. $\frac{10}{41}$

7. $\frac{144}{12}$

8. $\frac{50}{35}$

Lesson 5.2 Start Thinking!

For use before Lesson 5.2

You go to the grocery store to buy cereal. How do you determine which cereal brand is the better buy?

Lesson 5.2 Warm Up

For use before Lesson 5.2

Tell whether the two rates form a proportion.

1. 5 feet in 4 hours; 15 feet in 12 hours

2. 8 pages in 40 minutes; 15 pages in 70 minutes

3. 3 pounds for $3.75; 5 pounds for $6.50

4. 2 cups in 4 servings; 5 cups in 10 servings

Name______________________________ Date__________

5.2 Practice A

Tell whether the ratios form a proportion.

1. $\frac{1}{4}, \frac{3}{12}$

2. $\frac{1}{7}, \frac{4}{28}$

3. $\frac{2}{5}, \frac{30}{80}$

4. $\frac{18}{24}, \frac{15}{20}$

5. $\frac{35}{16}, \frac{5}{2}$

6. $\frac{5}{7}, \frac{35}{49}$

7. $\frac{15}{21}, \frac{40}{56}$

8. $\frac{33}{63}, \frac{26}{42}$

9. $\frac{54}{10}, \frac{81}{15}$

Tell whether the two rates form a proportion.

10. 8 feet in 15 seconds; 16 feet in 40 seconds

11. 28 people in 4 rooms; 63 people in 9 rooms

12. 14 girls to 6 boys; 35 girls to 15 boys

13. 45 marbles in 9 bags; 150 marbles in 36 bags

14. You can run 4 laps in 10 minutes. Your friend can run 6 laps in 15 minutes. Are these rates proportional? Explain.

Tell whether the ratios form a proportion.

15. $\frac{7}{4}, \frac{17.5}{10}$

16. $\frac{1.5}{6}, \frac{2}{8}$

17. $\frac{8}{5}, \frac{68}{45}$

18. You get $27 to spend at the mall for doing 6 chores. Your friend gets $36 for doing 8 chores.

a. What is your pay rate?

b. What is your friend's pay rate?

c. Are the pay rates equivalent? Explain.

19. You can buy 4 tickets for $75 or 5 tickets for $94. Are the costs proportional? If not, rewrite one of the rates so the costs are proportional.

20. A recipe requires a ratio of 4 potatoes to 6 carrots. You accidentally use 5 potatoes with 6 carrots. What is the least number of potatoes and carrots that you can add to get the correct ratio of potatoes to carrots?

Name ______________________________ Date __________

5.2 Practice B

Tell whether the ratios form a proportion.

1. $\frac{25}{16}, \frac{65}{56}$

2. $\frac{30}{75}, \frac{24}{60}$

3. $\frac{27}{48}, \frac{108}{192}$

Tell whether the two rates form a proportion.

4. \$24 for 16 burgers; \$15 for 10 burgers

5. 10 used books for \$4.50; 15 used books for \$6.00

6. 125 horsepower motor for an 18-foot boat; 225 horsepower motor for a 32-foot boat

Tell whether the ratios form a proportion.

7. $\frac{28.5}{42}, \frac{19}{28}$

8. $\frac{3.5}{4}, \frac{11.9}{13.6}$

9. $\frac{124}{98}, \frac{315}{225}$

10. The seventh-grade band has 15 drummers and 12 trumpet players. The eighth-grade band has 10 drummers and 8 trumpet players. Do the ratios form a proportion? Explain.

11. One mixture contains 6 fluid ounces of water and 10 fluid ounces of vinegar. A second mixture contains 9 fluid ounces of water and 12 fluid ounces of vinegar. Are the mixtures proportional? If not, how much water or vinegar would you add to the second mixture so that they are proportional?

12. A wholesale warehouse buys pairs of sandals to sell.

a. The warehouse can purchase 5 pairs of sandals for \$65. What is the cost rate?

b. The warehouse can purchase 8 pairs of sandals for \$96. What is the cost rate?

c. The warehouse can purchase 10 pairs of sandals for \$126.50 and will get one free pair. What is the cost rate?

d. Are any of the cost rates proportional? Explain.

e. Your buyer is to purchase 40 pairs of sandals. Use any combination of parts (a), (b), and (c) for your buyer to purchase the 40 pairs of sandals at the lowest possible cost.

Find the value of *x* so that the ratios form a proportion.

13. $\frac{3}{7}, \frac{x}{21}$

14. $\frac{16}{12}, \frac{20}{x}$

Name____ Date__________

5.2 Enrichment and Extension

The Big Paint "Mix Up"

Oh no! The paint mixing machine has gone crazy! Can you help figure out how to fix these batches by adding the least amount of paint possible? Otherwise all the paint will go to waste. For each situation, tell how much of one or more colors should be added to each batch. The machine can only measure in cups, pints, quarts, and gallons.

1. *Brick Red* is supposed to be 7 parts red to 2 parts blue. Today, the machine mixed 35 quarts of red with 5 quarts of blue.

2. *Ocean Blue* is supposed to be 8 parts blue to 3 parts yellow. Today, the machine mixed 55 quarts of blue with 15 quarts of yellow.

3. *Sour Apple* is supposed to be 9 parts yellow to 2 parts blue. Today, the machine mixed 2 gallons of yellow with 2 quarts of blue.

4. *Midnight Navy* is supposed to be 7 parts blue to 3 parts red. Today, the machine mixed 3 quarts of blue with 1 pint of red.

5. *Sunset Yellow* is supposed to be 6 parts yellow to 1 part red. Today, the machine mixed 70 pints of yellow with 11 pints of red.

6. The last correct batch of *Burnt Orange* had 15 quarts of yellow and 30 quarts of red. Today, the machine mixed 42 gallons of yellow with 72 gallons of red.

7. The last correct batch of *Perfectly Grape* had 20 pints of red and 24 pints of blue. Today, the machine mixed 28 quarts of red with 32 quarts of blue.

8. The last correct batch of *Mustard Brown* had 96 pints of yellow, 72 pints of red, and 24 pints of blue. Today the machine mixed 45 cups of yellow, 20 cups of red, and 8 cups of blue.

Name ______________________________ Date __________

5.2 Puzzle Time

What Can You Hold Without Ever Touching?

For each exercise, circle the letter in the columns under Yes or No to indicate the correct answer. The circled letters will spell the answer to the riddle.

	Yes	No
1.	Y	T
2.	A	O
3.	U	H
4.	R	S
5.	C	B
6.	O	R
7.	E	O
8.	A	L
9.	T	M
10.	S	H

Tell whether the ratios form a proportion.

1. $\frac{2}{5}, \frac{8}{20}$

2. $\frac{3}{7}, \frac{6}{13}$

3. $\frac{5}{6}, \frac{15}{18}$

4. $\frac{18}{24}, \frac{12}{16}$

Tell whether the two rates form a proportion.

5. 55 miles in 1 hour; 450 miles in 8 hours

6. $3.00 for 32 ounces of strawberries; $1.75 for 24 ounces of strawberries

7. 45 baskets in 85 shots; 54 baskets in 102 shots

8. 18 push-ups in 60 seconds; 27 push-ups in 90 seconds

9. One type of cereal has 2 grams of protein per 1-cup serving. Another cereal has 1 gram of protein per half-cup serving. Do these rates form a proportion?

10. A 50-fluid ounce bottle of laundry detergent washes 32 loads of laundry. A 100-fluid ounce bottle washes 60 loads of laundry. Are they proportional? Do these rates form a proportion?

Extension 5.2 Start Thinking!

For use before Extension 5.2

Your friend spent \$7 in 2 days on lunch. He says that he will need \$21 for 6 days to spend on lunch. Is your friend correct? Explain your reasoning.

Extension 5.2 Warm Up

For use before Extension 5.2

Tell whether *x* and *y* are proportional.

1.

x	1	2	3	4
y	4	8	12	16

2.

x	2	3	4	5
y	2	4	6	8

Name ________________________________ Date ________

Extension 5.2 Practice

Interpret each plotted point in the graph of the proportional relationship.

1. 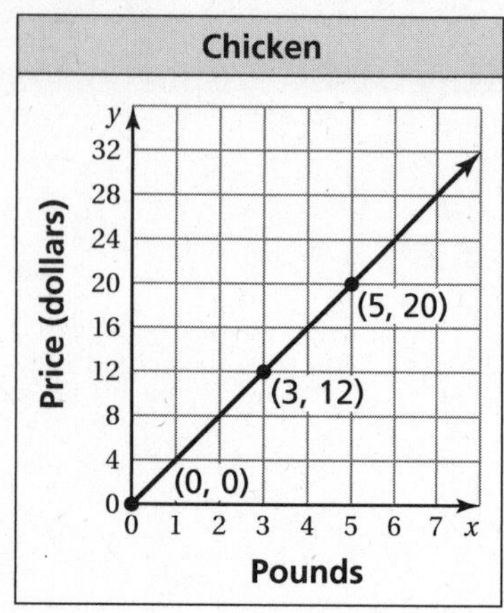

2. 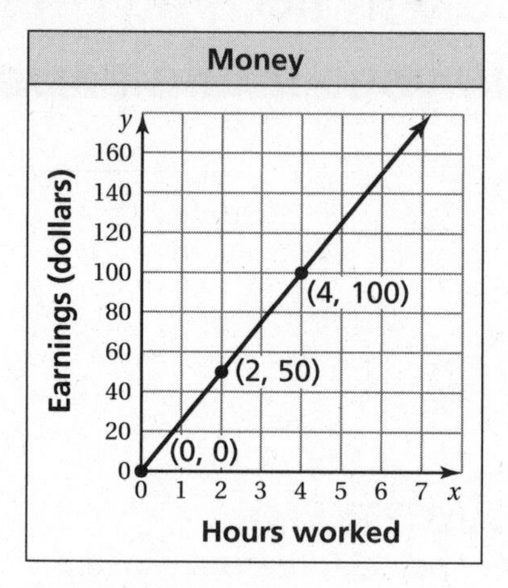

The graph of a proportional relationship passes through the given points. Find *y*.

3. $(4, 8), (1, y)$

4. $(3, 21), (1, y)$

5. $(1.5, 9), (1, y)$

6. $(3.5, 14), (1, y)$

7. Two classes have car washes to raise money for class trips. A portion of the earnings will pay for using the two locations for the car washes. The graph shows that the trip earnings of the two classes are proportional to the car wash earnings.

   a. Express the trip earnings rate for each class as a percent.

   b. What trip earnings does Class A receive for earning $75 from the car wash?

   c. How much less does Class B receive than Class A for earning $75 from the car wash?

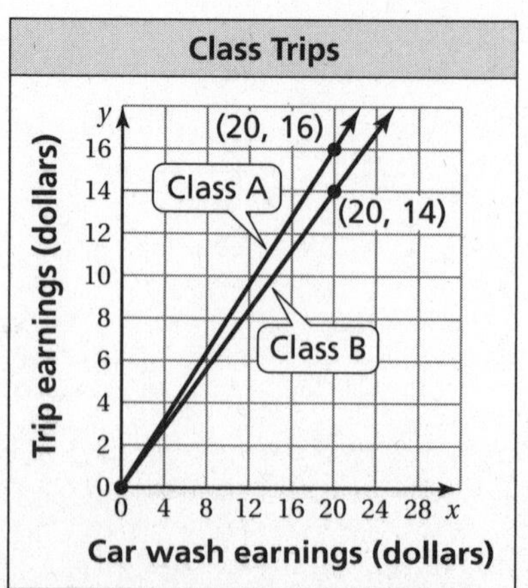

Activity 5.3 Start Thinking!

For use before Activity 5.3

How do you determine if two ratios are proportional? Explain giving an example of ratios that are proportional.

Activity 5.3 Warm Up

For use before Activity 5.3

Write two equivalent ratios for the ratio given.

1. $\frac{10}{15}$

2. $\frac{11}{20}$

3. $\frac{6}{14}$

4. $\frac{4}{10}$

5. $\frac{12}{20}$

6. $\frac{6}{15}$

7. $\frac{8}{9}$

8. $\frac{14}{24}$

Lesson 5.3 Start Thinking!

For use before Lesson 5.3

You scored 15 points in 3 minutes. Your cousin scored 25 points in 5 minutes.

Are the scores proportional? Explain.

Lesson 5.3 Warm Up

For use before Lesson 5.3

Write a proportion to find how many points a student needs to have on the test to get the given score.

1. test worth 100 points; test score of 85%
2. test worth 50 points; test score of 74%
3. test worth 25 points; test score of 80%
4. test worth 110 points; test score of 90%

Name____ Date__________

5.3 Practice A

Write a proportion to find how many points a student needs to earn on the test to get the given score.

1. test worth 70 points; test score of 90%

2. test worth 30 points; test score of 72%

Write a proportion to find how many free throws a player needs to get the given score.

3. 15 free-throw attempts; free-throw score of 60%

4. 24 free-throw attempts; free-throw score of 75%

Use the table to write a proportion.

5.

	August	September
Hurricanes	2	1
Storms	6	n

6.

	Day 1	Day 2
Wins	w	8
Races	21	12

7. The county requires 2 teachers for every 45 students. Write a proportion that gives the number t of teachers needed for 315 students.

Solve the proportion.

8. $\frac{2}{3} = \frac{a}{15}$

9. $\frac{4}{7} = \frac{44}{m}$

10. $\frac{d}{6} = \frac{72}{48}$

11. A paint color requires the ratio of green paint to yellow paint to be 4 : 9.

a. A container of this paint has 36 pints of yellow paint. Write a proportion that gives the number g of pints of green paint in the container.

b. How many pints of green paint are in the container?

c. How many *gallons* of paint are in the container altogether?

12. An orchestra has 10 cellists.

a. There are 3 violin players for every cellist in the orchestra. How many violin players are there?

b. There are 6 viola players for every 5 cellists in the orchestra. How many viola players are there?

c. What is the ratio of viola players to violin players? Give your answer in simplest form.

13. Give two possible pairs of values for p and q: $\frac{2}{5} = \frac{p}{q}$.

Name ______________________________ Date __________

5.3 Practice B

In Exercises 1 and 2, write a proportion to find how many strikes a bowler needs to get the given score.

1. 32 strike attempts; strike score of 75%

2. 80 strike attempts; strike score of 95%

3. Describe and correct the error in writing the proportion.

X

	Day 1	Day 2
Length	3.1	15.5
Height	h	45

$$\frac{15.5}{h} = \frac{3.1}{45}$$

4. There are 3 referees for every 16 players. Write a proportion that gives the number of referees r for 128 players.

Solve the proportion.

5. $\frac{5}{12} = \frac{x}{36}$

6. $\frac{20}{3.4} = \frac{800}{y}$

7. $\frac{2.8}{r} = \frac{70}{3}$

8. $\frac{48}{15} = \frac{k}{37.5}$

9. $\frac{21}{p} = \frac{252}{78}$

10. $\frac{2.3}{1.6} = \frac{46}{w}$

11. A recipe calls for $\frac{3}{4}$ cup of sugar and $\frac{1}{2}$ cup of brown sugar. You are reducing the recipe. You will use $\frac{1}{6}$ cup of brown sugar. How much sugar will you use?

12. A calculator has 50 keys in five colors: gray, black, blue, yellow, and green.

a. There are 6 gray keys for every 7 blue keys. Write the possible ratios for gray to blue keys.

b. There are 6 gray keys for every 11 black keys. Write the possible ratios for gray to black keys.

c. There are 6 gray keys for every 11 black keys. Also, the number of black keys is 2 less than twice the number of gray keys. Use your answer to part (b) to determine how many gray keys and how many black keys there are.

d. There is 1 yellow key for every 1 green key. How many keys of each color are there?

Name____ Date__________

5.3 Enrichment and Extension

Using Slope to Design Handicap Ramps

According to the Americans with Disabilities Act, the maximum rise to run ratio for all handicap ramps is 1 : 12. Also, there must be a landing, or flat rest area, at the top and bottom of all ramps. These landings must be at least 60 inches in length. If there is a change in direction on the ramp, there must be a flat landing that is at least 60 inches by 60 inches.

1. A ramp with a rise of 5 inches must have a run of at least how many inches? how many feet?

2. Why do you think there is a maximum slope for handicap ramps?

3. What would be the advantages and disadvantages of making a ramp that is less steep?

4. Your school is being renovated and a new handicap ramp must be built. The bottom of the new front door is 20 inches above the sidewalk. What is the minimum length (in feet) of a straight ramp to this door, including the landings at the top and bottom?

5. An existing ramp on the old part of the school has a rise of 10.5 inches and a run of 12 feet 3 inches, not including the landings. Does it meet the requirements or will it have to be changed? Explain.

6. At the main bus entrance, there is a ramp with a change in direction. It has a sloping part with a rise of 5 inches and a run of 6 feet 8 inches, a landing where there is a 90 degree turn, and another sloping part with a rise of 7.5 inches and a run of 13 feet 9 inches. Which part of the ramp is steeper? Why might the ramp have been designed in this way?

7. Another ramp with an overall rise of 30 inches has to be built to replace the ramp described in Exercise 6. It must have two sloping parts that should have the same slopes as the two sloping parts of the ramp in Exercise 6. Both parts should have a rise of 15 inches, with a landing in between. Find the lengths of both sloping parts of the ramp.

8. Find a place in your school, home, neighborhood, or elsewhere that could use a ramp. Measure the rise of that area. Then design a ramp, and draw a picture or describe it in detail. Be sure to tell how long each portion of the ramp should be as well as the overall length.

Name ______________________________ Date __________

5.3 Puzzle Time

Who Do Whales Go To See When Their Teeth Need To Be Fixed?

Write the letter of each answer in the box containing the exercise number.

Solve the proportion.

1. $\frac{3}{5} = \frac{h}{20}$

2. $\frac{4}{7} = \frac{24}{a}$

3. $\frac{1}{x} = \frac{5}{45}$

4. $\frac{2}{13} = \frac{m}{39}$

5. $\frac{t}{28} = \frac{3}{4}$

6. $\frac{18}{21} = \frac{6}{w}$

7. $\frac{3.4}{4.2} = \frac{r}{21}$

8. $\frac{1.5}{2.5} = \frac{6}{s}$

9. $\frac{q}{1.7} = \frac{16}{17}$

10. $\frac{2.2}{n} = \frac{44}{66}$

11. You need 3 tickets for one go-kart ride. How many tickets do you need for five go-kart rides?

12. Yesterday you downloaded 3 songs for $2.97. How many songs did you download today for $3.96?

13. There are 32 students in the school play. The ratio of girls to all students in the play is 5 : 8. How many girls are in the play?

14. Two out of three vehicles in a parking lot are SUVs. There are 18 SUVs in the parking lot. How many vehicles are in the parking lot?

Answers

I. 27	**T.** 10
O. 12	**H.** 9
T. 1.6	**N.** 17
D. 4	**C.** 20
E. 7	**T.** 3.3
A. 42	**R.** 21
S. 6	**O.** 15

9	3	6		11	5	13	2	-	12	1	7	10	14	4	8

Start Thinking!

For use before Activity 5.4

Give a real-world example of things that are proportional.

Warm Up

For use before Activity 5.4

Find the unit rate.

1. 140 students in 4 buses

2. 605 miles in 5 hours

3. 312 pages in 4 hours

4. 54 seats in 2 rows

5. 1023 miles in 3 hours

6. 544 pages in 4 days

Lesson 5.4 Start Thinking!

For use before Lesson 5.4

Describe how someone who works in a bakery might use proportions.

Lesson 5.4 Warm Up

For use before Lesson 5.4

Use the Cross Products Property to solve the proportion.

1. $\frac{a}{4} = \frac{15}{12}$

2. $\frac{11}{4} = \frac{m}{16}$

3. $\frac{2}{3} = \frac{v}{18}$

4. $\frac{8}{n} = \frac{11}{22}$

5. $\frac{t}{10} = \frac{15}{25}$

6. $\frac{10}{3} = \frac{8}{k}$

Name____ Date__________

5.4 Practice A

Use multiplication to solve the proportion.

1. $\frac{7}{4} = \frac{y}{28}$

2. $\frac{d}{48} = \frac{3}{4}$

3. $\frac{j}{8} = \frac{35}{56}$

Use the Cross Products Property to solve the proportion.

4. $\frac{14}{21} = \frac{b}{9}$

5. $\frac{10}{p} = \frac{6}{9}$

6. $\frac{55}{4} = \frac{h}{6}$

7. Eighteen oranges are packaged in 3 containers. How many oranges are packaged in 7 containers?

8. It costs $270 for 3 people to go on a fishing trip. How much does it cost for 10 people to go on the fishing trip?

Solve the proportion.

9. $\frac{3x}{10} = \frac{9}{4}$

10. $\frac{5x}{3} = \frac{80}{12}$

11. $\frac{7}{2} = \frac{x+1}{6}$

12. Tell whether the statement is *true* or *false*. Explain.

$$\text{If } \frac{p}{q} = \frac{3}{5}, \text{ then } \frac{5}{p} = \frac{3}{q}.$$

13. The dimensions of a miniature model are proportional to the dimensions of the actual building.

a. A wall that is 12 feet high on the building is 36 centimeters high on the model. Find the height on the model of a door that is 9 feet high on the building.

b. Use a different method than the one you used in part (a) to find the number of centimeters on the model for a window that is 3 feet wide.

14. The ratio of men to women at a lecture is 2 to 5. A total of 63 people are at the lecture. How many are men? Explain how you found your answer.

15. The distance traveled (in feet) is proportional to the number of seconds. Find the values of x, y, and z.

Feet	3	x	15	z
Seconds	5	65	y	3.5

16. You train for a race by running at a speed of 6 miles per hour.

a. At this speed, how many *minutes* does it take you to run 3.2 miles?

b. On race day, you run 3.2 miles in 30 minutes. What is your speed in miles per hour?

Name ______________________________ Date __________

5.4 Practice B

Use either multiplication or the Cross Products Property to solve the proportion.

1. $\frac{16}{9} = \frac{q}{36}$

2. $\frac{57}{r} = \frac{38}{18}$

3. $\frac{50}{14} = \frac{w}{98}$

4. $\frac{n}{121} = \frac{52}{22}$

5. $\frac{96}{45} = \frac{b}{20}$

6. $\frac{24}{f} = \frac{15}{36}$

7. Three shirts cost \$9.99. How much does it cost for 8 shirts?

Solve the proportion.

8. $\frac{8x}{13} = \frac{64}{52}$

9. $\frac{c - 3}{6} = \frac{7}{3}$

10. $\frac{20}{9} = \frac{10}{s + 2}$

11. The number of grams of protein is proportional to the number of servings.

a. How many servings provide 32.5 grams of protein?

b. Use a different method than part (a) to find how many servings provide 52 grams of protein.

c. How many grams of protein will 7 servings provide?

d. If 1 serving is equal to $\frac{3}{4}$ cup, how many cups does it take to get 19.5 grams of protein?

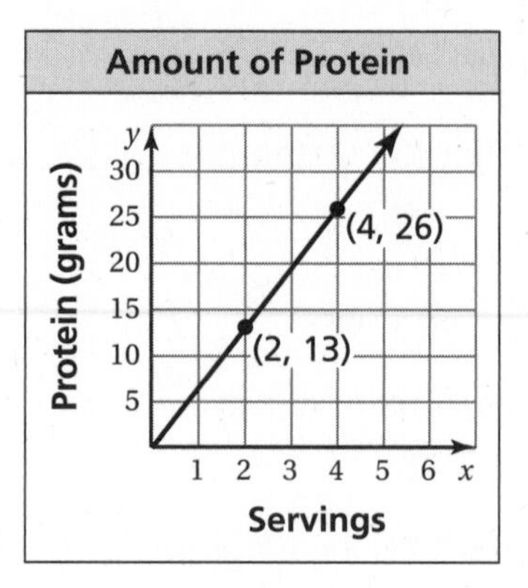

12. One day 176 people visited a small art museum. The ratio of members to nonmembers that day was 5 to 11. How many people who visited the museum that day were nonmembers?

13. Solve the proportion: $\frac{|h|}{3} = \frac{4}{9}$.

14. One gallon of water weighs about 8.34 pounds.

a. How much does 3.5 gallons of water weigh?

b. One inch of rain on a square foot of land weighs 5.2 pounds. About how many gallons of rain are there? Round your answer to the nearest tenth.

Name______________________________ Date__________

5.4 Enrichment and Extension

Sports Statistics and Proportions

In baseball, WHIP is a pitching statistic used to find the average number of walks and hits allowed per inning. It is calculated by adding the walks and hits that a pitcher allows and then dividing that by the number of innings pitched.

$$\text{WHIP} = \frac{\text{Walks} + \text{Hits}}{\text{Innings Pitched}}$$

In hockey, a goalie's save percentage is the number of saves divided by the number of shots on goal.

$$\text{Save percentage} = \frac{\text{Saves}}{\text{Shots on goal}}$$

1. A pitcher allowed 26 walks and 104 hits in 85 innings pitched. What is his WHIP? Round your answer to the nearest hundredth.

2. The pitcher in Exercise 1 wants to lower his WHIP to 1.3. How many more innings would he have to pitch without any walks or hits? Is this a reasonable goal? Explain your reasoning.

3. Another pitcher has a WHIP of 1.28 with 92 hits in 82 innings pitched. How many walks has he allowed? Explain how you used proportions to get your answer.

4. The pitcher from Exercise 3 wants to keep his WHIP under 1.3. How many more hits and walks can he allow in the next 9 innings? Is this a reasonable goal? Explain your reasoning.

5. A goalie has a save percentage of 0.914 after playing his first 6 games. He has made 191 saves. How many shots on goal has he faced?

6. Another goalie has a save percentage of 0.907 with 204 saves in his first 6 games. Your friend says that you could find their combined save percentage by adding 0.907 and 0.914 and then dividing by 2. Explain why this is incorrect. Then show how you would correctly calculate their combined save percentage. Round your answer to the nearest thousandth.

7. The goalie from Exercise 5 wants to finish the next game with an overall save percentage of 0.925. How many shots on goal would he have to face without allowing any goals? Based on the goalie's first 6 games, is it reasonable for him to achieve this goal in one game? Explain your reasoning.

Name ______________________________ Date __________

5.4 Puzzle Time

Did You Hear About...

A	B	C	D	E	F
G	H	I	J	K	L
M	N	O	P	Q	R

Complete each exercise. Find the answer in the answer column. Write the word under the answer in the box containing the exercise letter.

Answer
44 THE
20 BECAUSE
18 IN
16 DAY
3.8 MINUTES
45 SECONDS
9 FOR
5.2 LUNCH
$12\frac{3}{4}$ TIME
$18\frac{3}{4}$ CAFETERIA

Use multiplication to solve the proportion.

A. $\frac{m}{8} = \frac{3}{4}$

B. $\frac{7}{9} = \frac{y}{18}$

C. $\frac{6}{13} = \frac{r}{39}$

D. $\frac{g}{48} = \frac{11}{12}$

E. $\frac{z}{24} = \frac{25}{32}$

F. $\frac{b}{21} = \frac{5}{7}$

G. $\frac{11}{27} = \frac{n}{18}$

H. $\frac{9}{28} = \frac{s}{42}$

Use the Cross Products Property to solve the proportion.

I. $\frac{c}{12} = \frac{5}{3}$

J. $\frac{9}{4} = \frac{x}{16}$

K. $\frac{7}{8} = \frac{14}{p}$

L. $\frac{12}{7} = \frac{36}{n}$

M. $\frac{k}{20} = \frac{13}{50}$

N. $\frac{15}{a} = \frac{25}{14}$

O. $\frac{6.6}{1.2} = \frac{w}{4.2}$

P. $\frac{1.6}{3.2} = \frac{2.8}{t}$

Q. $\frac{5}{\$13.75} = \frac{p}{\$24.75}$

R. $\frac{230 \text{ cal}}{30 \text{ min}} = \frac{345 \text{ cal}}{x \text{ min}}$

Answer
36 EVERY
14 CLOCK
5.6 BACK
15 THAT
6 THE
21 AT
23.1 WENT
8.4 IT
$13\frac{1}{2}$ SLOW
$7\frac{1}{3}$ WAS

Activity 5.5

Start Thinking!

For use before Activity 5.5

Explain how judges use math to determine who receives the first place prize at a dance competition.

Activity 5.5

Warm Up

For use before Activity 5.5

Plot and label the points in a coordinate plane.

1. $A(2, -4)$
2. $B(-3, 1)$
3. $C(0, -3)$
4. $D(2, 0)$
5. $E(-1, -1)$
6. $F(-3, 0)$

Lesson 5.5 Start Thinking!

For use before Lesson 5.5

What math concepts are related to auto racing?

Lesson 5.5 Warm Up

For use before Lesson 5.5

Find the missing distances. Use the table to complete a line graph for each distance. Which graph is steeper?

Time (seconds)	Distance 1 (feet)	Distance 2 (feet)
0	0	0
1	12	6
2		
3		
4		

Name____ Date__________

5.5 Practice A

Find the slope of the line.

1.

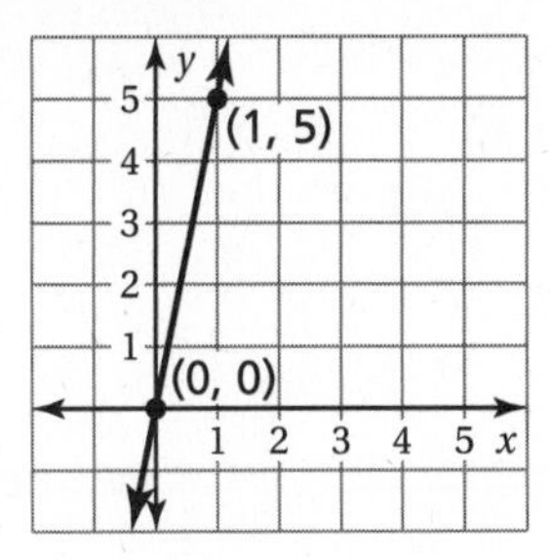

2.

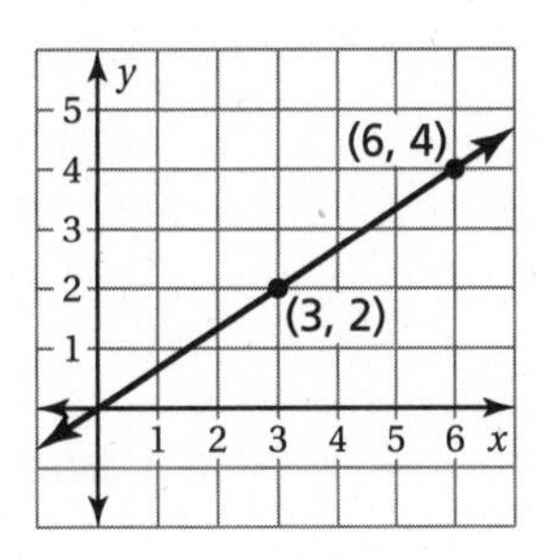

3.

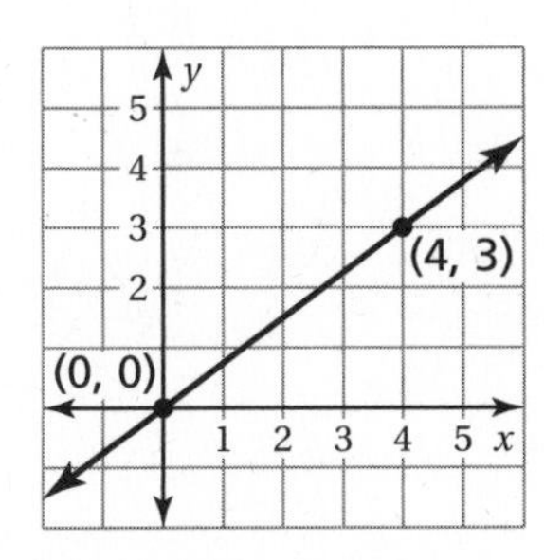

4.

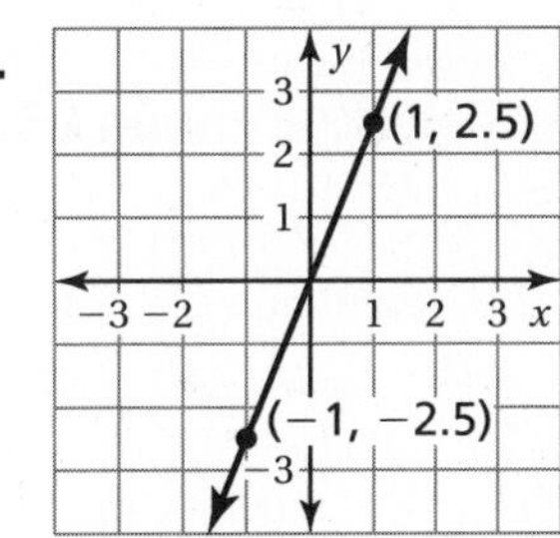

Graph the data. Then find and interpret the slope of the line through the points.

5.

Days, x	2	4	6	8
Pages, y	80	160	240	320

6.

Seconds, x	10	20	30	40
Feet, y	22	44	66	88

Graph the line that passes through the two points. Then find the slope of the line.

7. $(0, 0), (4, 3)$

8. $(-1, -2), (2, 4)$

9. $(-4, -1), (8, 2)$

10. The graph shows the amounts that you are collecting for selling calendars and boxes of greeting cards to raise money for the school band.

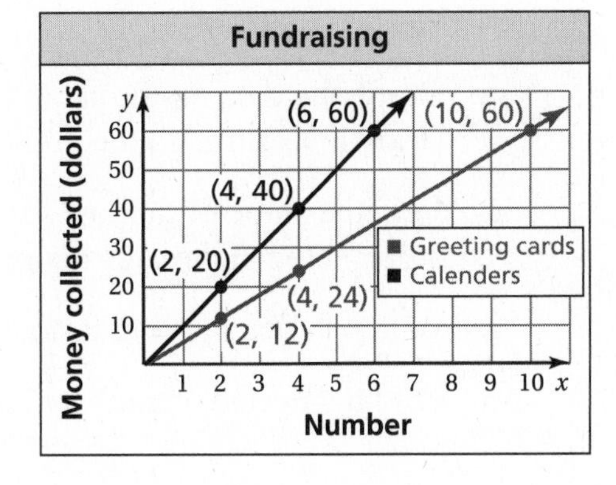

a. Compare the steepness of the lines. What does this mean in the context of the problem?

b. Find the slope of each line. What does each slope mean in the context of the problem?

c. How much more does it cost to buy 3 calendars than 4 boxes of greeting cards?

d. Find two different ways that you could collect exactly \$36.

Name ______________________________ Date __________

5.5 Practice B

Find the slope of the line.

1.

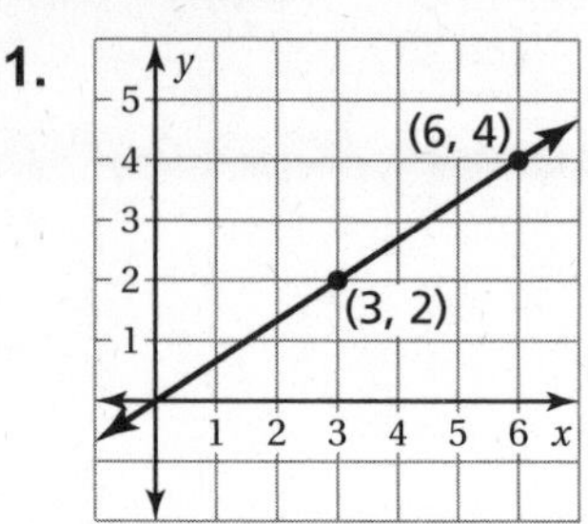

2.

Which line has the greater slope? Explain your reasoning.

3.

4.

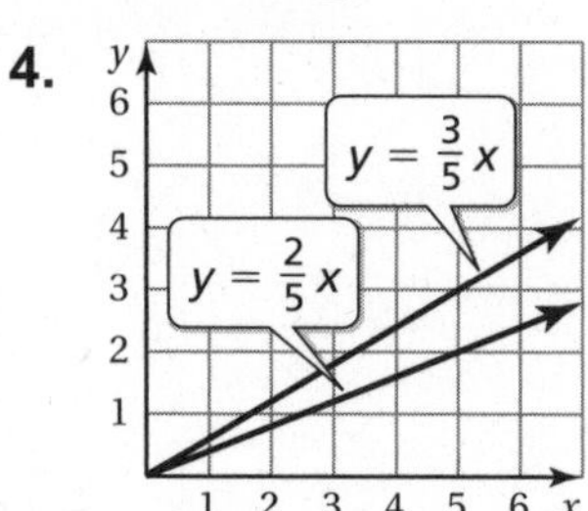

Graph the line that passes through the two points. Then find the slope of the line.

5. $(-2, -3), (3, 4.5)$

6. $(-8, -10), (-2, -2.5)$

7. Graph the line that passes through the points $(0, 1)$ and $(0, 4)$. Explain why the slope of this line is undefined.

8. You and a friend throw tennis balls up in the air at the same time. The table shows the height (in feet) of each ball.

Seconds, x	2	4	5
You, y	12	24	30
Friend, y	11	22	27.5

a. Graph the data on the same coordinate axes. Draw lines through the points. Label each graph.

b. Find the slope of the line for each tennis ball. What does each slope mean in the context of the problem?

c. Which ball is moving faster? How is this indicated in the slope?

d. Find the height of each ball 3.5 seconds after being thrown.

9. A line has a slope of $\frac{3}{5}$. It passes through the points $(5, 3)$ and $(x, 9)$. What is the value of x?

Name______________________________ Date__________

5.5 Enrichment and Extension

Finding Slope Without Drawing a Graph

You can find the slope of a line through two points without drawing a graph.
To find the change in y without a graph, subtract the y-values of two points.
To find the change in x without a graph, subtract the x-values of two points.
Be sure to put the coordinates of the points in the same order when subtracting.

Example: Find the slope of the line that passes through the points $(2, 7)$ and $(3, 10)$.

Let (x_1, y_1) be $(2, 7)$ and (x_2, y_2) be $(3, 10)$.

$$\frac{\text{change in } y}{\text{change in } x} = \frac{y_2 - y_1}{x_2 - x_1}$$ Write coordinates in the same order.

$$= \frac{10 - 7}{3 - 2}$$ Substitute.

$$= \frac{3}{1} = 3$$ Simplify.

The slope of the line is 3.

Without graphing, find the slope of the line through the points.

1. $(2, 5), (5, 6)$

2. $(-1, -2), (-5, -7)$

3. $(-6, 0), (2, 8)$

4. $(4, 1), (-8, 10)$

5. In the example above, let (x_1, y_1) be $(3, 10)$ and (x_2, y_2) be $(2, 7)$. Would the slope still be the same? Explain your reasoning.

6. A line has a slope of $\frac{3}{2}$. It passes through the point $(4, 5)$. Give the coordinates of three other points on this line.

7. A line has a slope of $\frac{4}{7}$. It passes through the points $(2, -4)$ and $(-5, y)$. What is the value of y?

8. A line has a slope of 5. It passes through the points $(-5, -1)$ and $(x, 9)$. What is the value of x?

Name ______________________ Date __________

Puzzle Time

What Is The Invisible Man's Favorite Drink?

Circle the letter of each correct answer in the boxes below. The circled letters will spell out the answer to the riddle.

Graph the line that passes through the two points. Then find the slope of the line.

1. (0, 0), (6, 7) **2.** (0, 0), (−3, −5) **3.** (1, 2), (4, 8)

4. (2, 2), (5, 5) **5.** (−4, −12), (2, 6) **6.** (−9, −2), (18, 4)

7. (−6, −2), (6, 2) **8.** (−2, −8), (5, 20) **9.** (10, 12), (20, 24)

10. (−12, −8), (12, 8) **11.** (−12, −2), (6, 1) **12.** (8, 1), (24, 3)

In Exercises 13 and 14, use the table below for the price of admission to a water park.

Water Park Admission				
Number of Persons	2	4	6	8
Child	\$46	\$92	\$138	\$184
Adult	\$62	\$124	\$186	\$248

13. Find the slope of the line for the price of a child's admission to the water park.

14. Find the slope of the line for the price of an adult's admission to the water park.

C	A	E	R	V	L	A	T	D	P	G	I	O	N	E	R
8	$\frac{1}{2}$	3	$\frac{2}{5}$	$\frac{1}{6}$	$\frac{3}{4}$	$\frac{1}{8}$	5	$\frac{1}{4}$	$\frac{2}{9}$	$\frac{6}{7}$	$\frac{5}{4}$	1	7	$\frac{1}{12}$	2
A	**M**	**K**	**T**	**L**	**E**	**D**	**I**	**M**	**A**	**R**	**I**	**T**	**L**	**K**	**S**
23	$\frac{7}{8}$	6	$\frac{7}{6}$	$\frac{3}{5}$	$\frac{1}{3}$	$\frac{2}{3}$	12	4	9	25	$\frac{5}{3}$	90	31	$\frac{6}{5}$	0

Activity 5.6 Start Thinking!

For use before Activity 5.6

You made a sketch of the school mascot. Your sketch has a height of 2 inches and a width of 4 inches. Describe the proportions you would use to make a mural of your sketch on a wall in your school. What would be the dimensions of your mural?

Plot the point in a coordinate plane. Describe the location of the point.

1. $A(-4, -2)$
2. $B(-2, -1)$
3. $C(0, 0)$
4. $D(2, 1)$
5. $E(4, 2)$
6. $F(6, 3)$

Lesson 5.6 Start Thinking!
For use before Lesson 5.6

You receive \$20 every time you mow your neighbor's lawn. How many times do you need to mow the lawn so that you can buy a digital camera that costs \$189? Write an equation to represent the situation. Does the equation show direct variation? Why or why not?

Lesson 5.6 Warm Up
For use before Lesson 5.6

Graph the ordered pairs in a coordinate plane. Do you think that the graphs show that the quantities vary directly? Explain your reasoning.

1. $(0, 0), (1, 6), (2, 12), (3, 18)$

2. $(-3, 6), (4, 8), (-5, 10), (6, 12)$

3. $(4, 6), (8, 12), (12, 18), (18, 27)$

Name____ Date__________

5.6 Practice A

Graph the ordered pairs in a coordinate plane. Do you think that graph shows that the quantities vary directly? Explain your reasoning.

1. $(-2, -2), (0, 0), (2, 2), (4, 4)$

2. $(-1, -4), (0, -1), (1, 2), (2, 5)$

Tell whether *x* and *y* show direct variation. Explain your reasoning. If so, find *k*.

3.

x	−1	0	1	2
y	2	0	2	4

4.

x	2	4	6	8
y	1	2	3	4

5. $y - 2 = 3x - 2$

6. $y + 3 = x$

7. $xy = 5$

8. The table shows the grams of fiber y for the grams of protein x. Graph the data. Tell whether x and y show direct variation. If so, write an equation that represents the line.

Grams of protein, *x*	3	6	9	12
Grams of fiber, *y*	2	4	6	8

The variables *x* and *y* vary directly. Use the values to find the constant of proportionality and write an equation that relates *x* and *y*.

9. $y = 6; x = 2$

10. $y = 15; x = 3$

11. $y = 40; x = 10$

12. To prepare an aquarium for use, you can clean it with a saltwater solution. The amount of salt varies directly with the volume of the water. The solution has 2 teaspoons of aquarium salt for every gallon of water.

a. How many teaspoons of aquarium salt are needed for 5 gallons of water?

b. Write an equation that relates x gallons of water to y teaspoons of salt.

c. Use the equation to find the number of gallons of water to use for 12 teaspoons of salt.

13. The total cost of football tickets varies directly with the number of tickets purchased. Four tickets cost $32. How many tickets can you buy for $56?

14. One quart is equivalent to 0.95 liter.

a. Write a direct variation equation that relates x quarts to y liters.

b. Write a direct variation equation that relates x gallons to y liters.

c. Write a direct variation equation that relates x liters to y quarts.

d. What is the relationship between the values of k in the direct variation equations in parts (a) and (c)?

Name ______________________________ Date __________

5.6 Practice B

Graph the ordered pairs in a coordinate plane. Do you think that graph shows that the quantities vary directly? Explain your reasoning.

1. $(-3, -1), (3, 1), (6, 2), (9, 3)$

2. $\left(-2, -\frac{5}{2}\right), (4, 5), \left(6, \frac{15}{2}\right), (12, 15)$

Tell whether *x* and *y* show direct variation. Explain your reasoning. If so, find *k*.

3. $y - 2x = x$

4. $y - 5 = 5x + 2$

5.

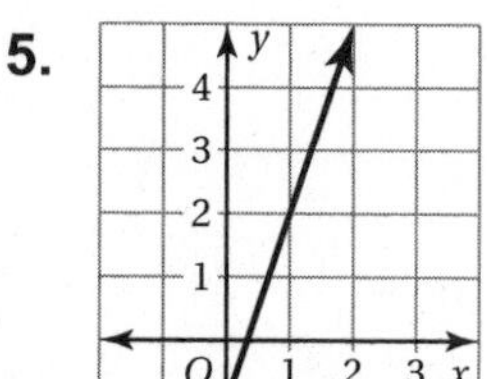

6.

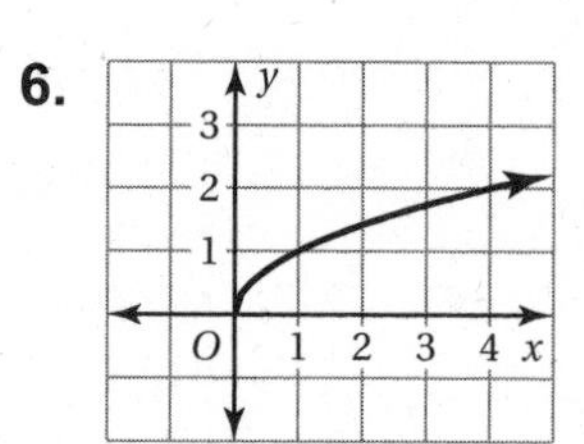

7. The percent y of correct answers on a test varies directly with the points x earned on the test.

Points earned, *x*	?	48	?	64	72	?
Percent, *y*	50	60	70	80	90	100

a. Copy and complete the table.

b. Write an equation that relates x and y.

c. Graph the points. Draw a line through the points. Does the graph confirm that x and y show direct variation? Explain your reasoning.

The variables *x* and *y* vary directly. Use the values to find the constant of proportionality and write an equation that relates *x* and *y*.

8. $y = 36;\ x = 18$

9. $y = 51;\ x = 34$

10. $y = 55;\ x = 10$

11. The table shows the heights of the vertical supports for two skateboard ramps. Which ramp(s) show direct variation between distance and height?

Distance from base (feet)	1	2	3	4
Height for your ramp (feet)	1	1.5	2	2.5
Height for friend's ramp (feet)	0.5	1	1.5	2

12. Does the graph of every direct variation equation pass through the origin? Is every relationship whose graph passes through the origin direct variation? Explain your reasoning.

Name______________________________ Date__________

5.6 Enrichment and Extension

More on Direct Variation

1. A store's profit P varies directly with the number of items n that they purchase from a warehouse. Greg's Gadgets buys 75 Wacky Widgets from the Gadget Supply Warehouse. The store's profit for this purchase is $-\$190.50$.

   a. Write a direct variation equation. Then, explain what the equation means.

   b. Explain why the profit is negative. Predict what a graph of the equation will look like.

   c. Fill in the table and make a graph of the data. Then compare the graph with your prediction.

Items purchased, n	0	5	10	15	20
Profit, P					

2. The volume V of blood that your heart pumps varies directly with your pulse rate p. Tomiqua's heart pumps 2.6 liters of blood in one minute when she is at her resting heart rate of 52 beats per minute.

   a. Write a direct variation equation. Then, explain what the equation means.

   b. After Tomiqua has been running on a treadmill, she measures her pulse for 10 seconds and gets 15 beats. How much blood is her heart pumping per minute?

3. The weight of an object on the moon varies directly with its weight on Earth. For example, the Apollo 13 Command Module weighed approximately 6 tons on Earth and about 1 ton on the moon.

   a. Write a direct variation equation. Then, explain what the equation means.

   b. Find your approximate weight on the moon.

Name ______________________________ Date __________

5.6 Puzzle Time

How Do Bees Get To School?

For each exercise, circle the letter in the columns under Yes or No to indicate the correct answer. The circled letters will spell the answer to the riddle.

	Yes	No
1.	T	S
2.	H	T
3.	A	E
4.	Y	R
5.	M	T
6.	E	A
7.	L	K
8.	E	R
9.	T	S
10.	O	H
11.	E	K
12.	A	B
13.	U	N
14.	Z	T
15.	Y	Z

Tell whether *x* and *y* show direct variation.

1. $(1, 1), (2, 2), (3, 3), (4, 4)$
2. $(1, 3), (2, 6), (3, 9), (4, 12)$
3. $(-2, 3), (0, 0), (2, 3), (4, 12)$
4. $(-4, -1), (4, 1), (8, 2), (12, 3)$
5. $(-1, -1), (0, 1), (1, 3), (2, 5)$
6. $(1, 1), (2, 4), (3, 9), (4, 16)$
7. $(1, 3), (2, 3), (3, 3), (4, 3)$
8. $y = 12x$
9. $y = \frac{1}{7}x$
10. $y = 4x^2$
11. $y = -2x$
12. $y = 7x - 3$
13. $x = 5y$
14. $6 = \frac{x}{y}$
15. $x^3 = 12y$

Name_______________________________ Date__________

Chapter 5

Technology Connection

For use after Section 5.5

Exploring Slope

Dynamic geometry software allows you to explore the slope of a line or line segment. The directions below might not correspond exactly to your software, but the same actions can be performed using similar commands.

Open a **New Sketch**. Use the **Segment** tool on the left side of the screen to draw a short segment that slants upward.

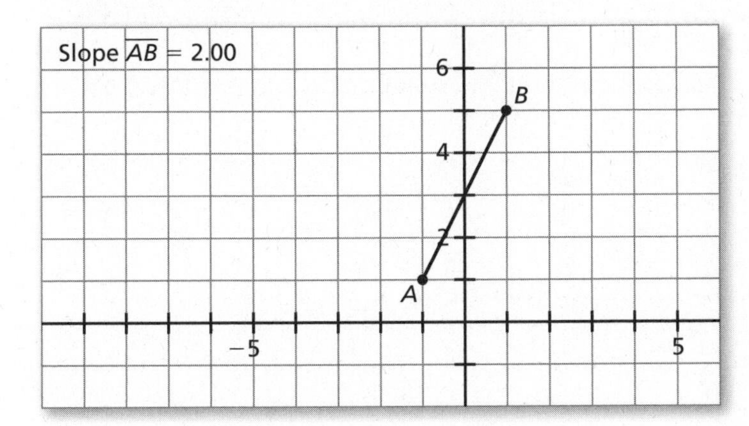

Click on the middle of the segment you made. In the **Measure** menu, select **Slope**. A grid and the slope of the segment will appear on the screen. In the **Graph** menu, select **Snap Points**. This will make it so that the segment will always have integer coefficients.

Move the endpoints of the segment you made so that it has the given slope. Sketch the line segment on a piece of paper and label the slope.

1. 1 **2.** $\frac{1}{2}$ **3.** $\frac{2}{3}$ **4.** $-\frac{4}{5}$

5. 0 **6.** -0.75 **7.** $1.\overline{3}$ **8.** -2.5

9. Which line segments are the steepest? Which are the least steep? Describe their slopes.

10. What is the slope of a horizontal line segment?

11. What happens to the slope when you make the line segment vertical? Why?

12. Can you move the line segment without changing the slope? How?

Chapter 6

Name____ Date__________

Percents

Dear Family,

Family vacations provide a relaxing break from the usual routine. Whether you visit family, go camping, or plan a "stay-cation" at home, you usually need to make a budget for your expenses.

Budgeting for a family vacation may take some research. The rates advertised for hotel rooms usually do not include taxes or resort fees. When dining out, the tax and tip can add quite a bit to the total bill. Vehicle rentals often have extra charges for taxes, insurance, fuel, and mileage that are not usually included in advertised rates. All of these "hidden" costs can increase the cost of your vacation significantly.

Ask your student to help create a budget for your next family vacation. Have him or her research common fees, taxes, and tips for each part of your plan. For example:

- Hotels sometimes have multiple tax rates—for state, county, and local sales taxes, and sometimes even an extra "room tax." Find out what the tax rates are and determine how much this will add to the room charges. Are there other fees, such as resort fees or parking charges?
- Employees in restaurants and hotels usually earn most of their money from tips, so find out who usually receives a tip and what amount is appropriate.
- Rental cars can have many taxes and fees—sometimes these cost as much as the rental fee itself. Try to find out each rate or fee so you can plan accordingly.

Work with your student to make a budget that includes the extra charges for each item. You might want to include other categories in your budget—such as categories for shopping, entertainment, fuel, and parking. Having your student help with the budgeting will involve him or her in planning the vacation—and help your student realize you can't always afford everything you want to do.

Don't forget to write!

Nombre ____ Fecha ________

Porcentajes

Estimada Familia:

Las vacaciones familiares ofrecen un descanso relajante de la rutina cotidiana. Ya sea que se visite a la familia, se vaya de campamento o se quede en casa descansando, generalmente se necesita hacer un presupuesto para los gastos.

Hacer el presupuesto para una vacación familiar puede demandar algo de investigación. Por lo general, las tarifas publicadas de habitaciones de hotel no incluyen impuestos ni costos del complejo.

Al salir a cenar, el impuesto y la propina aumentan de manera regular la cuenta total. Los alquileres de vehículos a menuda cobran extra por impuestos, seguro, gasolina y millaje, lo que generalmente no está incluido en las tarifas publicadas. Todos estos costos "ocultos" pueden aumentar el costo de las vacaciones de manera significativa.

Pida a su estudiante que lo ayude a crear un presupuesto para sus próximas vacaciones familiares. Haga que investigue acerca de precios, impuestos y propinas comunes para cada parte de su plan. Por ejemplo:

- Los hoteles a veces cobran impuestos múltiples—impuestos de ventas estatales, nacionales y locales, y a veces incluso cobran un "impuesto por habitación" adicional. Averigüen las tarifas de impuestos y determinen en cuánto aumentará la cuenta de la habitación. ¿Hay otros gastos, como los costos del complejo o del estacionamiento?
- Los empleados de restaurantes y hoteles generalmente ganan la mayoría de su dinero en propinas, así que averigüen quién recibe generalmente una propina y qué cantidad es la apropiada.
- Los autos de alquiler pueden incluir muchos impuestos y gastos—a veces estos cuestan tanto como el alquiler del coche mismo. Intenten encontrar cada tarifa o gasto para que puedan planear como corresponde.

Trabaje con su estudiante para hacer un presupuesto que incluya los gastos extras de cada artículo. Querrán incluir otras categorías en su presupuesto—como por ejemplo: compras, entretenimiento, gasolina y estacionamiento. El permitir que su estudiante ayude con el presupuesto, lo animará a que planee las vacaciones—y ayude a su estudiante a darse cuenta que no siempre se puede conseguir todo lo que se desea.

¡No se olviden de escribir!

Activity 6.1 Start Thinking!
For use before Activity 6.1

Study the decimal and percent equivalents below. What do you notice?

$0.35 = 35\%$ $\qquad$ $0.27 = 27\%$ $\qquad$ $0.03 = 3\%$

$0.2 = 20\%$ $\qquad$ $0.5 = 50\%$ $\qquad$ $1.38 = 138\%$

$2.1 = 210\%$ $\qquad$ $6 = 600\%$ $\qquad$ $0.886 = 88.6\%$

Andrew says that $0.98 = 9.8\%$. Is he correct? Why or why not?

Activity 6.1 Warm Up
For use before Activity 6.1

Write the fraction represented by the model.

1. 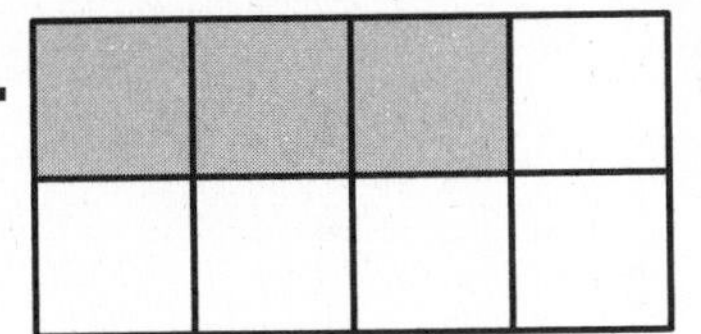

2. 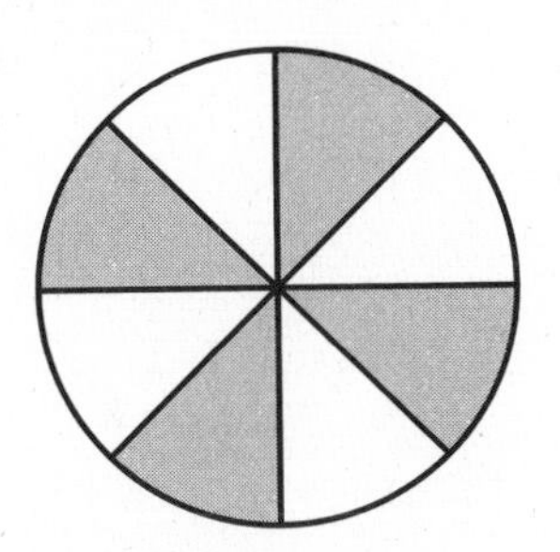

3. 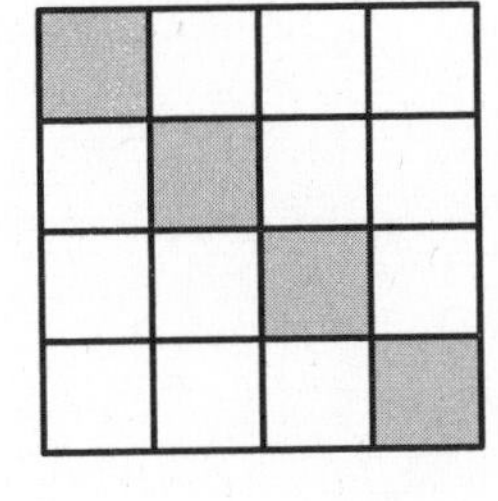

4. 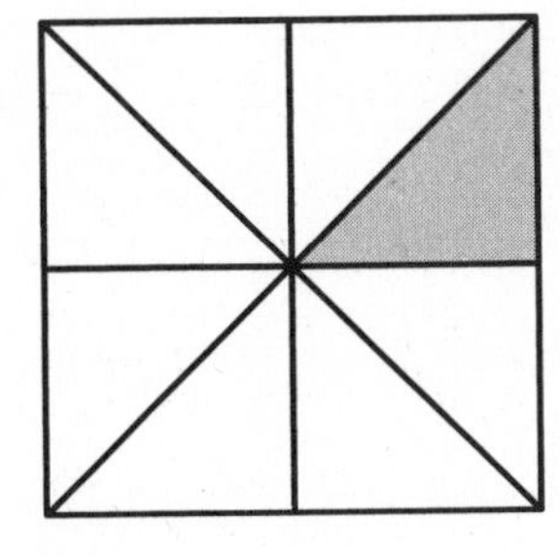

5. 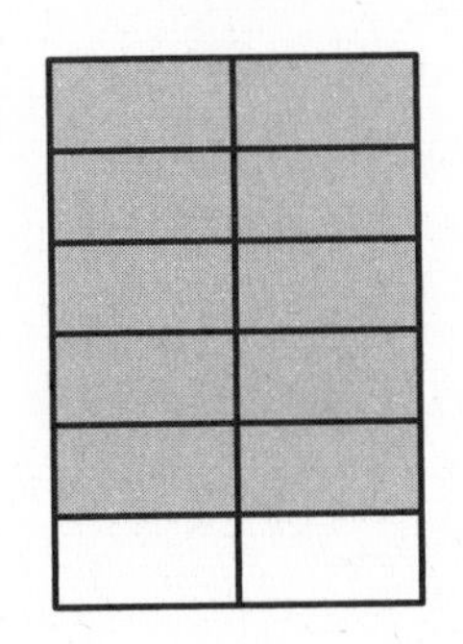

6. 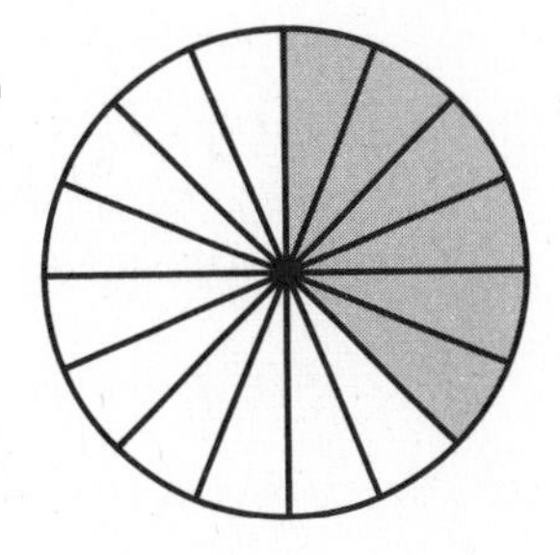

Start Thinking!

For use before Lesson 6.1

Describe a situation in sports when decimals and percents are used interchangeably.

Warm Up

For use before Lesson 6.1

Write the percent as a decimal.

1. 19% **2.** 2% **3.** 7.5%

Write the decimal as a percent.

4. 0.89 **5.** 0.54 **6.** 0.1

Name ____________________ Date ________

6.1 Practice A

Write the percent as a decimal.

1. 81%
2. 78%
3. 5%
4. 8%
5. 40%
6. 60%
7. 23.7%
8. 16.75%
9. 150%
10. 210%
11. 186%
12. 416%
13. 100.8%
14. 5.17%
15. 0.4%
16. 0.04%
17. Describe and correct the error in writing 1.475% as a decimal.

   ✗ $1.475\% = 1.475\% = 147.5$

Write the decimal as a percent.

18. 0.66
19. 0.32
20. 0.51
21. 0.97
22. 0.01
23. 0.04
24. 0.312
25. 0.468
26. 0.5
27. 1.2
28. 1.08
29. 1.16
30. 0.003
31. 0.025
32. 0.0245
33. 2.025
34. Describe and correct the error in writing 1.8 as a percent.

   ✗ $1.8 = 1.8 = 18\%$

35. Fifty-four percent of the students in your class have moved at least one time. Write this percent as a decimal.
36. Only 0.15 of the total number of vehicles in your school parking lot are buses. What percent of the vehicles are buses?
37. You spent 0.88 of your allowance this week. What percent of your allowance did you spend?
38. On a history test, you get 86 out of a possible 100 points. Write a decimal and a percent that represent a score of 86 out of 100.
39. Of the fluids that you drink on a typical day, $\frac{1}{10}$ is milk and 50% is water. How many times more water do you drink than milk?

Write the percent as a fraction in simplest form and as a decimal.

40. 21%
41. 75%
42. 64%
43. 85%

Name ____ Date __________

6.1 Practice B

Write the decimal as a percent.

1. 0.54 **2.** 0.37 **3.** 0.222 **4.** 0.929

5. 1.4 **6.** 2.5 **7.** 20 **8.** 0.005

Write the percent as a fraction in simplest form and as a decimal.

9. 68% **10.** 9% **11.** 55% **12.** 26%

13. 42.4% **14.** 73.6% **15.** 31.25% **16.** 44.65%

17. About 36% of the students at a middle school are seventh graders. What percent are *not* in seventh grade?

18. The percents of three types of tickets collected at the gate for a high school football game are shown.

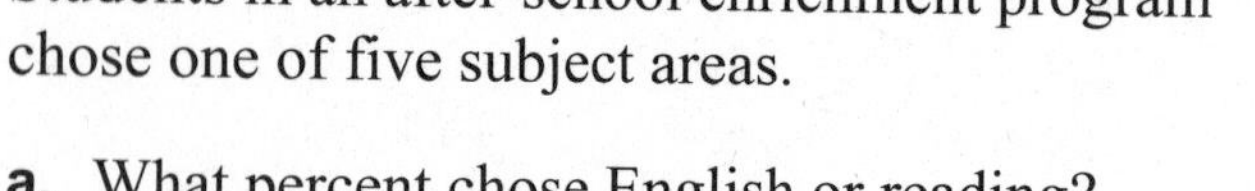

Ticket type	Student	Adult	Senior (65 and older)
Percent	48%	28%	14%

a. Write the percents as decimals and as fractions.

b. There is one other type of ticket that is not shown. It is a ticket for a child under 5. What percent of the tickets were of this type?

c. Make a bar graph to represent the percents for all four ticket types.

19. Students in an after-school enrichment program chose one of five subject areas.

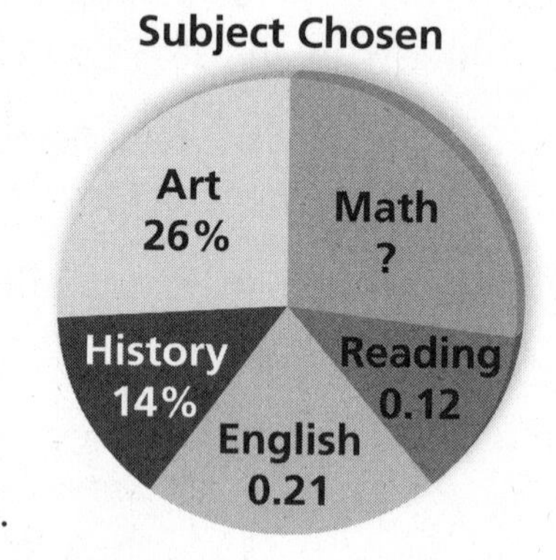

a. What percent chose English or reading?

b. What percent chose English or history?

c. How many times more students chose English than reading?

d. What percent chose math? Write the percent as a decimal.

20. At one school, half of the students live within 1 mile, 78% live within 2 miles, and 0.1 of the students live between 2 and 3 miles from the school.

Make a table to show the percent of students who live at each distance from the school.

a. within 1 mile **b.** between 1 and 2 miles

c. between 2 and 3 miles **d.** more than 3 miles

Name____ Date__________

6.1 Enrichment and Extension

Playing Checkers

You set up a game of checkers as shown. Write a decimal for each percent you find.

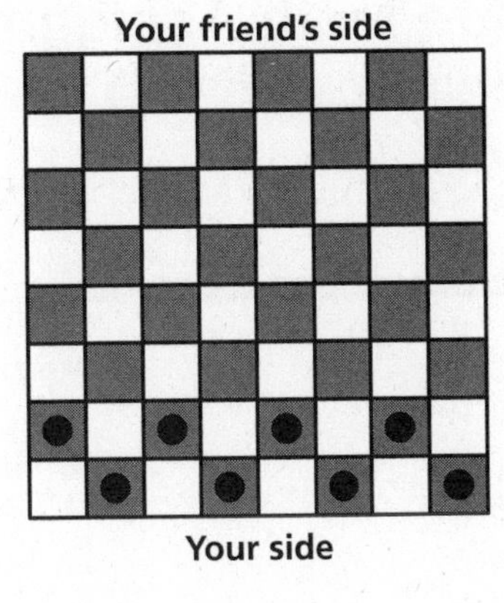

1. What percent of the checker board is shaded?

2. What percent of the checker board is *not* shaded?

3. Your friend arrives and places pieces appropriately on the board.

   a. What percent of the checker board squares have pieces?

   b. What percent of the checker board squares do *not* have pieces?

   c. What percent of the shaded checker board squares have pieces?

   d. What percent of the shaded checker board squares do *not* have pieces?

4. At one point during the game, 4 of your pieces have been removed and 5 of your friend's pieces have been removed.

   a. What percent of the original number of pieces have been removed?

   b. What percent of the original number of pieces are remaining?

5. The checker piece near the middle of the board can move diagonally in any direction between shaded squares.

   a. On how many different squares can the piece be located at the end of on turn?

   b. On how many different squares can the piece be located at the end of two turns? What percent of the shaded checker board is this? Write this percent as a decimal.

Name ______________________ Date ________

6.1 Puzzle Time

Why Are Math Assignments Like The Water That Is Found On The Ground In The Early Morning?

Write the letter of each answer in the box containing the exercise number.

Write the percent as a decimal.

1. 67%

2. 44%

3. 29.6%

4. 46.3%

5. 22%

6. 8%

7. 58.74%

8. 80.14%

9. 277%

10. 106%

11. 0.05%

12. 0.045%

Write the decimal as a percent.

13. 0.85

14. 0.41

15. 0.98

16. 0.657

17. 0.77

18. 0.51

19. 0.376

20. 0.239

21. 2.57

22. 0.0482

Answers for Exercises 1–12

T. 2.77

E. 0.00045

S. 0.296

O. 0.22

B. 0.67

H. 1.06

R. 0.8014

P. 0.5874

O. 0.463

D. 0.44

E. 0.0005

U. 0.08

Answers for Exercises 13–22

B. 37.6%

P. 65.7%

S. 4.82%

A. 23.9%

T. 85%

E. 98%

U. 257%

D. 51%

O. 77%

E. 41%

19	17	9	10		20	8	12		3	6	7	16	5	22	15	2		13	4		1	11		18	21	14

Activity 6.2 Start Thinking!

For use before Activity 6.2

What are two methods that you can use to compare two fractions?

Use one of the methods you described to tell whether $\frac{21}{40}$ or $\frac{7}{12}$ is greater. Which method did you use? Why?

Activity 6.2 Warm Up

For use before Activity 6.2

Order the decimals from least to greatest.

1. 1.4, 1.44, 4.1, 4.3, 3.3

2. 0.5, 1.75, 0.7, 1.57, 1.5

3. 0.44, 0.02, 0.2, 0.04, 0.24

4. 6.6, 0.6, 6.63, 6.06, 0.06

5. 0.2, 0.17, 0.02, 0.72, 2.27

6. 8.1, 0.81, 0.18, 1.8, 1.88

Lesson 6.2 Start Thinking!

For use before Lesson 6.2

Design a survey with three yes or no questions to ask the students in your class. Take turns in class asking the survey questions. Record your results in a table like the one below. You can use fractions, percents, or decimals. Which of your questions had the greatest percent of students answering yes?

Question	Yes	No
Do you play a sport?	$\frac{19}{28}$	$\frac{9}{28}$
Do you play a musical instrument?	20%	80%
Do you have any siblings?	0.85	0.15

Lesson 6.2 Warm Up

For use before Lesson 6.2

Tell which number is greater.

1. $\frac{5}{13}$, 45%

2. 30%, 0.4

3. $\frac{9}{17}$, 50%

4. 0.6, $\frac{4}{7}$

5. $\frac{2}{5}$, 0.225

6. 240%, 0.24

Name________________________ Date__________

6.2 Practice A

Tell which number is greater.

1. $\frac{3}{4}$, 70%

2. $\frac{1}{2}$, 0.54

3. 0.21, 21%

4. $\frac{2}{3}$, 66%

5. 0.482, 49%

6. 16%, 0.108

7. $\frac{12}{25}$, 48%

8. $\frac{1}{10}$, 12%

9. 1.2, 11%

10. 58%, $\frac{31}{50}$

11. 5020%, $50\frac{1}{4}$

12. 12.25%, $\frac{1}{8}$

13. Describe and correct the error in comparing 0.7% and $\frac{17}{25}$.

> ✗ $\frac{17}{25} \overset{\times 4}{=} \frac{68}{100} = 0.68\%$
>
> 0.7% is greater than 0.68%, so 0.7% is the greater number.

Use a number line to order the numbers from least to greatest.

14. 0.64, $\frac{13}{20}$, 63%

15. 45%, 0.46, $\frac{11}{25}$

16. 0.12, $\frac{1}{8}$, 0.135, 13%

17. $\frac{15}{16}$, 90%, 0.925, $\frac{7}{8}$, 0.93

18. $3\frac{2}{3}$, 362%, 3.66, $3\frac{3}{5}$, 36

19. 0.3, 27.3%, $\frac{11}{40}$, 28%, 0.27

20. You use 8 fluid ounces of fruit juice in a recipe to make 64 fluid ounces of fruit punch. A fruit punch you can buy at the store has 10% real fruit juice. Which has a higher percent of fruit juice?

21. While shooting baskets at a basketball hoop, you make 36 out of 80 shots. Your friend makes 43.75% of the shots. Who made a higher percent?

22. To earn a bonus in a video game, you must find at least 60% of the hidden gems. You find 25 out of 40 gems. Do you get the bonus? Explain.

23. The table shows the portion of students at a middle school that are in each grade. Order the grades from the least to the greatest number of students.

Grade	6	7	8
Portion of students	$33\frac{1}{3}\%$	0.3125	$\frac{17}{48}$

Name ______________________________ Date __________

6.2 Practice B

Tell which number is greater.

1. $\frac{1}{4}$, 22% **2.** $\frac{5}{9}$, 55% **3.** 3.2, 32% **4.** 99.9%, 1

Use a number line to order the numbers from least to greatest.

5. $\frac{1}{3}$, 0.3, 33%, $\frac{8}{25}$, 33.6% **6.** 210%, 2.2, $2.\overline{2}$, $\frac{43}{20}$

Tell which letter shows the graph of the number.

7. 0.884 **8.** $\frac{8}{9}$ **9.** $\frac{22}{25}$ **10.** 0.89

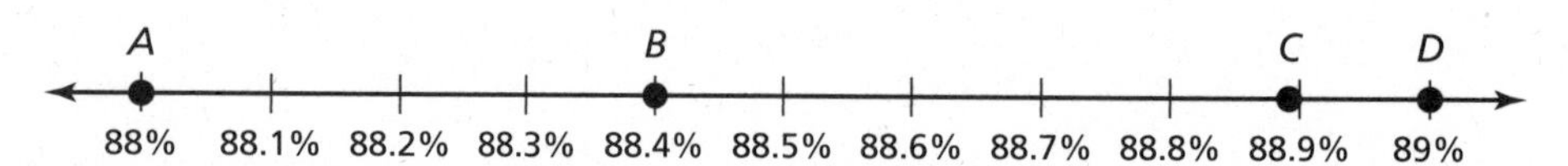

11. Describe a process that you can use to find a decimal whose value is between 31% and 32%.

12. Is 6 centimeters greater than 5% of a meter? Explain.

13. Does 6% of a pound weigh more than an ounce? Explain.

14. Order the periods of time from least to greatest.

1% of an hour $\frac{2}{3}$ of a minute 0.0004 of a day

15. The table shows the portions of the U.S. population that lived in Florida in certain years.

Year	1860	1910	1960	2010
Portion of U.S. Population in Florida	0.45%	0.0082	$\frac{1}{36}$	$\frac{1}{16}$

a. Order the portions from least to greatest.

b. Since 1860, how has the population of Florida increased compared to the population of the United States? Why do you think this happened?

c. Do you think this will always happen? Explain your reasoning.

16. Arsenic is toxic to humans. The greatest amount of arsenic that is allowed in drinking water is 10 parts per billion. A test shows that a source of drinking water contains 0.000002% arsenic. Is this an allowable amount? Explain.

Name____ Date__________

6.2 Enrichment and Extension

Treasure Map

You want to find the treasure shown on the map. To get to the treasure, start at 1% and move to 100%. The only way you can move on the map is if a number is greater than the number you are currently on. List the numbers in order that make up your path to the treasure.

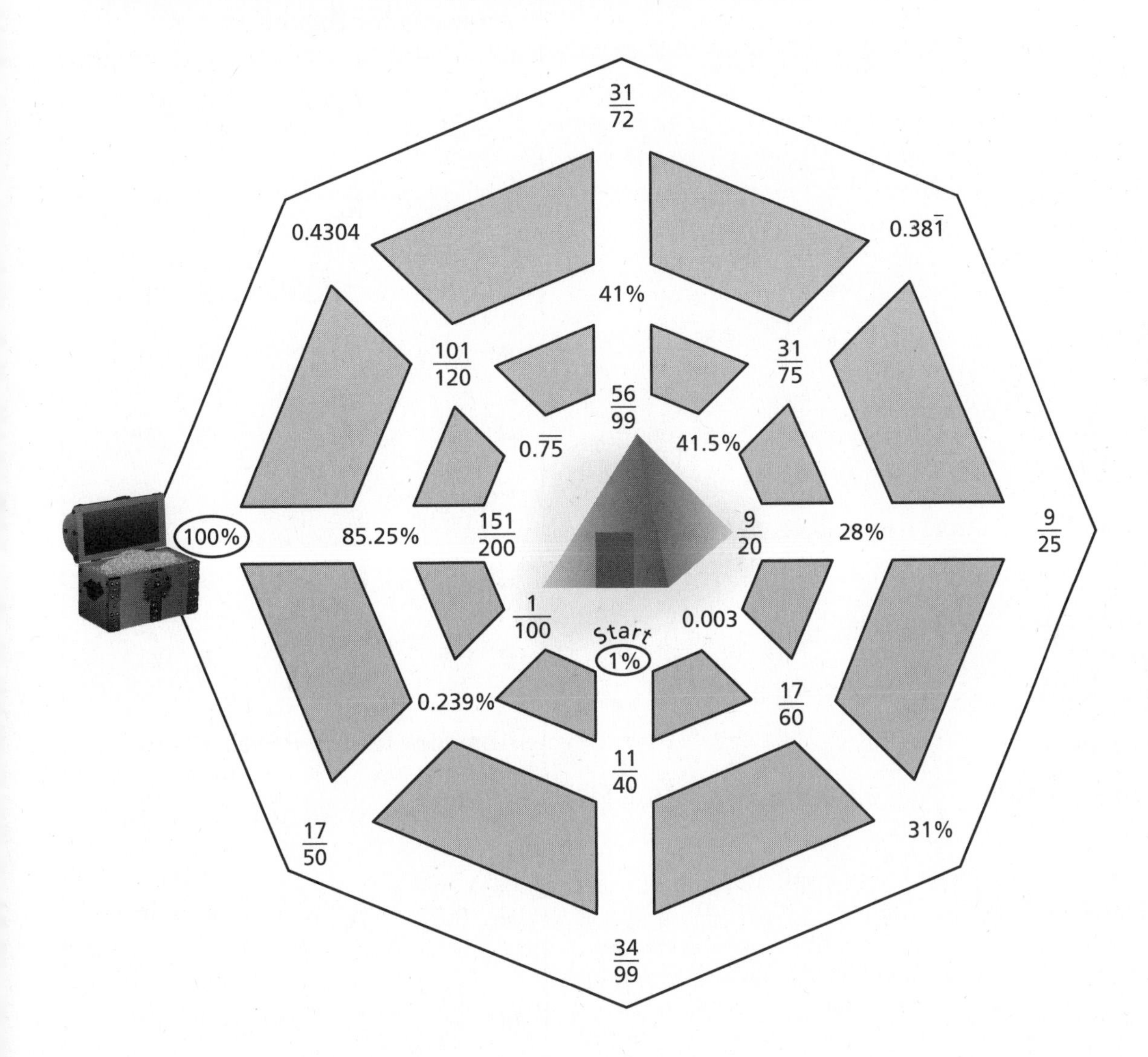

Name ____________________ Date __________

6.2 Puzzle Time

Why Did The Peach Need To Hire An Attorney?

Write the letter of each answer in the box containing the exercise number.

Tell which number is greater.

1. $\frac{17}{20}$, 95%

2. 30%, 0.03

3. $\frac{27}{50}$, 27%

4. $\frac{14}{25}$, 60%

5. 0.097, 97%

6. 65%, 0.56

7. 80%, $\frac{7}{8}$

8. 0.13, 13.5%

9. 11%, $\frac{3}{25}$

10. 150%, 0.15

11. $\frac{1}{4}$, 20%

12. 66%, $\frac{2}{3}$

Tell which number is the greatest.

13. 28%, $\frac{3}{5}$, 0.31

14. 79%, 0.52, $\frac{17}{20}$

15. $\frac{17}{25}$, 0.81, $\frac{5}{8}$, 64%

16. 0.14%, $\frac{7}{20}$, 0.014, 1.4

Answers for Exercises 1–12

W. $\frac{3}{25}$	**D.** $\frac{14}{25}$	**G.** 80%
E. 66%	**R.** $\frac{1}{4}$	**L.** $\frac{27}{50}$
Y. 97%	**I.** 20%	**E.** 11%
A. 95%	**T.** 150%	**C.** 0.15
J. $\frac{2}{3}$	**N.** $\frac{7}{8}$	**O.** 27%
I. 60%	**S.** 30%	**K.** 0.56
U. 13.5%	**M.** 0.03	**A.** 65%
P. 0.097	**B.** $\frac{17}{20}$	**F.** 0.13

Answers for Exercises 13–16

E. $\frac{5}{8}$	**D** 0.014	**T.** 0.81
W. 0.52	**A.** $\frac{3}{5}$	**S.** 0.31
M. $\frac{17}{20}$	**Y.** 79%	**I.** 1.4
M. 28%	**R.** 0.14%	**B.** 64%

4	15	■	9	13	2	■	10	11	8	3	5	■	16	7	■	6	■	12	1	14

Activity 6.3 Start Thinking!

For use before Activity 6.3

Discuss with a partner how percents are used in a real-life situation.

Activity 6.3 Warm Up

For use before Activity 6.3

Estimate the sum or difference.

1. $162 + 98$

2. $148 - (-69)$

3. $-239 + 102$

Estimate the product or quotient.

4. $32(-43)$

5. $\dfrac{-187}{12}$

6. $(-49)(-12)$

Lesson 6.3 Start Thinking!

For use before Lesson 6.3

At a soccer game, your team scored 6 goals during regular play and 4 goals on penalty kicks. Use a model to show the percent of goals that were not scored by penalty kicks.

Lesson 6.3 Warm Up

For use before Lesson 6.3

Use a model to estimate the answer to the question. Use a ratio table to check your answer.

1. What number is 20% of 60?
2. 35 is what percent of 70?
3. 51 is 102% of what number?
4. What number is 80% of 130?
5. 72 is what percent of 100?
6. 44 is 55% of what number?

Name____ Date__________

6.3 Practice A

Use a model to estimate the answer to the question. Use a ratio table to check your answer.

1. What number is 20% of 40?
2. 12 is what percent of 50?
3. 42 is 60% of what number?
4. What number is 150% of 92?

Write and solve a proportion to answer the question.

5. 40% of what number is 15?
6. 24 is 0.6% of what number?
7. What percent of 75 is 27?
8. 17 is what percent of 68?
9. Of the 60 seeds that you plant, 80% germinate. How many seeds germinate?
10. You are charged 6% sales tax. You purchase a new bicycle and pay $27 in sales tax. What is the purchase price of the bicycle?

Write and solve a proportion to answer the question.

11. 0.2 is what percent of 16?
12. 19.6 is 24.5% of what number?
13. $\frac{3}{5}$ is 30% of what number?
14. What number is 45% of $\frac{5}{9}$?
15. You are making 28 name badges for a committee. You complete 75% of these on Monday. How many do you have left to complete on Tuesday?
16. You and your friend are selling tickets for the orchestra concert. On Thursday, you sold 15 tickets and your friend sold 10 tickets.
    a. What percent of the tickets sold on Thursday did you sell?
    b. On Friday, you sold 9 tickets and your friend sold 16 tickets. What percent of the tickets sold on Friday did you sell?
    c. What percent of the total tickets sold on Thursday and Friday did you sell?

Name ____________________ Date ________

6.3 Practice B

Write and solve a proportion to answer the question.

1. 55% of what number is 33?

2. What percent of 120 is 42?

3. 36 is 0.8% of what number?

4. 48 is what percent of 64?

5. Of the 360 runners at a 5-kilometer race, 20% are in the 35–39 age bracket. How many runners at the 5-kilometer race are in the 35–39 age bracket?

6. You pay $3.69 for a gallon of gasoline. This is 90% of the price of a gallon of gasoline one year ago. What was the price of a gallon of gasoline one year ago?

7. Describe and correct the error in using the percent proportion to answer the question below.

"6 is 6.25% of what number?"

✗

$$\frac{a}{w} = \frac{p}{100}$$

$$\frac{6}{w} = \frac{0.0625}{100}$$

$$w = 9600$$

Write and solve a proportion to answer the question.

8. $\frac{7}{8}$ is 70% of what number?

9. 7.2 is 250% of what number?

10. What number is 72% of $\frac{3}{8}$?

11. 1.4 is what percent of 1.12?

12. You earn a score of 86.8 on a standardized exam. Your score is 140% higher than your friend's score on the standardized exam. What is your friend's score?

13. 80% of a number is x. What is 40% of the number?

14. Answer each question.

a. What is 35% of $90x$?

b. What percent of $16x$ is $9x$?

Name____ Date__________

6.3 Enrichment and Extension

Proportions and Similar Triangles

Triangles that have the same shape but not necessarily the same size are called **similar triangles**. Two triangles are similar when corresponding side lengths are proportional and corresponding angles are congruent.

Example: The triangles are similar. Find x.

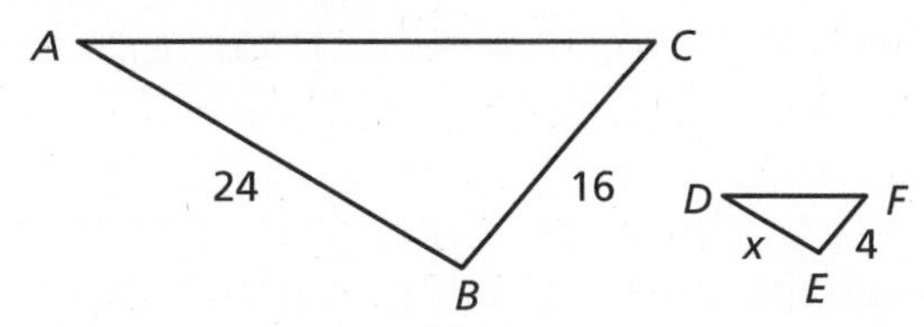

Triangle ABC Triangle DEF

$\frac{AB}{BC} = \frac{DE}{EF}$ Set up a proportion.

$\frac{24}{16} = \frac{x}{4}$ Substitute.

$24 \bullet 4 = 16 \bullet x$ Cross multiply.

$96 = 16x$ Simplify.

$6 = x$ Divide each side by 16.

So, $x = 6$.

The triangles are similar. Find *x*.

1.

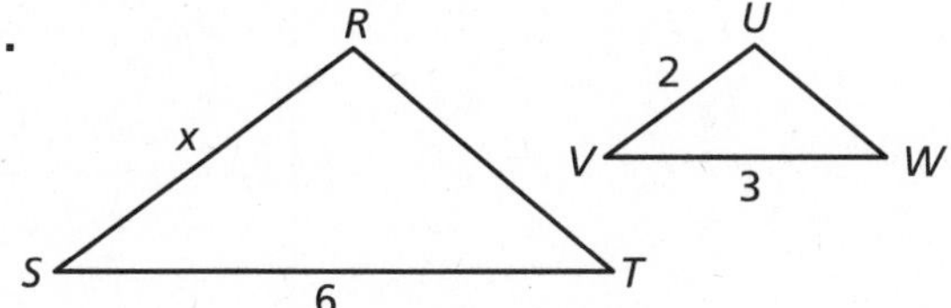

2.

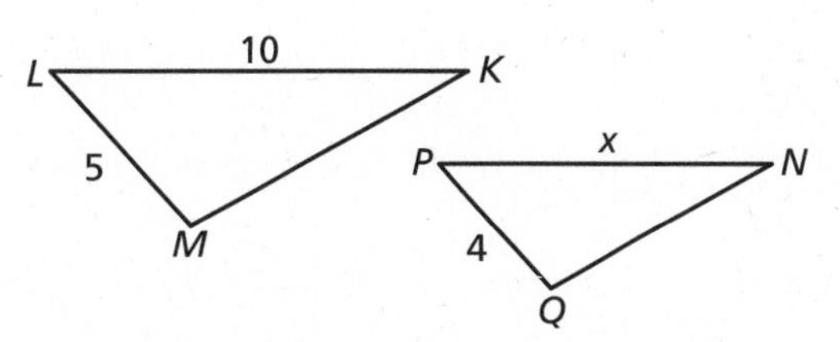

3.

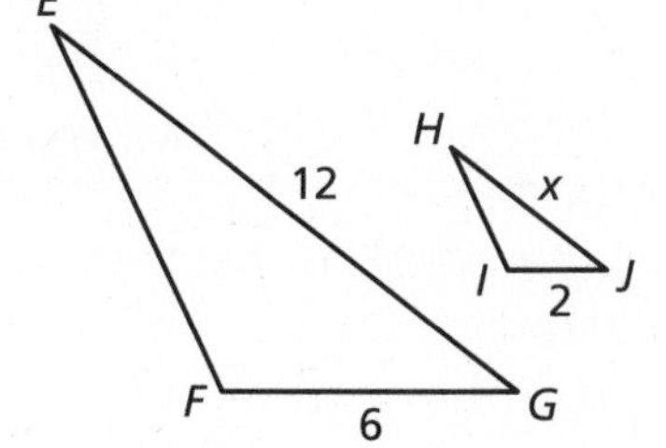

4.

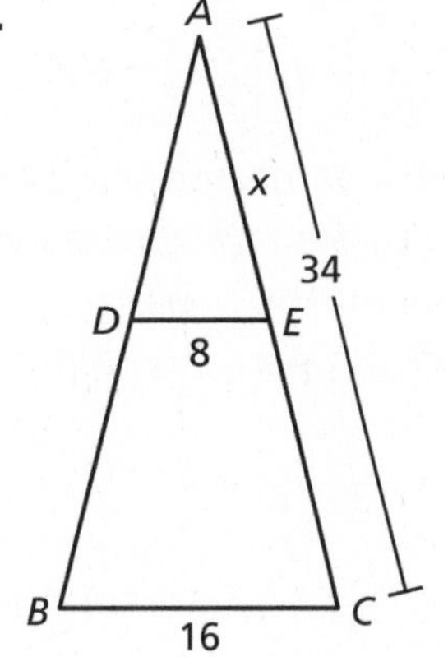

Name ________________________________ Date __________

Puzzle Time

Did You Hear About The...

A	B	C	D	E	F
G	H	I	J	K	L
M					

Complete each exercise. Find the answer in the answer column. Write the word under the answer in the box containing the exercise letter.

525 UP
15 HOLDUP
10% WIND
96 PANTS
22 SOCKS
25% WHEN
33 PAIR
200 HELD
37.5% IN
625 ROPE

Write and solve a proportion to answer the question.

A. What number is 25% of 60?

B. 30 is what percent of 80?

C. 20 is 40% of what number?

D. What number is 110% of 70?

E. 32 is what percent of 128?

F. What percent of 48 is 36?

G. 12 is what percent of 60?

H. 8% of what number is 16?

I. 42 is 8% of what number?

J. 30% of 140 is what number?

K. 150% of 22 is what number?

L. What number is 0.6% of 30?

M. The seventh-grade class needs to earn money for a trip to the amusement park. Of the 160 seventh-grade students, 60% participate in the fundraiser. How many students participate in the fundraiser?

20% CLOTHESPINS
50 THE
36% SUNSHINE
0.18 OF
120 CLOTHESLINE
75% TWO
100% RAINBOW
42 A
77 YARD
325 WRINKLES

Activity 6.4

Start Thinking!

For use before Activity 6.4

Explain to a partner how to use the percent proportion to determine the number of points you need to get a score of 80% on a 45-point test.

Activity 6.4

Warm Up

For use before Activity 6.4

Write and solve a proportion to answer the question.

1. What percent of 15 is 6?
2. 22 is what percent of 40?
3. What number is 60% of 55?
4. 30% of 63 is what number?
5. 0.5% of what number is 0.425?
6. $\frac{3}{4}$ is 75% of what number?

Lesson 6.4 Start Thinking!

For use before Lesson 6.4

Two students are working together to answer the question “what number is 5% of 30?”.

One student says to use the percent proportion $\frac{30}{x} = \frac{5}{100}$ and the other student says to use the percent equation $a = 0.05(30)$.

Who is correct? Explain your reasoning. Find the number.

Lesson 6.4 Warm Up

For use before Lesson 6.4

Answer the question. Explain the method you chose.

1. What number is 30% of 60?

2. 25 is what percent of 75?

3. 42 is 105% of what number?

4. What number is 80% of 120?

5. 45 is what percent of 100?

6. 91 is 65% of what number?

Name____ Date__________

6.4 Practice A

Answer the question. Explain the method you chose.

1. 24 is what percent of 60?
2. 8 is 40% of what number?

Write and solve an equation to answer the question.

3. What number is 70% of 120?
4. 30 is what percent of 120?
5. 112 is 56% of what number?
6. 128 is what percent of 80?
7. What number is 140% of 45?
8. 15 is 6% of what number?
9. There are 35 competitors in a marathon. Sixty percent of these finished the race in under four hours. How many competitors finished the race in under four hours?
10. Your class is going on a field trip. Twenty-four students have turned in their permission slips so far. This is 80% of the students in the class. How many students are in the class?
11. You take a test with 32 questions on it. You answer 24 questions correctly. What percent of the questions do you answer correctly?
12. You have r rare coins, consisting of p pennies and n nickels.
    a. p is 20% of 190. How many pennies do you have?
    b. 190 is 200% of r. How many rare coins do you have?
    c. n is 60% of r. How many nickels do you have?
13. The table shows the sales receipt for your purchase.
    a. The items with a "T" next to the price are subject to sales tax. What percent sales tax did you pay?
    b. Calculate the price of the top.
    c. The price you paid for the top was 60% of the original price. What was the original price of the top?

Item	Price
top	p
earrings	\$ 3.00 T
socks	\$ 2.00
granola bar	\$ 0.50 T
Subtotal	\$13.00
Tax	\$ 0.21
Total	\$13.21

Tell whether the following statement is *true* or *false*. Explain your reasoning.

14. 120% of a whole number is always greater than the number.
15. You can find 0.5% of a number by multiplying the number by $\frac{5}{100}$.

Name ______________________________ Date ________

6.4 Practice B

Answer the question. Explain the method you chose.

1. 27 is what percent of 90?

2. 7 is 5% of what number?

Write and solve an equation to answer the question.

3. 27 is 0.5% of what number?

4. What number is 125% of 240?

5. 1.4% of what number is 28?

6. 27 is what percent of 72?

7. During a given month, there was a total of 23.6 inches of rain. This was 250% of the average rainfall for that month. What is the average rainfall for that month?

8. To maintain an acceptable level of chlorine in your pool, you add 1.4 gallons of chlorine. This is 0.007% of the amount of water in your pool. How many gallons of water are in your pool?

9. You must attend a minimum of 85% of the practices in order to play in the playoffs. You have made 37 of the 42 practices. Will you be able to play in the playoffs?

10. You are in charge of the seventh grade graduation dinner. The table shows the results of a survey of students' meal preferences.

Choice	Percent
Chicken Nuggets	25%
Spaghetti	?
Pizza	45%
Fish Sticks	?

   a. 144 students chose pizza. How many students responded to the survey?

   b. How many students chose chicken nuggets?

   c. The number of students choosing fish sticks was 50% of the number of students choosing spaghetti. How many students chose fish sticks?

   d. How many students chose spaghetti?

11. What is 15% of 40% of $180?

12. There are 15 copies of a popular CD left to be sold in a store. This is between 1% and 1.5% of the original number of copies of the CD in the store. The original number of CDs was between what two numbers?

13. Tell whether the statement is *true* or *false*. Explain your reasoning.

If A is 45% of B, then the ratio $A : B$ is $9 : 20$.

Find the percent to the nearest hundredth.

14. 16 is what percent of 38?

15. 50 is what percent of 38?

Name______________________________ Date__________

6.4 Enrichment and Extension

The Percent is Right: Closest Without Going Over

Preparation:

- Cut index cards to make 20 playing cards. Write each number from 1 to 20 on the cards.

To play:

- On your turn, choose a card and record that value. Your current score is the percent of the value chosen out of the total possible, which is 20.
- With each consecutive turn, add your new card value to your previous total. Also add 20 to the total possible. Then, calculate your new score.
- You may choose to draw as many as 4 times, replacing the cards each time. You may end your turn at any time. But, you may not go back and draw again after other players have gone.
- The player who comes closest to 50% without going over, wins. If players tie, the one who drew the fewest number of cards is the winner.

A sample score card is shown.

Turn	Card Value	Total Card Value	Total Possible	Score
1	7	7	20	35%
2	15	22	40	55%
3	6	28	60	46.6%

Complete the following exercises after playing a few rounds.

1. You draw a 5 on your first turn and a 17 on your second turn. Should you draw again? Explain your reasoning.
2. On your first 3 draws, you draw a 5, a 6, and a 10. Your opponent has already finished with a score of 42%. What is the lowest card you can draw and still win? What is the highest card you can draw and still win? What percent of the cards could you draw and still win the game?
3. What is the most you can get on your first three draws and still have a chance to win? Explain.
4. You are the first player to take a turn and you draw a 7. Should you draw again? Explain your reasoning. What if your first draw is an 8 or a 9?
5. If you have time, you can play multiple rounds with a different target percent score each round. Explain how a higher or lower target percent score would change the game.

Name ______________________________ Date __________

Puzzle Time

Did You Hear About...

A	B	C	D	E	F
G	H	I	J	K	L
M	N				

Complete each exercise. Find the answer in the answer column. Write the word under the answer in the box containing the exercise letter.

Answer	Word
42	HIS
150	THAT
52	TO
48	DOWN
15	THE
300	TEAM
75	MATCH
55	SET
145	ENDED
175	PLAYED
65	EAR
35	CHEER
93	FAN
82	TOO

Solve an equation to answer the question.

A. What percent of 60 is 9?

B. 12 is what percent of 30?

C. 62% of 150 is what number?

D. 8% of what number is 2?

E. What percent of 130 is 91?

F. 13% of 400 is what number?

G. 72 is what percent of 80?

H. 64% of what number is 48?

I. 175% of 8 is what number?

J. What percent of 200 is 290?

K. 156 is what percent of 60?

L. 0.5% of what number is 3?

M. Of the 60 students in the seventh grade, 70% own a pet. How many of the seventh grade students own a pet?

N. A pair of sunglasses is on sale for $26. The original price of the sunglasses was $40. What percent of the original price is the sale price?

Answer	Word
17	ARE
25	WHO
225	THIS
260	UP
125	GAME
90	A
45	COACH
40	SPORTS
97	STARTED
70	LISTENED
20	RUNNING
600	BURNING
400	PLAYING
14	AND

Explain how a meteorologist uses percents.

Activity 6.5 Warm Up

For use before Activity 6.5

Write the decimal as a percent.

1. 0.45 **2.** 1.34 **3.** 0.549

4. 1.08 **5.** 0.985 **6.** 0.3225

Start Thinking!

For use before Lesson 6.5

Work with a partner to write and solve a real-life problem for the following situation: $250 decreased by 35%.

Warm Up

For use before Lesson 6.5

Find the new amount.

1. 15 inches increased by 20%

2. 145 gallons decreased by 60%

3. 70 meters increased by 80%

4. 150 grams decreased by 74%

5. 120 pounds decreased 5%

6. 40 liters increased by 25%

Name____ Date__________

6.5 Practice A

Find the new amount.

1. 12 dogs decreased by 25%
2. 140 fluid ounces increased by 45%
3. 100 textbooks increased by 99%
4. 75 students decreased by 80%

Identify the percent of change as an *increase* or a *decrease*. Then find the percent of change. Round to the nearest tenth of a percent, if necessary.

5. 5 cups to 8 cups
6. 150 pounds to 135 pounds
7. 14 dollars to 10 dollars
8. 28 seconds to 23 seconds
9. $\frac{1}{3}$ to $\frac{2}{3}$
10. $\frac{1}{3}$ to $\frac{1}{6}$
11. Yesterday your bus ride to school took 10 minutes. Today your bus ride took 12 minutes. What is the percent of change?
12. Yesterday 270 concert tickets were sold. Today 216 tickets were sold.
    a. Find the percent of change in the number of tickets sold from yesterday to today.
    b. Use the percent of change from part (a) to predict the number of tickets sold tomorrow. Round to the nearest ticket, if necessary.
    c. Find the predicted percent of change in the number of tickets sold from yesterday to tomorrow. Round to the nearest tenth of a percent, if necessary.
13. This month a band has 6 musicians. This is a 50% increase from the number of musicians in the band last month. How many musicians were in the band last month?
14. The sides of a square garden are 8 feet long.
    a. You enlarge the garden to create a 25% increase in the length of each side. Find the new length of the sides.
    b. Find the percent of change in the perimeter of the garden. Round to the nearest tenth of a percent, if necessary.
    c. Find the percent of change in the area of the garden. Round to the nearest tenth of a percent, if necessary.

Name ______________________ Date __________

6.5 Practice B

Find the new amount.

1. 55 employees increased by 20%
2. 25° decreased by 60%
3. 15 customers increased by 200%
4. 4200 fans increased by 0.5%

Identify the percent of change as an *increase* or a *decrease*. Then find the percent of change. Round to the nearest tenth of a percent, if necessary.

5. 3.2 kilograms to 2.4 kilograms
6. 41 euros to 85 euros
7. $\frac{2}{7}$ to $\frac{4}{7}$
8. $\frac{5}{6}$ to $\frac{1}{3}$
9. Last month you swam the 50-meter freestyle in 28.38 seconds. Today you swam it in 27.33 seconds. What is your percent of change? Round to the nearest tenth of a percent, if necessary.
10. Last week 1200 burgers were served at the Burger Barn.
    a. This week 1176 burgers were served. What is the percent of change?
    b. Use the percent of change from part (a) to predict the number of burgers served next week. Round to the nearest whole number, if necessary.
11. The price of a share of a stock was $37.50 yesterday.
    a. Today there was a price decrease of 4%. What is today's price?
    b. Based on today's price in part (a), what percent of change is needed to bring the price back up to $37.50? Round to the nearest tenth of a percent, if necessary.
12. The table shows the membership of two scout troops.

Year	Troop A	Troop B
2010	14	21
2011	16	24

    a. What is the percent of change in membership from 2010 to 2011 for Troop A? Round to the nearest tenth of a percent, if necessary.
    b. What is the percent of change in membership from 2010 to 2011 for Troop B? Round to the nearest tenth of a percent, if necessary.
    c. Which troop has the better record in terms of the number of new members?
    d. Which troop has the better record in terms of the percent of change in membership?

Name______________________________ Date__________

6.5 Enrichment and Extension

Selling Super Slick Shoes

You work at Super Slick Shoes and are planning a new advertising campaign. You have been given information about the shoe sales of your company and your competitor. You want to be sure that you are making your company sound as good as possible.

1. Your athletic shoe department sold 135 thousand pairs of Super Sweet Sneakers this year, which is a 125% increase over last year's sales.

   a. How many pairs of Super Sweet Sneakers did you sell last year?

   b. By how many thousand pairs did the sales increase?

2. The athletic shoe department of your competitors sold 258 thousand pairs of Super Smooth Sneakers this year, which is a 50% increase over last year's sales.

   a. How many pairs of Super Smooth Sneakers did they sell last year?

   b. By how many thousand pairs did their sales increase?

3. For sneakers, would you use the *amount* of increase in sales or the *percent* of increase in sales to make your company sound better than your competitor? Explain your reasoning.

4. Your designer shoe department sold 175 thousand pairs of Super Sleek Sandals this year and anticipates a 70% increase in sales next year.

   a. How many pairs of Super Sleek Sandals do you expect to sell next year?

   b. By how many thousand pairs do you expect the sales to increase?

5. Your competitors sold 150 thousand pairs of Super Snazzy Sandals this year and anticipate an 80% increase in sales next year.

   a. How many pairs of Super Snazzy Sandals do they expect to sell next year?

   b. By how many thousand pairs do they expect their sales to increase?

6. For sandals, would you use the anticipated *amount* of increase in sales or the anticipated *percent* of increase in sales to make your company sound better than your competitor? Explain your reasoning.

Name ______________________________ Date __________

Puzzle Time

Why Did the Picture Go To Jail?

Write the letter of each answer in the box containing the exercise number.

Find the new amount.

1. 42 customers increased by 50%
2. 70 pennies decreased by 20%
3. 30 inches increased by 30%
4. 125 pounds decreased by 4%
5. 15 acres decreased by 80%
6. 12 stamps increased by 125%
7. 68 days increased by 75%
8. 440 miles decreased by 95%

Identify the percent of change as an *increase* or a *decrease*. Then find the percent of change.

9. 20 ounces to 25 ounces
10. 10 hours to 6 hours
11. 15 feet to 33 feet
12. 80 books to 64 books
13. 64 minutes to 144 minutes
14. 16 skateboards to 2 skateboards
15. 55 videos to 77 videos
16. 13.6 kilometers to 23.8 kilometers

Answers for Exercises 1–8

T. 3 **O.** 119

S. 120 **D.** 27

F. 22 **E.** 56

M. 39 **Y.** 63

Answers for Exercises 9–16

O. 40% increase

B. 87.5% decrease

I. 20% decrease

R. 25% increase

E. 75% increase

M. 120% increase

D. 40% decrease

A. 125% increase

4	15	11	2	14	7	10	1		8	9	13	3	16	6		12	5

Start Thinking!

For use before Activity 6.6

Why does a store mark up an item?

Why does a store then sometimes discount that item?

Warm Up

For use before Activity 6.6

Write and solve an equation to answer the question.

1. What is 40% of 238?
2. 28 is what percent of 70?
3. What is 34% of 240?
4. 5% of what number is 6?
5. What is 110% of 150?
6. 42 is 250% of what number?

Lesson 6.6

Start Thinking!

For use before Lesson 6.6

You go to a store to buy a new pair of jeans. You find 2 pairs of jeans each on sale for a different price. Explain which is a better bargain.

Regular Price: \$35; Discount: 30%

Regular Price: \$40; Discount: 35%

Warm Up

For use before Lesson 6.6

Find the cost to store, percent of markup, or selling price.

1. Cost to store: \$20
Markup: 15%
Selling price: ___?___

2. Cost to store: \$65
Markup: 30%
Selling price: ___?___

3. Cost to store: \$100
Markup: 12%
Selling price: ___?___

4. Cost to store: \$150
Markup: 5%
Selling price: ___?___

Name____ Date__________

6.6 Practice A

Copy and complete the table.

	Original Price	Percent of Discount	Sale Price
1.	$75	30%	
2.	$18	65%	
3.		30%	$42
4.		55%	$90
5.	$35		$28
6.	$55		$46.75

Find the cost to store or selling price.

7. Cost to store: $65

Markup: 25%

Selling price: __?__

8. Cost to store: __?__

Markup: 80%

Selling price: $122.40

9. The cost to a store for a box of cereal is $2.50. The store is selling the box of cereal for $3.50. What is the percent of markup?

10. A store pays $120 for a bicycle.

a. The store has a 60% markup policy. What is the selling price of the bicycle?

b. The store is now going out of business and is selling all of the bicycles at a 30% discount. What is the sale price of the bicycle?

c. Will the store make money or lose money on the bicycle? How much?

11. The selling price of a skateboard is $147. The store has a 75% markup policy. What is the cost of the skateboard to the store?

12. You buy a watch for $60.

a. There is a 6% sales tax. What is your total cost for the watch?

b. Your friend buys the same watch a month later. It is now sold at a discount of 15%. What is the new sale price?

c. What is your friend's total cost for the watch including tax?

d. What is the percent of change in the total cost?

Name ______________________________ Date ________

6.6 Practice B

Find the original price, discount, sale price, selling price, markup, or cost to store. Round to the nearest penny, if necessary.

1. Original price: $130

Discount: 45%

Sale price: ___?___

2. Original price: $500

Discount: ___?___

Sale price: $175

3. Original price: ___?___

Discount: 5%

Sale price: $68.40

4. Cost to store: $1600

Markup: 33%

Selling price: ___?___

5. Cost to store: $65

Markup: ___?___

Selling price: $91

6. Cost to store: ___?___

Markup: 25%

Selling price: $437.50

7. You are buying shoes online. The selling price is $29.99. Round to the nearest penny, if necessary.

a. The sales tax is 6.5%. What is the total cost?

b. The cost of shipping is 15% of the total cost. What is the total cost plus shipping?

c. If the total cost plus shipping is greater than $35, then you receive a 10% discount off the original selling price. Do you qualify? If so, what is the new total cost plus shipping?

8. You have a coupon for $15 off a video game. You can use it on 2 separate days.

a. On Monday, the discounted price of your video game is $22.99. What is the original price of the game?

b. What is the percent of discount to the nearest percent?

c. On Thursday, the discounted price of your video game is $12.99. What is the original price of the game?

d. What is the percent of discount to the nearest percent?

9. You buy a bracelet for $15. You sell it at a craft show for $25. What is the percent of markup to the nearest percent?

Name____ Date__________

6.6 Enrichment and Extension

Would You Get the Job?

You are applying for a job selling electronics. The interviewer asks you to answer the following questions in order to prove that you would be able to do your job effectively.

1. A customer is purchasing last year's version of the Football Mania video game, which was originally \$55. It is on a sale rack that says 60% off. The customer has a coupon for 30% off his entire purchase. He incorrectly says, "I can't believe I'm getting this for 90% off!"

   a. Find the cost of the video game after the coupon.

   b What would be the cost of the video game at 90% off?

   c. How would you explain to the customer that he is not getting 90% off?

2. Your boss has decided to make room for new models of digital cameras by discounting old models. In order to sell them quickly, she wants to sell them at the store's original cost before markup. She asks you to change the price tags and make signs with the percent of discount. Knowing that the store's percent of markup on digital cameras is 40%, a co-worker suggests that you just mark them down by 40%.

   a. How would you explain to your co-worker that the store would actually lose money if you did this?

   b. The original price of a camera is \$175.50. What was the store's original cost on this camera?

   c. What percent off will you write on the sign in order to sell this camera at cost? Round to the nearest percent.

3. If a customer brings an ad that shows a competitor is selling an item for less, you are permitted to match the competitor's price as long as the percent of discount is 15% or less.

   a. The newest Blu-ray disc player is listed at \$307.99. You notice on your way to work that the same player is for sale at another store for \$298.79. Would you be allowed to match this store's price if asked? Explain.

   b. The store sells a wireless Internet router for \$74.75. A customer has printed out a page from the Internet that shows the same router for \$59.97 plus 5% shipping and handling. Can you match this price? Explain.

Name ____ Date __________

6.6 Puzzle Time

What Does A Monster Say When Introduced?

Write the letter of each answer in the box containing the exercise number.

Find the sale price or the selling price.

1. $99 watch with a 40% discount
2. $32 earrings marked up 80%
3. $59 cell phone with a 10% discount
4. $65 digital camera marked up 50%
5. $35 soccer cleats marked up 95%

Find the original price.

6. swimsuit discounted 65% on sale for $20.30
7. tennis racket marked up 50% to $36.75
8. backpack discounted 10% on sale for $37.80
9. tickets marked up 35% to $59.40
10. perfume discounted 25% on sale for $27

Find the percent of discount or markup.

11. $52 sweater on sale for $31.20
12. $23 rollerblades marked up to $41.40
13. $30 board game on sale for $20.10
14. $0.60 bottle of iced tea marked up to $1.35
15. $3.50 nail polish marked up to $8.75

Answers for Exercises 1–10

T. $68.25	**U.** $53.10
E. $97.50	**A.** $42
L. $44	**O.** $57.60
A. $58	**D.** $24.50
S. $36	**E.** $59.40

Answers for Exercises 11–15

P. 125%	**Y.** 80%
E. 150%	**T.** 33%
O. 40%	

14	9	1	6	10	4	7		13	2		15	8	5		12	11	3

Start Thinking!

For use before Activity 6.7

Discuss the pros and cons of having a credit card.

Warm Up

For use before Activity 6.7

Find the sale price for the item. Round to the nearest cent.

1. Original price: $300; Discount: 25%
2. Original price: $75; Discount: 40%
3. Original price: $1300; Discount: 20%
4. Original price: $95; Discount: 15%
5. Original price: $725; Discount: 10%
6. Original price: $845; Discount: 35%

Start Thinking!

For use before Lesson 6.7

You earned $150 babysitting. You want to open a savings account. What factors must you consider before opening an account?

Warm Up

For use before Lesson 6.7

An account earns simple interest. (a) Find the interest earned. (b) Find the balance of the account.

1. $750 at 2% for 3 years

2. $300 at 6% for 2 years

3. $1400 at 4% for 5 years

4. $600 at 4.5% for 7 years

5. $550 at 8% for 6 months

6. $1200 at 3.5% for 6 months

Name____ Date__________

6.7 Practice A

An account earns simple interest. (a) Find the interest earned. (b) Find the balance of the account.

1. \$200 at 3% for 5 years

2. \$750 at 8% for 2 years

3. \$1600 at 5% for 1 year

4. \$500 at 12% for 6 months

Find the annual interest rate.

5. $I = \$18$, $P = \$150$, $t = 6$ years

6. $I = \$164.50$, $P = \$940$, $t = 2.5$ years

Find the amount of time.

7. $I = \$72$, $P = \$600$, $r = 4\%$

8. $I = \$174$, $P = \$1450$, $r = 8\%$

9. You deposit \$350 in a savings account. The account earns 2.5% simple interest per year. What is the balance after 2 years?

Find the amount paid for the loan.

10. \$1000 at 8% for 5 years

11. \$3500 at 10% for 2 years

12. You deposit \$2000 in a savings account earning 5% simple interest. How long will it take for the balance of the account to be \$3800?

13. Your parents charge a family ski trip of \$3000 on a credit card.

a. The simple interest rate is 20%. The charges are paid after 6 months. What is the amount of interest paid?

b. What is the total amount paid for the ski trip?

14. Your parents could have taken out a loan for the ski trip in Exercise 13.

a. The simple interest rate is 6% and the time for the loan is 2 years. What would have been the total amount paid for the \$3000 ski trip?

b. What would be the monthly payment, if there were equal monthly payments?

c. Which loan option costs less, the credit card or the loan?

15. You deposit \$1200 in an account earning 8% simple interest.

a. What is the account balance after 1 year?

b. At the end of the first year, you deposit the balance of the account in a CD (certificate of deposit) earning 8% simple interest. What is the account balance after another year?

Name ____________________ Date __________

6.7 Practice B

An account earns simple interest. (a) Find the interest earned. (b) Find the balance of the account.

1. \$2600 at 3.2% for 4 years
2. \$75,000 at 8.5% for 3 months

Find the annual interest rate.

3. $I = \$41.80$, $P = \$440$, $t = 2$ years
4. $I = \$893.75$, $P = \$5500$, $t = 30$ months

Find the amount of time.

5. $I = \$9.90$, $P = \$360$, $r = 5.5\%$
6. $I = \$2064$, $P = \$10,000$, $r = 6.88\%$

Find the amount paid for the loan.

7. \$20,000 at 7.5% for 10 years
8. \$6000 at 12% for 2.5 years
9. You deposit \$2000 in an account. The account earns \$120 simple interest in 8 months. What is the annual interest rate?
10. You put money in two different accounts for one year each. The total simple interest for the two accounts is \$140. You earn 6% interest on the first account, in which you deposited \$1000. You deposited \$800 in the second account. What is the annual interest rate for the second account?
11. You deposit \$1200 in an account.

    a. The account earns 2.7% simple interest rate. What is the balance of the account after 3 months?

    b. The interest rate changes, and your new balance now earns 2% simple interest rate. What is the balance of the account after the next 6 months? Round to the nearest penny, if necessary.

    c. The interest rate changes again, and your new balance now earns 2.6% simple interest rate. What is the balance of the account after an additional 3 months? Round to the nearest penny, if necessary.

    d. How much did the account earn in simple interest for the year?

    e. Based on the interest in part (d), what is the actual simple interest rate for the year? Round to the nearest tenth of a percent.

12. You purchase a new guitar and take out a loan for \$450. You have 18 equal monthly payments of \$28 each. What is the simple interest rate for the loan? Round to the nearest tenth of a percent, if necessary.

Name____ Date__________

6.7 Enrichment and Extension

Buying the Car of Your Dreams

People who sell cars are usually good negotiators. So, being an educated consumer is important. Most car loans are based on *compound interest*, which means that you pay interest on your interest. You will be better able to negotiate a good deal on a car if you understand how compound interest is calculated.

Example: A loan for $15,560 is taken out for 5 years at a yearly interest rate of 7.2% that is compounded annually. (a) What is the balance after 5 years? (b) What is the monthly payment?

a. $B = P(1 + r)^t$ — B = balance, P = principal, r = interest rate (in decimal form), t = number of times the interest is compounded

$= 15{,}560(1 + 0.072)^5$ — Substitute 15,560 for P, 0.072 for r, and 5 for t.

$= 15{,}560(1.072)^5$ — Add.

$= 22{,}028.43$ — Simplify.

The balance after 5 years is $22,028.43.

b. There are 60 months in 5 years. So, divide the balance by 60.

$$\frac{22{,}028.43}{60} \approx 367.14$$

The monthly payment is $367.14.

1. A salesman offers two discount options on the car in the example. He can decrease the initial cost of the car by $500 or decrease the interest rate by 0.5%. Find the final cost of both options. Which is the better deal?

2. Find the final cost of the loan in the example if the interest were compounded monthly instead of annually. (*Hint:* Divide the yearly interest rate by 12 to find the monthly interest rate.) Why are some loans compounded monthly and even daily?

3. Find the price of your dream car on the Internet, in the newspaper, or from another source. Then research the current interest rate at a dealership or other lender and how often their interest is compounded. Calculate the final cost and monthly payment for your dream car on a 5-year loan.

Name ______________________________ Date __________

6.7 Puzzle Time

What Question Can You Never Answer Yes To?

Write the letter of each answer in the box containing the exercise number.

Find the interest earned.

1. $500 at 4% for 3 years

D. $50 **E.** $60 **F.** $100

2. $1200 at 7% for 5 years

A. $420 **B.** $240 **C.** $400

3. $750 at 6% for 18 months

T. $58.50 **U.** $67.50 **V.** $56.75

4. $1500 at 8.5% for 6 months

J. $62.50 **K.** $65.50 **L.** $63.75

Find the annual interest rate.

5. $I = \$200, P = \$1000, t = 4$ years

E. 5% **F.** 6% **G.** 7%

6. $I = \$30, P = \$600, t = 2$ years

M. 1.5% **N.** 2.25% **O.** 2.5%

7. $I = \$150, P = \$2500, t = 9$ months

R. 6% **S.** 8% **T.** 9%

8. $I = \$75, P = \$800, t = 15$ months

D. 6.5% **E.** 7.5% **F.** 8.5%

Find the amount of time.

9. $I = \$144, P = \$400, r = 6\%$

P. 3 years **Q.** 4 years **R.** 6 years

10. $I = \$236.25, P = \$750, r = 4.5\%$

N. 3 years **O.** 5 years **P.** 7 years

11. $I = \$87.50, P = \$3500, r = 5\%$

W. 3 years **X.** 3 months **Y.** 6 months

12. $I = \$108.75, P = \$2000, r = 7.25\%$

A. 9 months **B.** 6 months **C.** 3 months

2	9	5		11	6	3		12	7	4	1	8	10

Name______________________________ Date__________

Technology Connection

For use after Section 6.4

Finding Percents with a Calculator

In Section 6.4, to solve the percent equation $a = p \bullet w$ you converted the percent p to fraction or decimal form. You could also use a calculator with a [%] key to solve the equation.

EXAMPLE What number is 55% of 176?

SOLUTION

Step 1 This problem can be modeled by the equation $a = 55\% \bullet 176$.

Step 2 On your calculator, enter 55 [%] [×] 176 [=].

ANSWER 96.8

EXAMPLE 42 is 15% of what number?

SOLUTION

Step 1 This problem can be modeled by the equation $42 = 15\% \bullet w$.

Step 2 On your calculator, enter 42 [÷] 15 [%] [=].

ANSWER 280

Use a calculator to solve the percent problem.

1. What number is 84% of 325?
2. What number is 250% of 4?
3. What number is 90% of 30?
4. What number is 15% of 450?
5. 75% of what number is 225?
6. 98 is 14% of what number?
7. 6.2 is 40% of what number?
8. 0.5% of what number is 2?
9. According to the NCAA, only 2% of high school athletes receive an athletic scholarship for college. Your school has a total of 250 athletes. How many athletes would you expect to receive a scholarship?
10. In a fundraiser, 15% of the money collected goes to your school. If your school earns $780, how much money is collected?

Chapter 7

Name______________________________ Date__________

Constructions and Scale Drawings

Dear Family,

Do you know someone who works with blueprints on a daily basis? One possible career that uses blueprints often is an architect. An architect designs buildings and structure. They must take in consideration not only the look but also safety, function, and cost when designing structures. Architects work with engineers to create blueprints for the builders.

How do architects and engineers create blueprints? How does a builder then read the blueprint to create buildings or other types of structures?

Spend some time with your student researching the questions above. If you are able, visit a local architecture or engineering firm. They will be able to show you examples of blueprints.

Have your student take note of how engineers create the blueprints of a structure an architect has designed.

What type of lines and angles do they use?

What type of polygons are formed when they create the drawings?

What information does the engineer need about the location of the structure to make accurate drawings?

Once the blueprints are complete, they are sent to the builder. Consider visiting a building site to watch how the drawings come to life. Here are some other questions to think about as you watch the building being built.

How does a builder read the blueprints?

What if the measurements are wrong? How does that affect the final product?

Enjoy your journey with your student as you watch a building come to life that once started out as a drawing in an architect's office.

Nombre ________________________________ Fecha ________

Construcciones y dibujos a escala

Estimada familia,

¿Conocen a alguien que trabaje con planos diariamente? Una carrera que usualmente requiere del uso de planos es la arquitectura. Un arquitecto diseña edificios y estructuras. Debe tener en cuenta no solo la apariencia sino también la seguridad, función y el costo al diseñar estructuras. Los arquitectos trabajan con los ingenieros para hacer planos para los constructores.

¿Cómo hacen sus planos los arquitectos e ingenieros? ¿De qué forma luego un constructor lee los planos para construir edificios u otro tipo de estructuras?

Dedique tiempo con su estudiante a investigar las preguntas anteriores. Si puede, visite una firma de arquitectos o ingenieros de su vecindario. Ellos le enseñarán muestras de planos.

Haga que su estudiante tome nota de la manera en que los ingenieros hacen los planos de una estructura que ha diseñado un arquitecto.

> ¿Qué tipo de líneas y ángulos usan?
>
> ¿Qué tipo de polígonos se forman cuando ellos hacen los dibujos?
>
> ¿Qué información necesita el ingeniero sobre la ubicación de la estructura para hacer dibujos exactos?

Una vez se terminan los planos, se envían al constructor. Considere la posibilidad de visitar una construcción para ver cómo los dibujos se hacen realidad. He aquí otras preguntas que pueden hacerse mientras ven cómo se construye el edificio.

> ¿Cómo lee los planos el constructor?
>
> ¿Qué pasaría si las medidas estuvieran equivocadas? ¿Cómo afectaría esto al producto final?

Disfrute de una aventura con su estudiante observando cómo un edificio que comienza como un dibujo en la oficina de un arquitecto, cobra vida.

Activity 7.1 Start Thinking!

For use before Activity 7.1

What is an acute angle?

What is an obtuse angle?

Use a dictionary to look up the non-math definitions of *acute* and *obtuse*. How do the definitions relate to the math definitions?

Activity 7.1 Warm Up

For use before Activity 7.1

Identify the angles as *acute*, *right*, *obtuse*, or *straight*.

1. 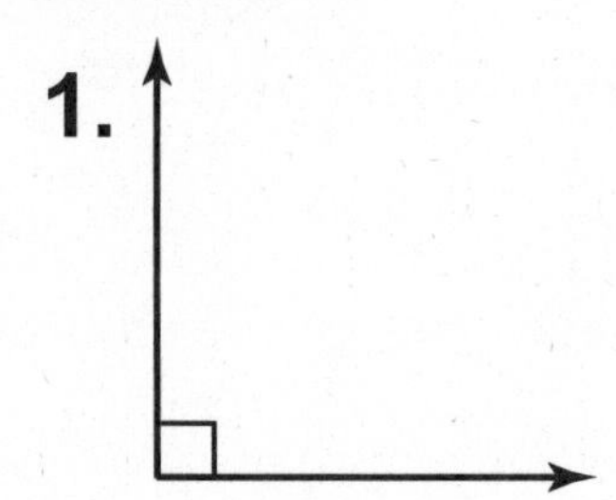

2. 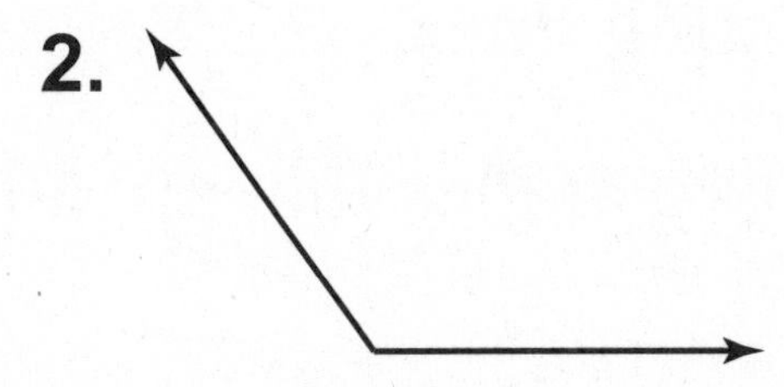

3. 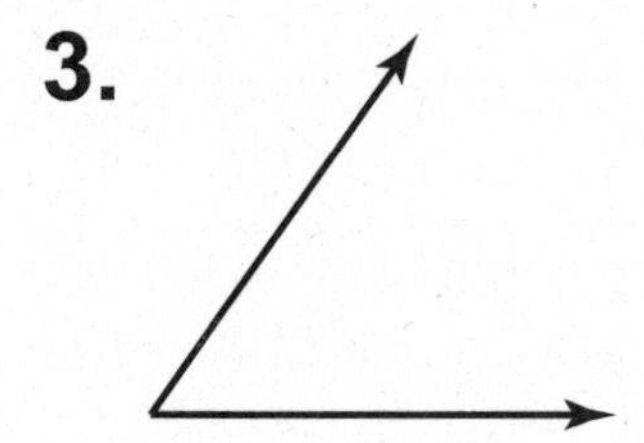

4. 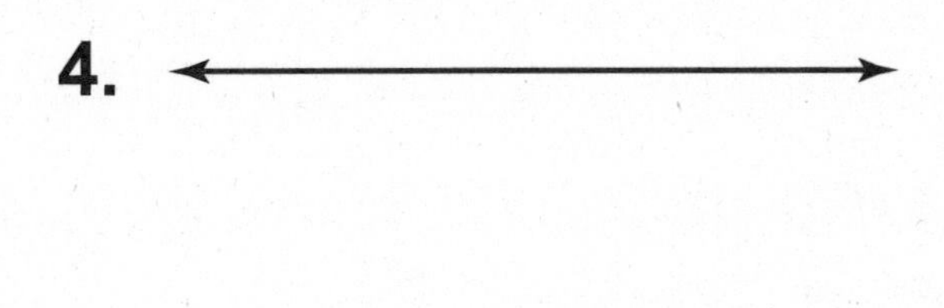

Lesson 7.1 Start Thinking!
For use before Lesson 7.1

Draw two lines that intersect. Explain to a partner how to locate a pair of adjacent angles.

Lesson 7.1 Warm Up
For use before Lesson 7.1

Use the figure below.

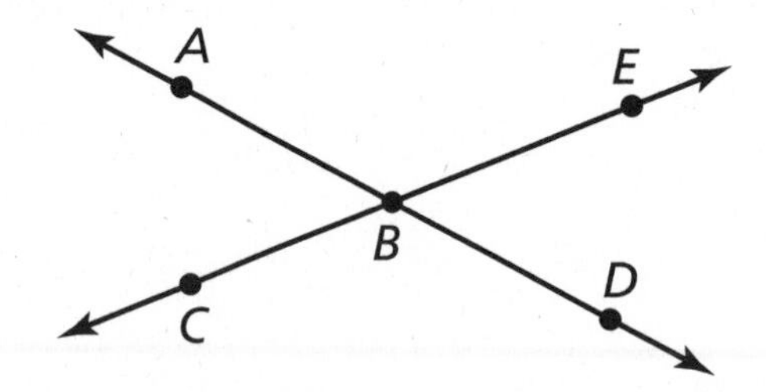

1. Measure each angle formed by the intersecting lines.

2. Name two angles that are adjacent to $\angle ABC$.

Name____ Date__________

7.1 Practice A

Name two pairs of adjacent angles and two pairs of vertical angles in the figure.

1.

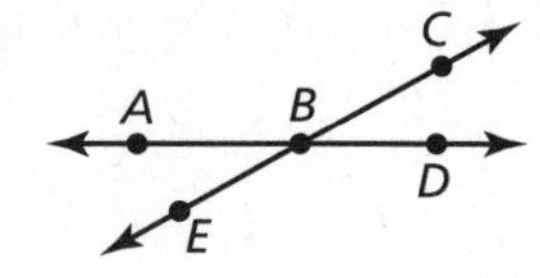

2.

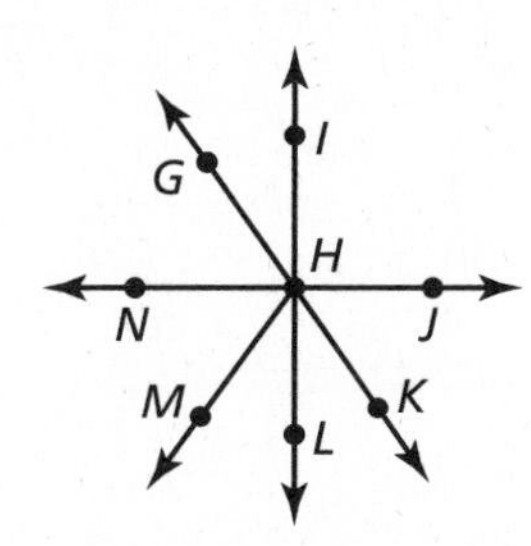

Tell whether the angles are *adjacent* or *vertical*. Then find the value of *x*.

3.

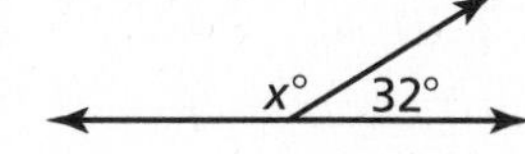

4.

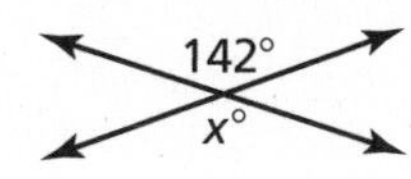

5.

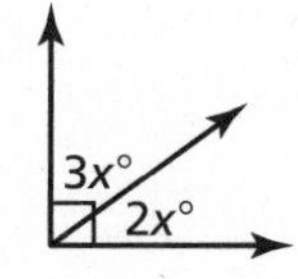

6.

$6x°$

$(x - 2)°$

Draw a pair of vertical angles with the given measure.

7. 40° **8.** 75° **9.** 120°

10. Draw a pair of adjacent angles with the given description.

a. Both angles are obtuse.

b. The sum of the angle measures is 180°.

c. The sum of the angles measures is 60°.

11. What are the measures of the other three angles formed by the intersection?

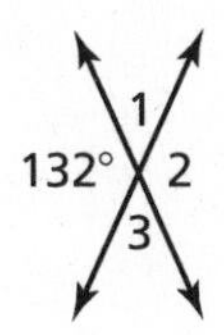

Name ______________________________ Date __________

7.1 Practice B

Name two pairs of adjacent angles and two pairs of vertical angles in the figure.

1.

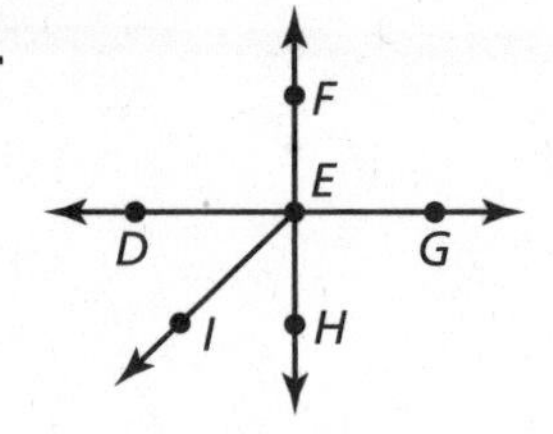

2.

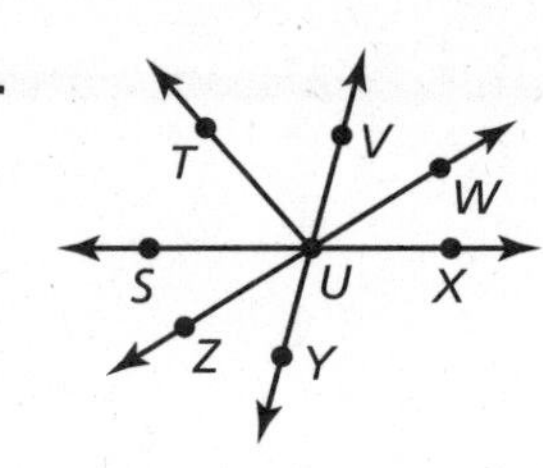

Tell whether the angles are *adjacent* or *vertical*. Then find the value of *x*.

3.

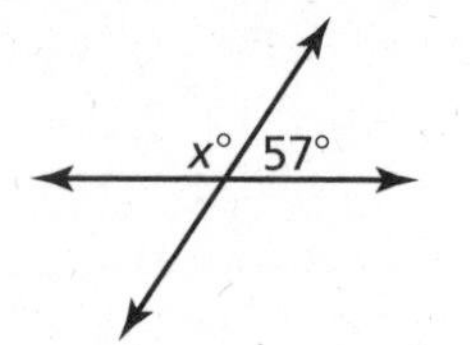

4.

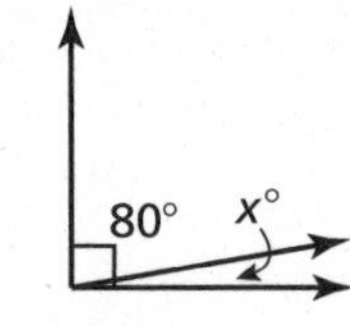

5.

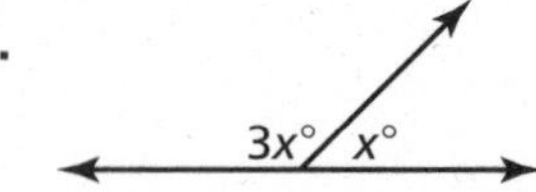

6.

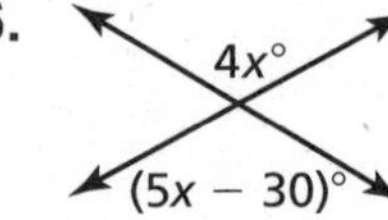

Draw a pair of vertical angles with the given measure.

7. 100° **8.** 15° **9.** 150°

10. Draw five angles so that ∠2 and ∠3 are acute vertical angles, ∠1 and ∠2 are supplementary, ∠2 and ∠5 are complementary, and ∠4 and ∠5 are adjacent.

11. The measures of two adjacent angles have a ratio of 3 : 5. The sum of the measures of the two adjacent angles is 120°. What is the measure of the larger angle?

Name____ Date__________

7.1 Enrichment and Extension

Compass Straightedge Construction

In addition to modern technology, a geometer's most important tools are a compass and a straightedge (ruler without marks). These instruments can be used for geometric drawings called constructions.

Bisecting the Angle

Bisecting the angle is the process of cutting a given angle in half so that both halves are equal. This can be done using only a compass and straightedge.

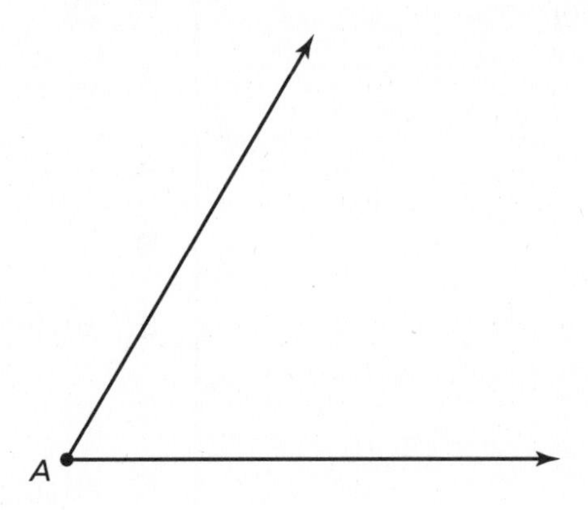

1. Classify angle A.
2. Open the compass. Place the point of the compass at A and draw an arc (a small curved line that is part of a circle) that intersects both sides of the angle. Label the points where the curved line intersects the angle as B and C.
3. Place the point of the compass on B. Draw an arc that is contained within the angle. Without changing the width of the compass, place the point of the compass on C and draw an arc that intersects the one just drawn.
4. Label the intersection point of the two arcs G. Use a straightedge to connect points A and G. The line segment AG bisects angle BAC.
5. Use a protractor to measure angle BAC. Divide this measure by two to determine what each smaller angle should measure. Then, use your protractor and measure the smaller angles. How do the numbers compare?

Name ______________________________ Date __________

Puzzle Time

What Runs Around All Day And Lies Under The Bed With Its Tongue Hanging Out?

Write the letter of each answer in the box containing the exercise number.

Tell whether the angles are *adjacent* or *vertical*.

1.

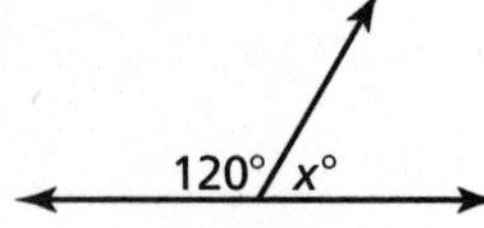

2.

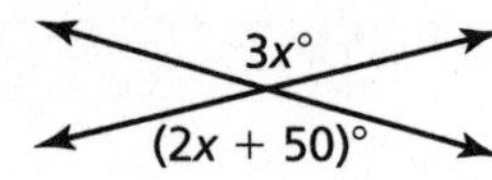

3.

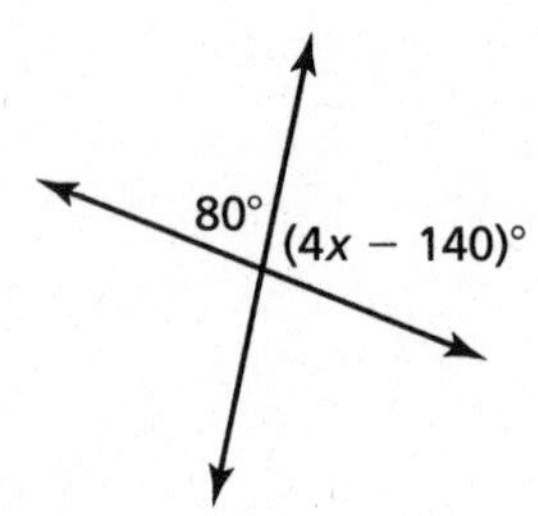

Find the value of *x*.

4.

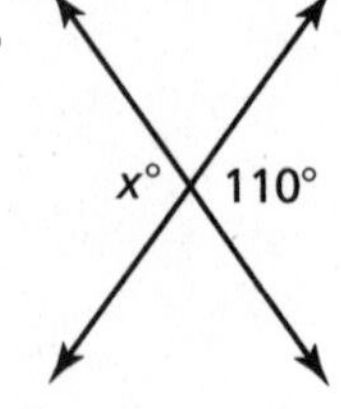

5.

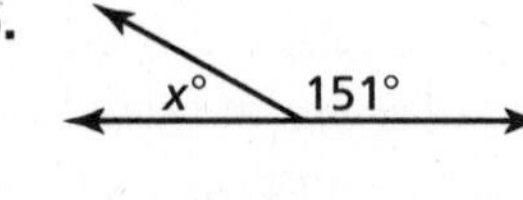

6.

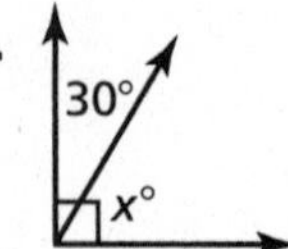

7.

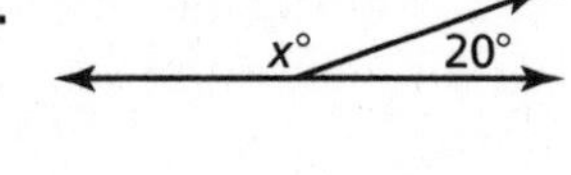

8. A road intersects another road at an angle of 45°. Find the value of x.

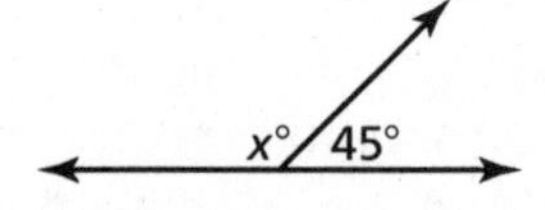

Answers

N. 29

K. vertical

A. 160

S. 60

E. adjacent

R. 135

E. adjacent

A. 110

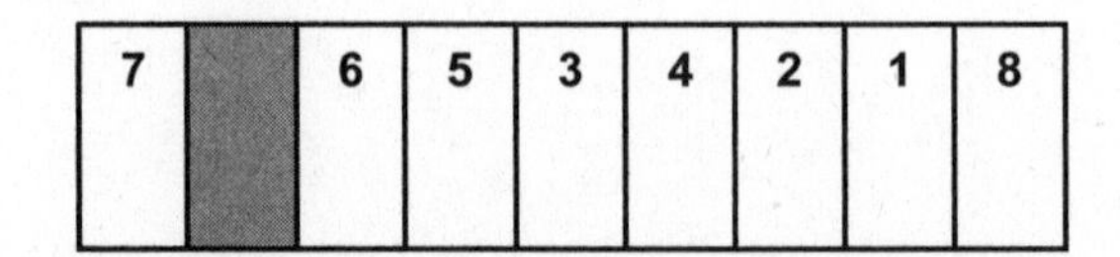

Activity 7.2 Start Thinking!

For use before Activity 7.2

Review with a partner what the difference is between adjacent angles and vertical angles.

Activity 7.2 Warm Up

For use before Activity 7.2

Draw a pair of vertical angles with the given measure.

1. $120°$

2. $45°$

3. $160°$

4. $30°$

Start Thinking!

For use before Lesson 7.2

Complete the statement.

Two angles are ___?___ if the sum of their measures is 90°.

Two angles are ___?___ if the sum of their measures is 180°.

People often have trouble remembering which is 90° and which is 180°. Make up your own way to help you remember the definitions.

Lesson 7.2

Warm Up

For use before Lesson 7.2

Tell whether the statement is *always*, *sometimes*, or *never* true. Explain.

1. If x and y are supplementary angles, then x is right.

2. If x and y and complementary angles, then y is acute.

3. If x is a right angle and y is an acute angle, then x and y are supplementary angles.

4. If x is acute and y is obtuse, then x and y are supplementary angles.

Name ____________________ Date __________

7.2 Practice A

Tell whether the statement is *always*, *sometimes*, or *never* true. Explain.

1. If x and y are supplementary angles, then y is acute.

2. If x and y are complementary angles, then x is obtuse.

Tell whether the angles are *complementary*, *supplementary*, or *neither*.

3.

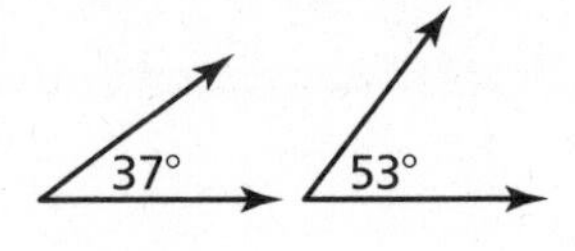

4.

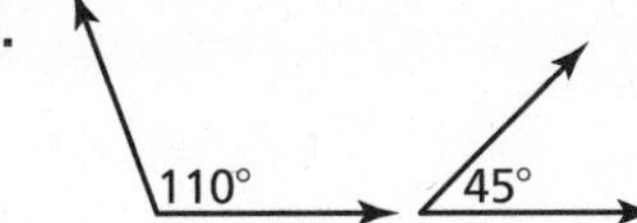

5.

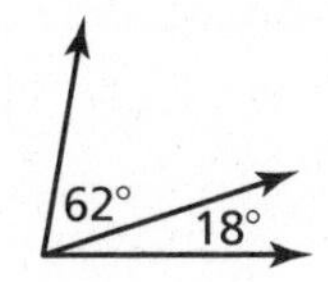

6.

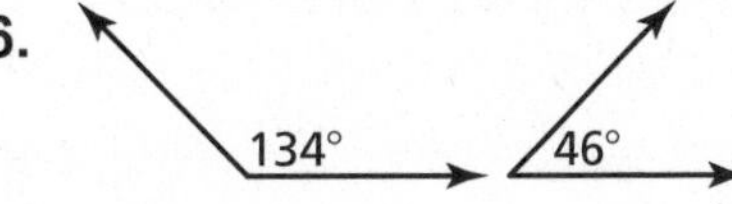

7. Angle x and angle y are complementary. Angle x is supplementary to a 128° angle. What are the measures of angle x and angle y?

Tell whether the angles are *complementary* or *supplementary*. Then find the value of x.

8.

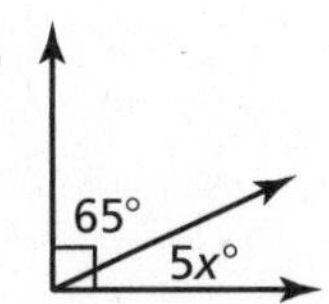

9.

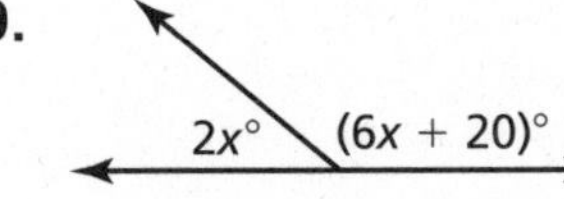

Draw a pair of adjacent supplementary angles so that one angle has the given measure.

10. 50° **11.** 110° **12.** 135°

13. Two angles have the same measure. What are their measures if they are also complementary angles? supplementary angles?

Name ______________________________ Date __________

7.2 Practice B

Tell whether the angles are *complementary*, *supplementary*, or *neither*.

1. 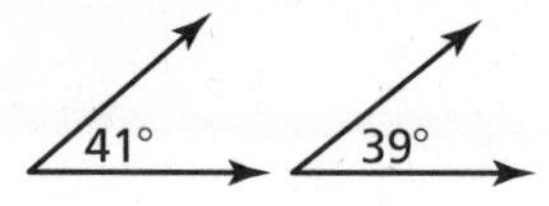

2. 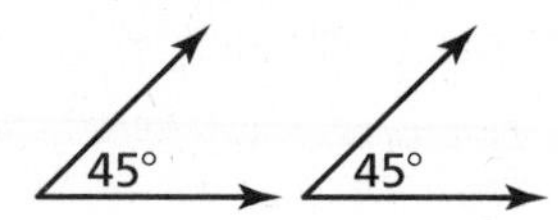

3. 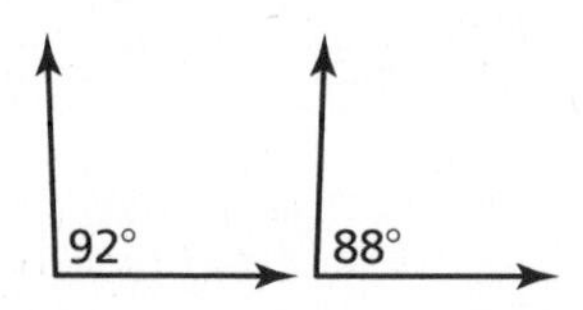

4. 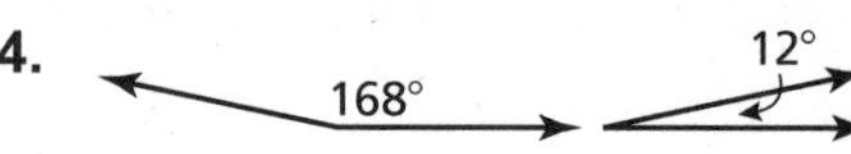

Tell whether the angles are *complementary* or *supplementary*. Then find the value of *x*.

5. 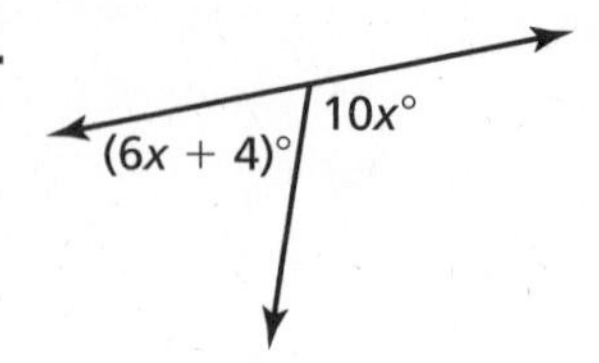

6. 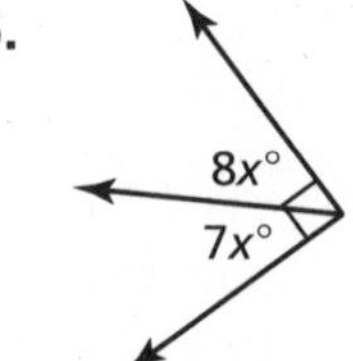

7. The measures of two supplementary angles have a ratio of 2 : 4. What is the measure of the smaller angle?

8. Find the values of x and y.

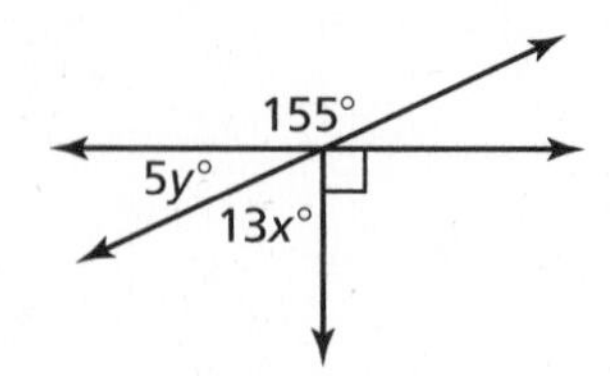

9. Let x be an angle measure. Let c be the measure of the complement of the angle and let s be the measure of the supplement of the angle.

a. Write an equation involving c and x.

b. Write an equation involving s and x.

Name____ Date__________

7.2 Enrichment and Extension

Finding Missing Angles

Use properties of shapes and angles to find the missing measures.

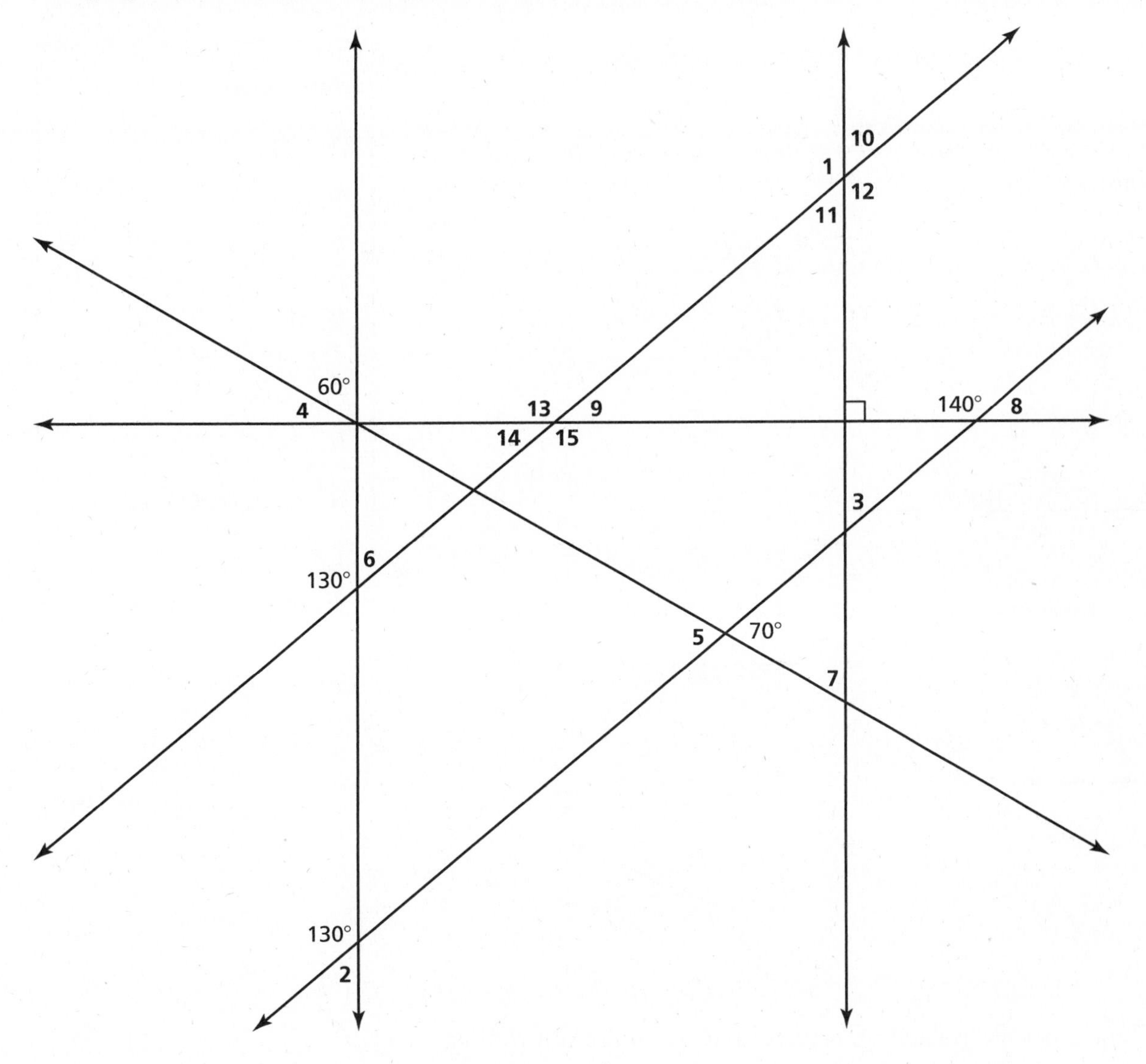

Name ______________________________ Date __________

7.2 Puzzle Time

What Is The Best Year For Grasshoppers?

Write the letter of each answer in the box containing the exercise number.

Answers

Y. 10

P. 20

A. neither

E. 4

A. 15

E. complementary

R. supplementary

L. 27

Tell whether the angles are *complementary*, *supplementary*, or *neither*.

1.

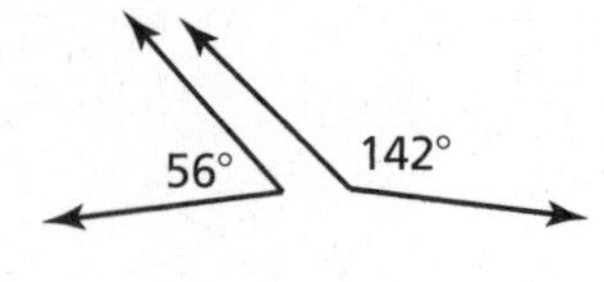

2.

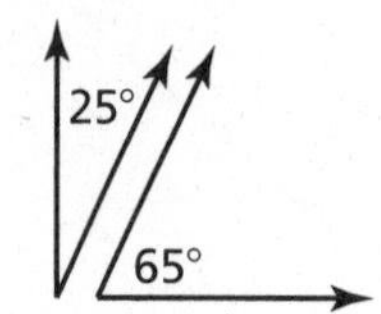

3.

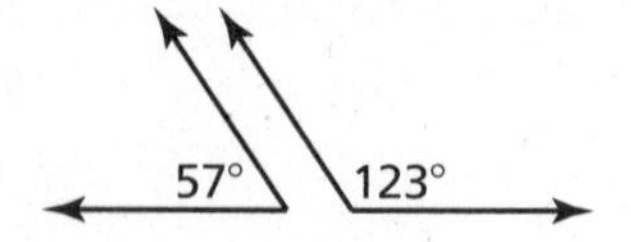

Find the value of *x*.

4.

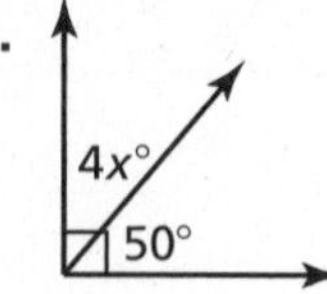

5.

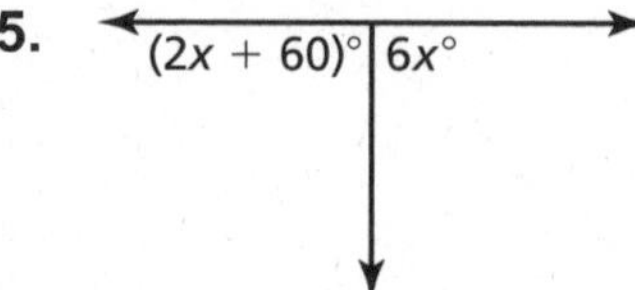

6.

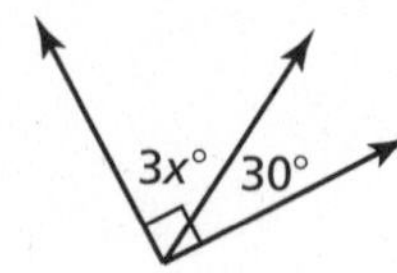

7.

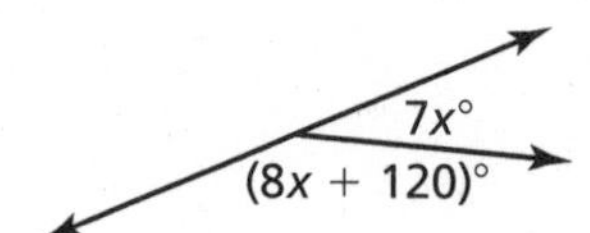

8. The crosswalk in front of a school intersects the sidewalk at an angle of 99°. Find the value of x.

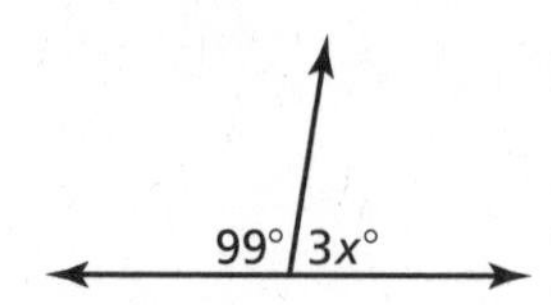

8	7	1	6		4	2	5	3

Activity 7.3 Start Thinking!

For use before Activity 7.3

Describe some real-life triangles. What kind of triangles are they?

Activity 7.3 Warm Up

For use before Activity 7.3

Construct the line segment.

1. Line segment: 7 in.

2. Line segment: 10 cm

3. Line segment: 3.5 in.

4. Line segment: 4 cm

5. Line segment: 1 in.

6. Line segment: 30 mm

Lesson 7.3 Start Thinking!

For use before Lesson 7.3

Given two of the three angles of a triangle, explain to a partner how to find the third angle. Use the angle measures 30° and 60°.

Lesson 7.3 Warm Up

For use before Lesson 7.3

Construct a triangle with the given description.

1. side lengths: 2 cm, 2 cm
2. side lengths: 7 cm, 10 cm
3. angles: 90°, 45°
4. angles: 60°, 60°

Name________________________________ Date__________

7.3 Practice A

Classify the triangle.

1.

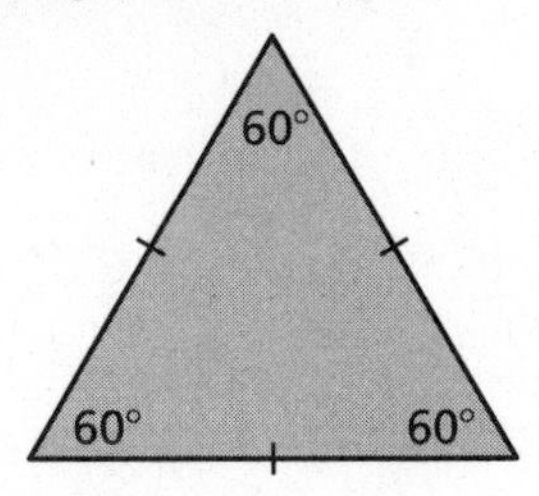

2.

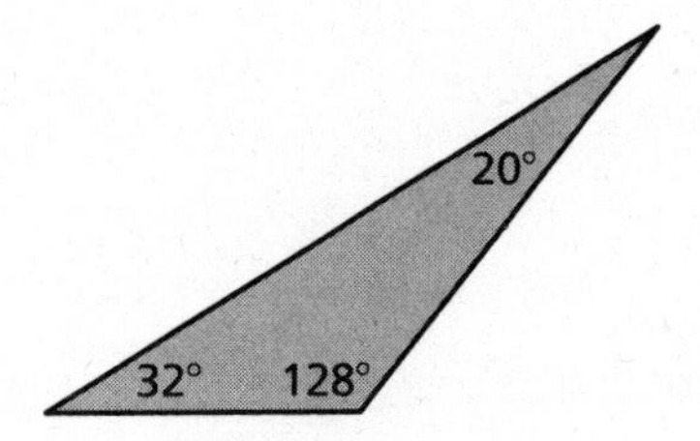

3.

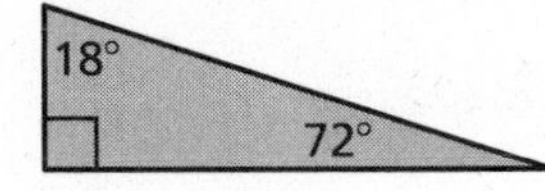

4.

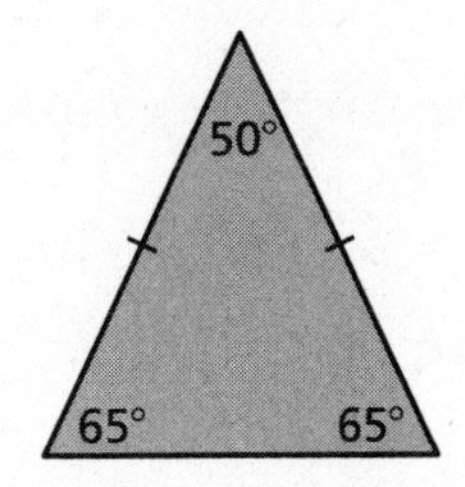

Draw a triangle with the given description.

5. a right triangle with two congruent sides

6. a scalene triangle with a 3-inch side and a 4-inch side that meet at a 110° angle

7. Consider the three isosceles right triangles.

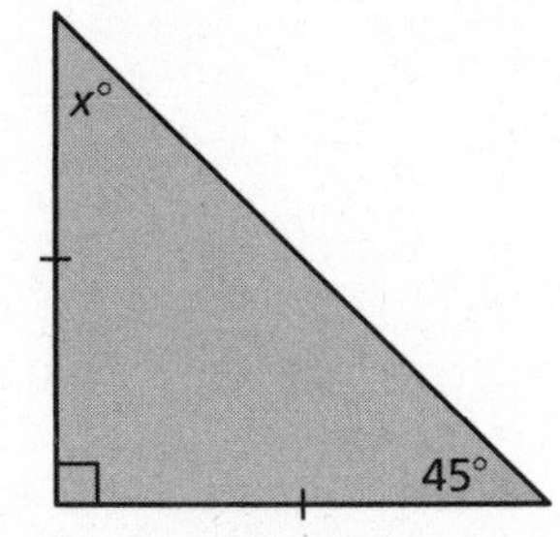

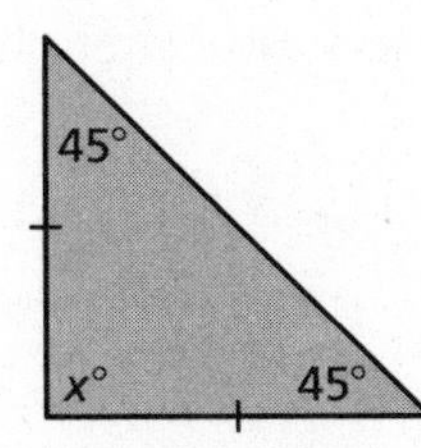

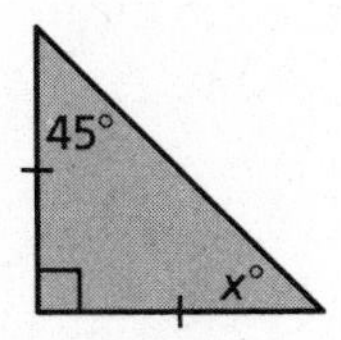

a. Find the value of x for each triangle.

b. What do you notice about the angle measures of each triangle?

c. Write a rule about the angle measures of an isosceles right triangle.

Name ______________________ Date ________

7.3 Practice B

Classify the triangle.

1. 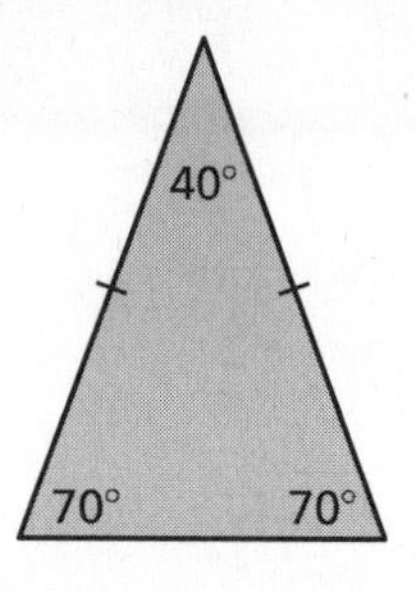

2. 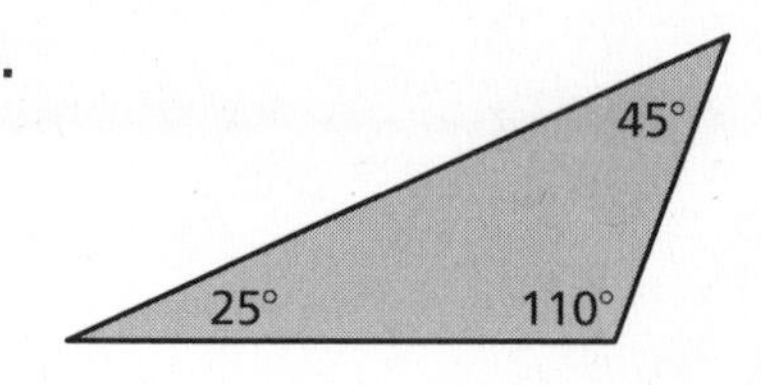

Draw a triangle with the given angle measures. Then classify the triangle.

3. 25°, 65°, 90°

4. 45°, 60°, 75°

Draw a triangle with the given description.

5. an obtuse scalene triangle

6. a triangle with a 110° angle connected to a 25° angle by a 6-inch side

Determine whether you can construct *many*, *one*, or *no* triangle(s) with the given description. Explain your reasoning.

7. a triangle with a 2-inch side, a 4-inch side, and a 5-inch side

8. a scalene triangle with two 7-centimeter sides

9. a triangle with one angle measure of 100° and one 6-inch side

10. Draw a circle. Draw a triangle with the given description such that all three vertices of the triangle touch the circle.

    a. Draw an obtuse triangle.

    b. Draw a right triangle.

    c. Draw an acute triangle.

Name____ Date____________

7.3 Enrichment and Extension

Writing Ratios

The sides of a right triangle can be named by their locations with respect to an angle of the triangle.

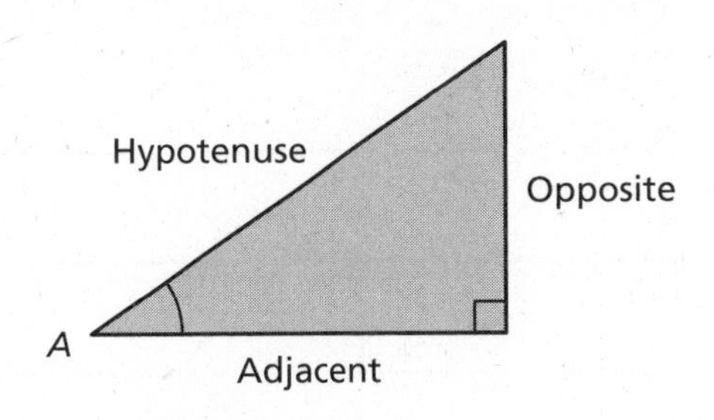

Trigonometry

It is possible to write ratios that compare the lengths of the sides in the triangle using special functions and a given angle. These ratios are called sine (sin), cosine (cos), and tangent (tan) and are studied in depth in a branch of mathematics called trigonometry.

$$\sin A = \frac{\text{Opposite}}{\text{Hypotenuse}} \qquad \cos A = \frac{\text{Adjacent}}{\text{Hypotenuse}} \qquad \tan A = \frac{\text{Opposite}}{\text{Adjacent}}$$

Write the ratios. Use your answers and the color key to shade the mosaic.

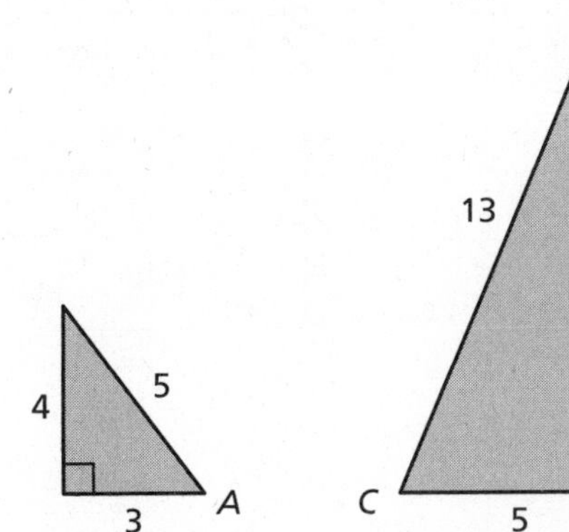

1. $\sin A$
2. $\tan C$
3. $\cos A$
4. $\tan A$
5. $\sin C$
6. $\cos C$

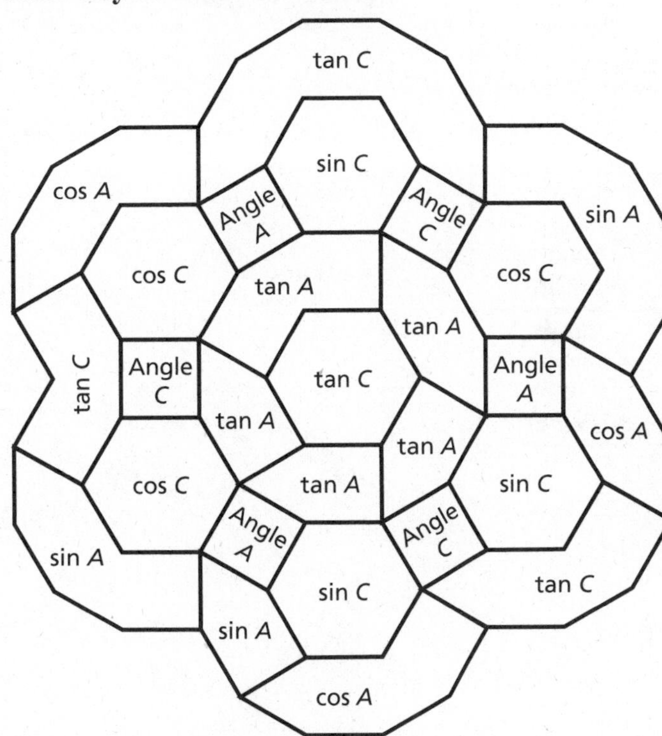

Key:
Angles = Blue
5 in the denominator = Red
13 in the denominator = Yellow
3 in the denominator = Purple

Name ____________________ Date __________

Puzzle Time

Why Did The Kindergartener Take Her Books To The Zoo?

A	B	C	D	E	F
G	H	I			

Complete each exercise. Find the answer in the answer column. Write the word under the answer in the box containing the exercise letter.

Answer
Equilateral and Equiangular Triangle **TO**
80 **LIONS**
Acute Scalene Triangle **READ**
Obtuse Scalene Triangle **WANTED**
Right Isosceles Triangle **BETWEEN**

Classify the triangle.

A.

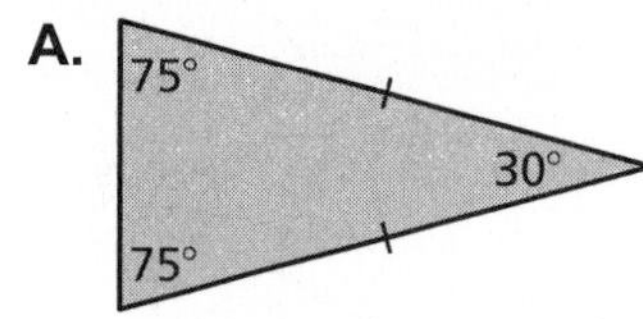

B.

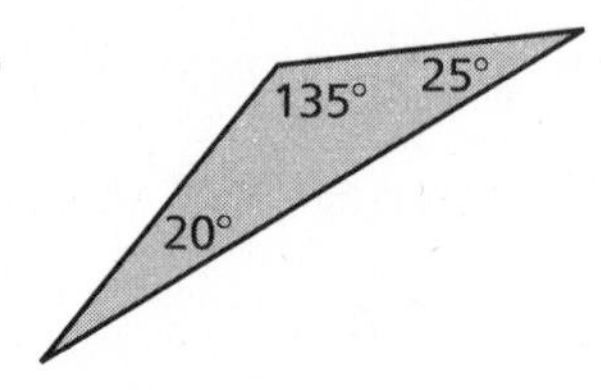

C.

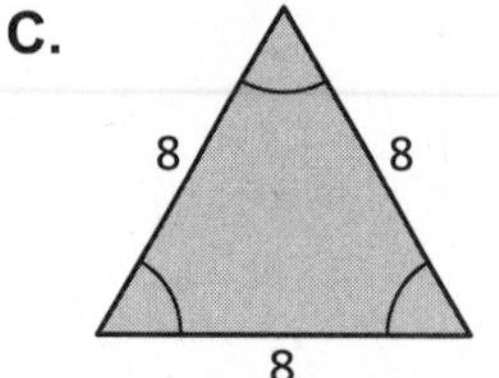

D.

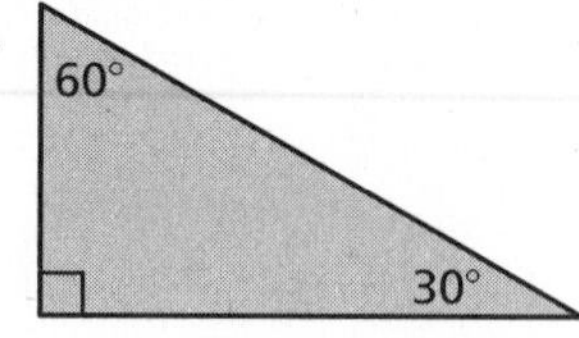

E. 100°, 40°, 40°

F.

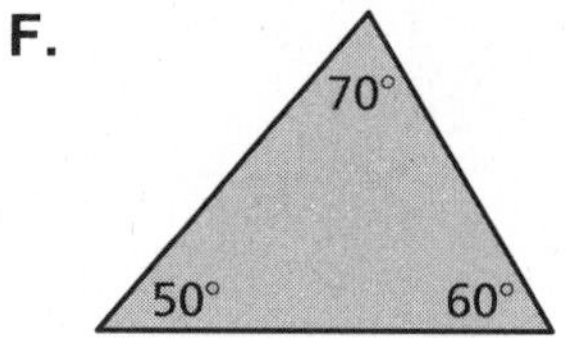

G.

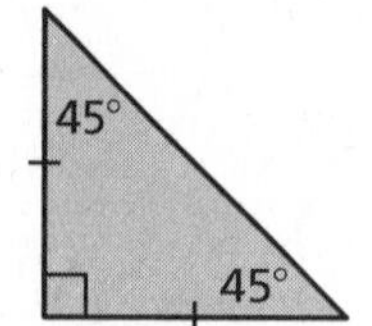

H. A triangle contains angles measuring 28° and 37°. How many degrees is the third angle of the triangle?

I. A triangle contains angles measuring 25° and 75°. How many degrees is the third angle of the triangle?

Answer
Obtuse Isosceles Triangle **TO**
Acute Isosceles Triangle **SHE**
Right Scalene Triangle **LEARN**
115 **THE**

In real life, when is it important to know the angle measures of a triangle?

Classify the triangle.

1.

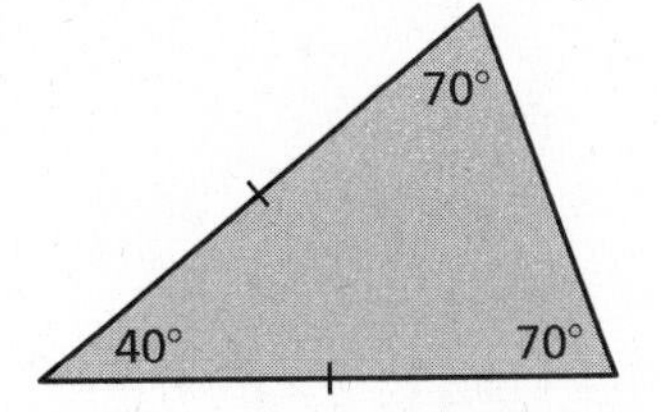

2.

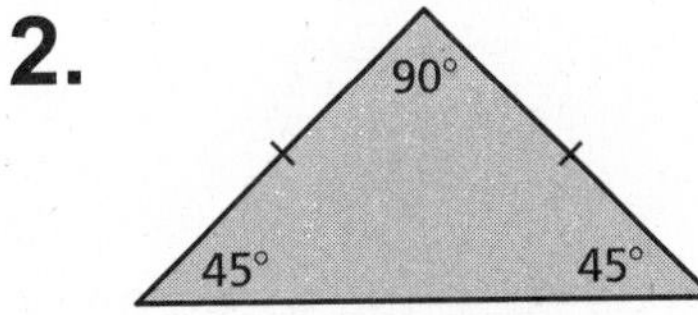

3.

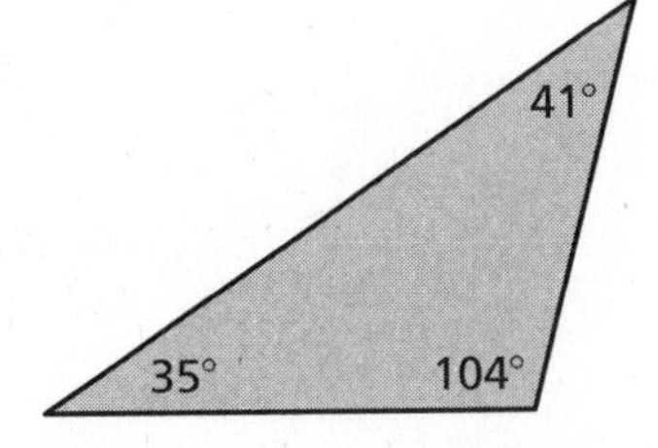

4.

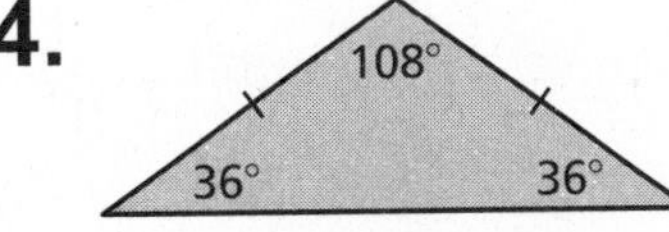

5.

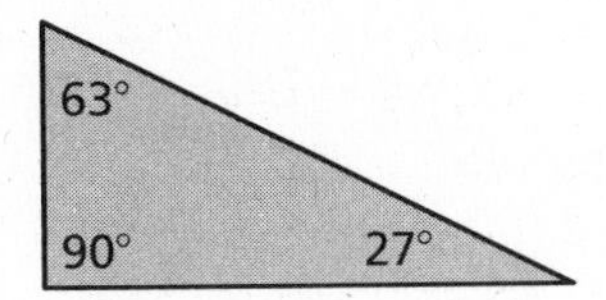

6.

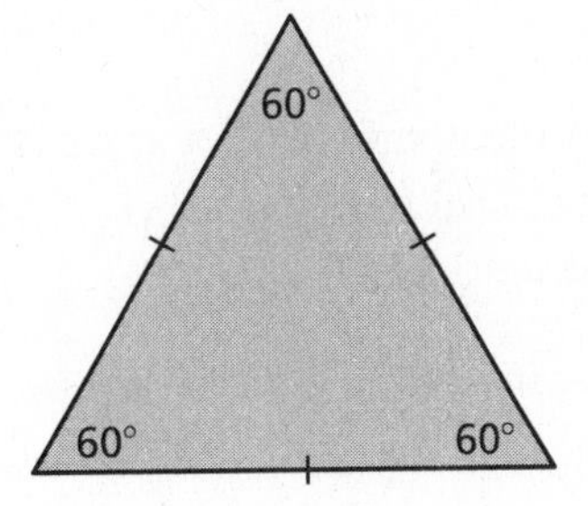

Name ______________________________ Date __________

Extension 7.3 Practice

Find the value of *x*. Then classify the triangle.

1.

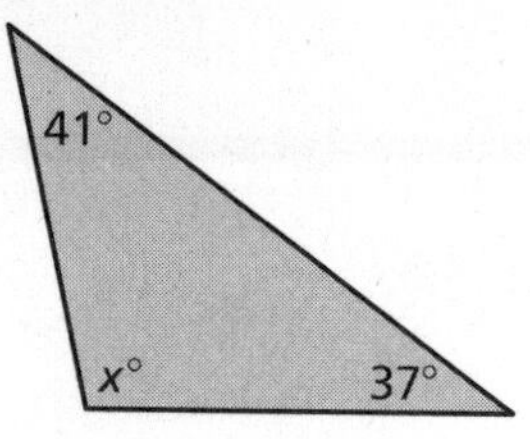

2.

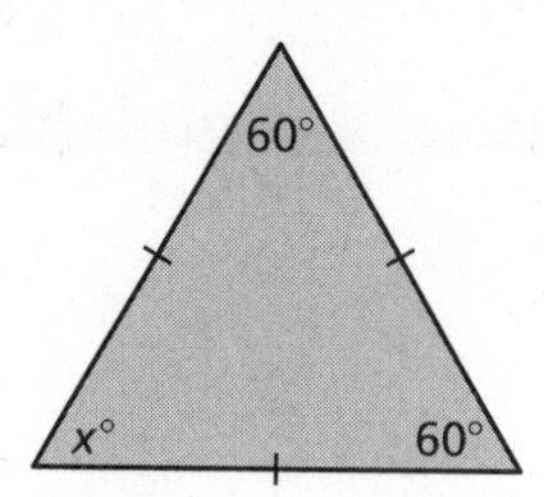

3.

4.

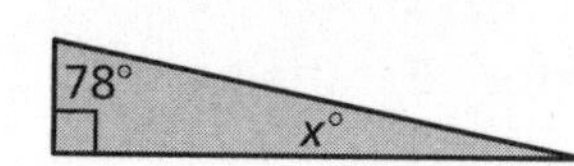

5.

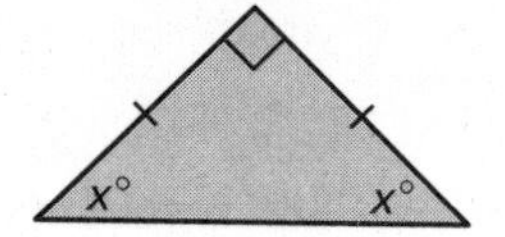

6.

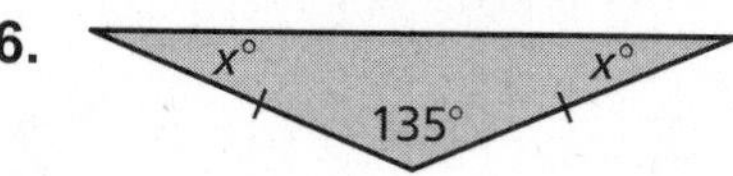

Tell whether a triangle can have the given angle measures. If not, change the first angle measure so that the angle measures form a triangle.

7. $46\frac{1}{3}°, 81\frac{1}{2}°, 52\frac{1}{6}°$

8. 36.9°, 121.4°, 33.7°

9. Using 3 equal-sized craft sticks, put the ends together to make a triangle.

a. Use a protractor to find the measure of each angle.

b. Classify the triangle.

c. Replace one of the sticks with either a longer stick or a longer pencil. Use a protractor to find the measure of each angle and classify this new triangle.

d. Replace the longest side with a stick or pencil that is shorter than the two other sides. Use a protractor to find the measure of each angle and classify this new triangle.

e. What do you notice about the triangle when two of its sides are equal in length?

Activity 7.4 Start Thinking!

For use before Activity 7.4

List some words that start with the prefix *quad-*.

What do those words mean?

What do you think *quad-* means?

Activity 7.4 Warm Up

For use before Activity 7.4

Identify the polygon.

1.

2.

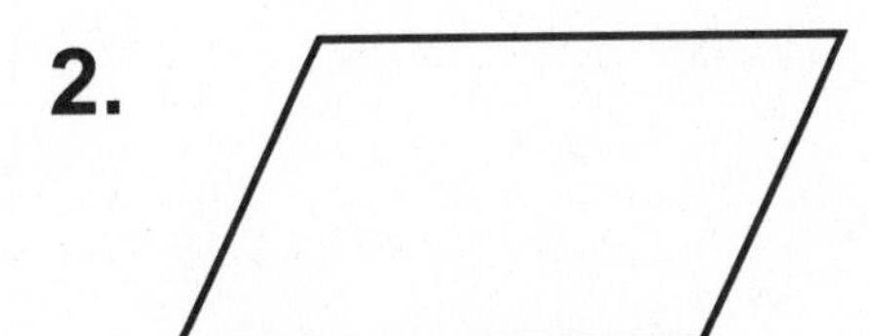

3.

4.

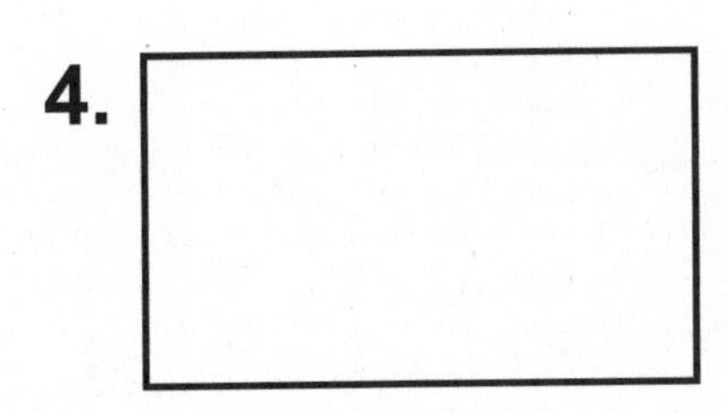

5.

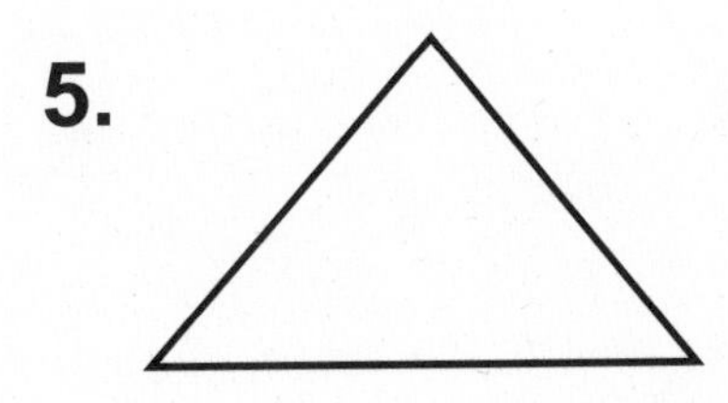

6.

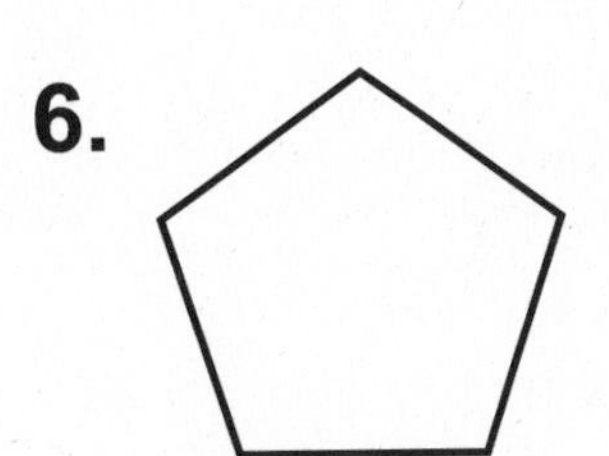

Name all the different kinds of four-sided figures that you know. Draw a sketch of each.

Classify the quadrilateral.

1. 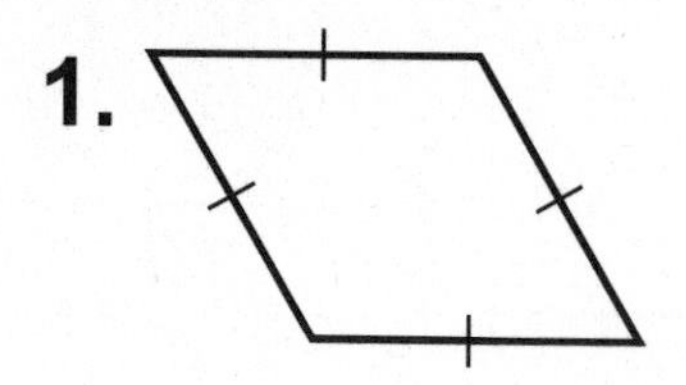

2. 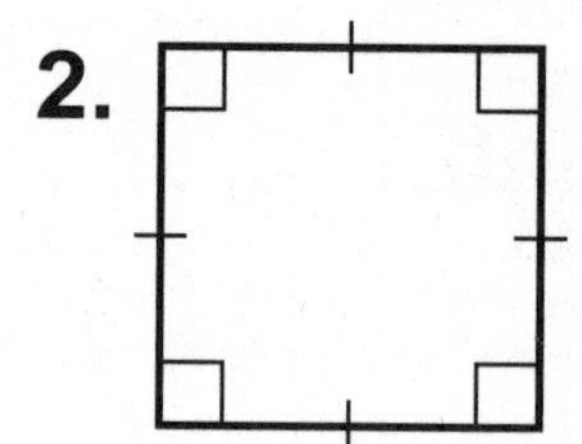

3. 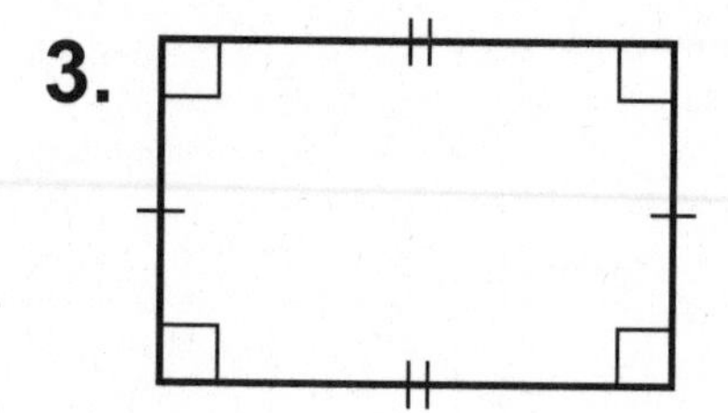

4. 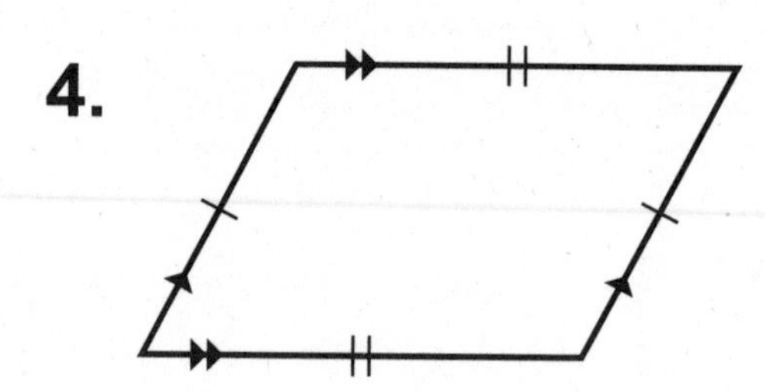

5. 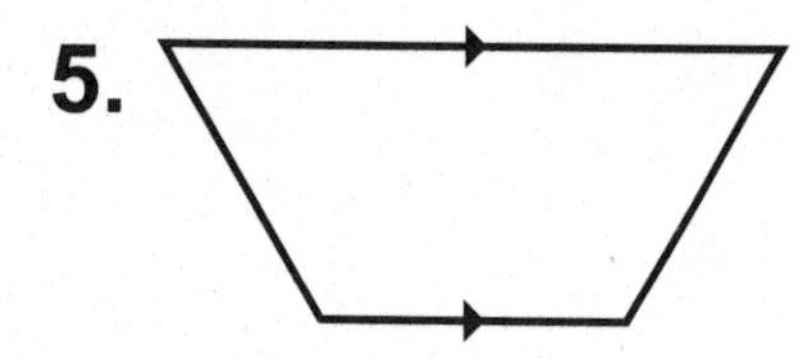

6. 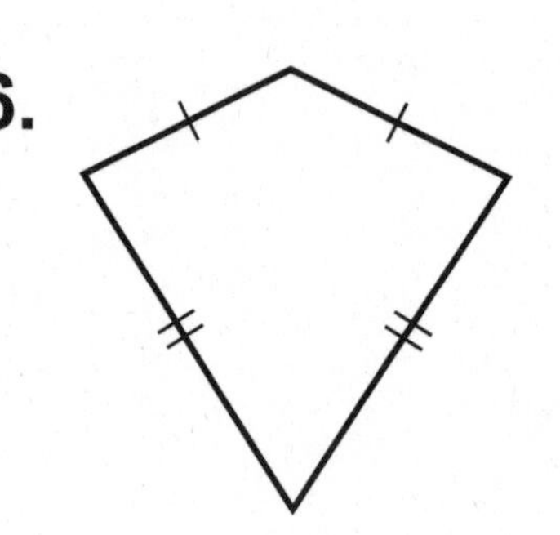

Name____ Date__________

7.4 Practice A

Classify the quadrilateral.

1.

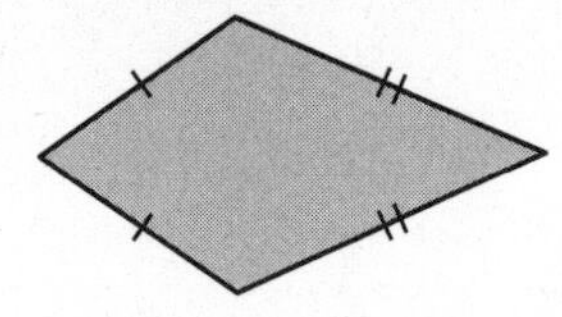

2.

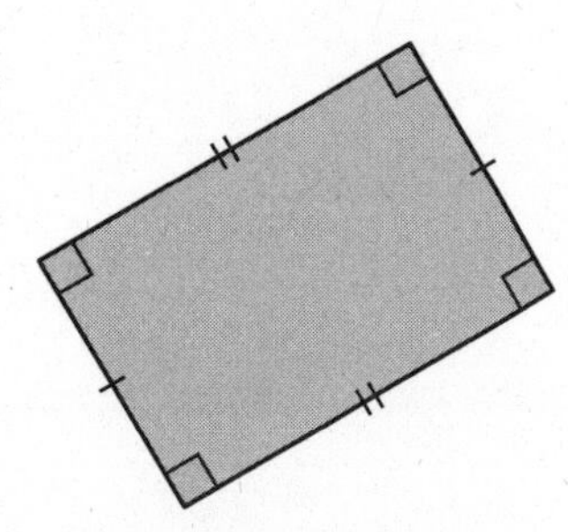

3.

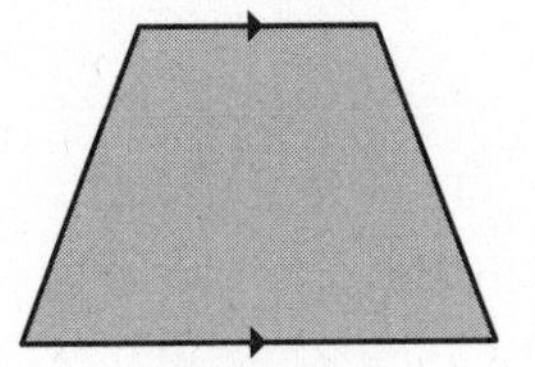

4.

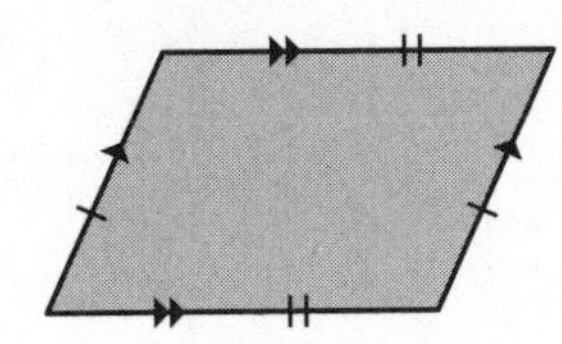

Find the value of *x*.

5.

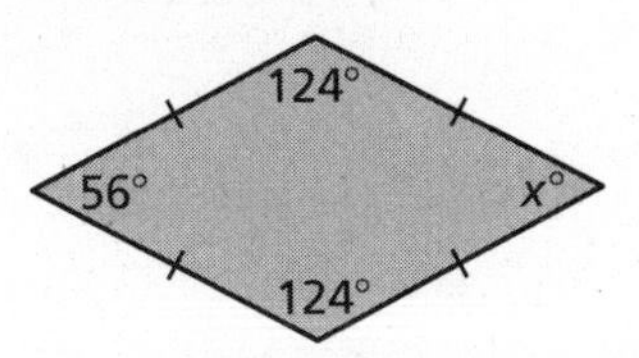

6.

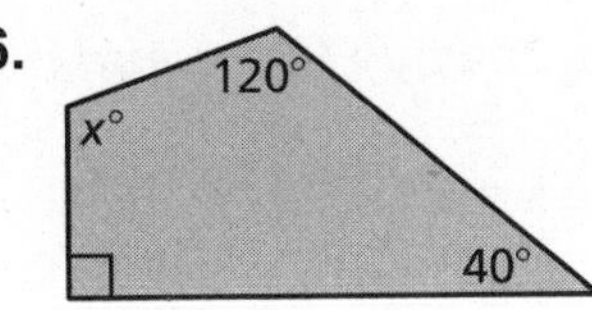

Copy and complete using *always*, *sometimes*, or *never*.

7. A square is __?__ a rhombus.

8. A parallelogram is __?__ a rectangle.

9. A kite is __?__ a square.

10. A trapezoid is __?__ a square.

11. Draw the following trapezoids. If it is not possible, explain why.

- **a.** a trapezoid with one right angle
- **b.** a trapezoid with two right angles
- **c.** a trapezoid with three right angles
- **d.** a trapezoid with four right angles

Name ______________________________ Date __________

7.4 Practice B

Classify the quadrilateral.

1.

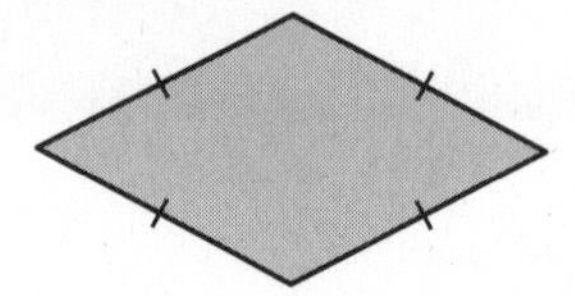

2.

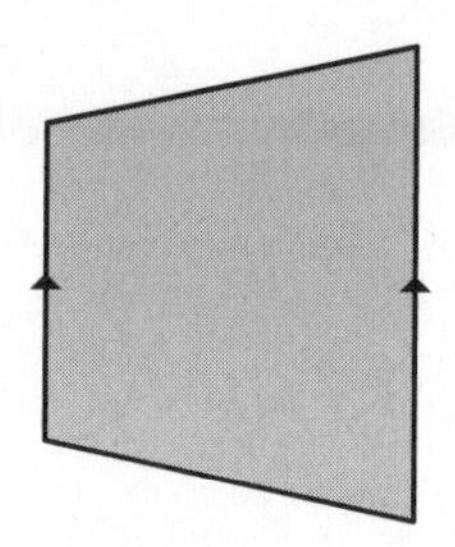

Find the value of *x*.

3.

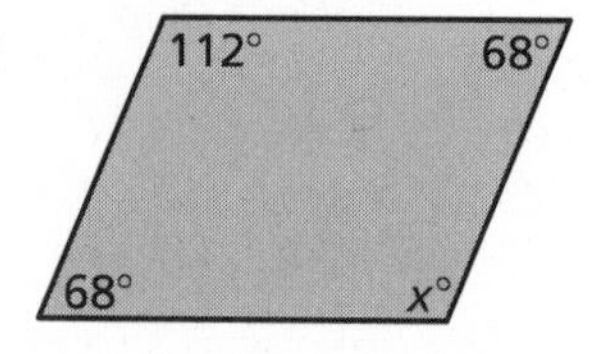

4.

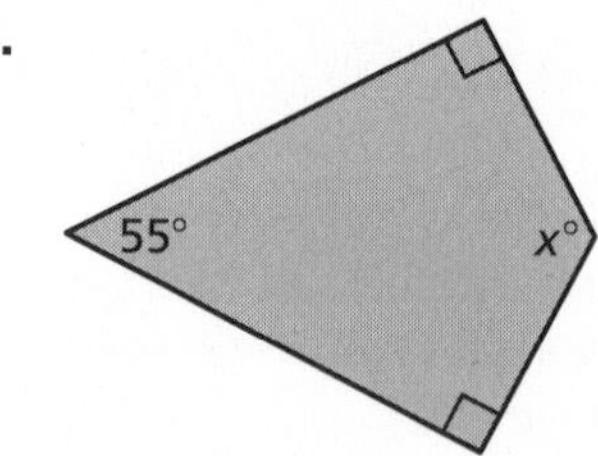

Copy and complete using *always*, *sometimes*, or *never*.

5. A rectangle is __?__ a square.

6. A rhombus is __?__ a parallelogram.

7. A trapezoid is __?__ a kite.

8. A parallelogram is __?__ a rhombus.

9. Determine whether the statement is *true* or *false*. Explain your reasoning. You may use diagrams to explain your reasoning.

a. A rectangle that is 30 inches long and 10 inches wide can be divided into two congruent squares.

b. A rectangle that is 30 inches long and 10 inches wide can be divided into three congruent squares.

c. A parallelogram with opposite congruent sides of 6 feet and 3 feet can be divided into two congruent rhombuses.

d. A rectangle that is 30 inches long and 10 inches wide can be divided into two congruent trapezoids.

e. A rhombus that has side length 8 meters can be divided into two congruent parallelograms.

Name______________________________ Date__________

7.4 Enrichment and Extension

Sum of Interior Angles

A regular polygon is a shape in which all sides are the same length and all angles have equal measure. For example, a regular quadrilateral is most often called a square.

Because all the angles in a regular polygon have equal measure, a formula can be used to calculate the measure of each angle:

Measure of one angle in a regular polygon $= \dfrac{(n-2)180^\circ}{n}$, where n is the number of sides in the polygon.

Identify which polygon the real-life object resembles. Then determine the measure of each angle using the formula.

1. 

2. 

3. 

4. 

5. 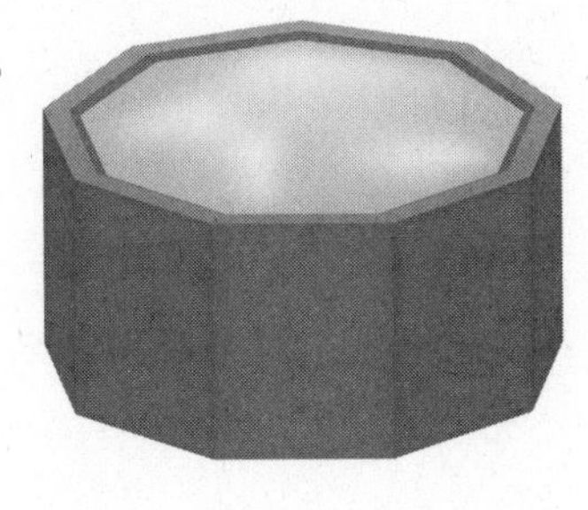

6. 

Name ______________________________ Date __________

7.4 Puzzle Time

What's The Healthiest Insect?

Write the letter of each answer in the box containing the exercise number.

Answers

N. rectangle

A. kite

E. rhombus

M. trapezoid

I. square

T. parallelogram

E. 90

B. 65

V. 38

I. 55

Classify the quadrilateral.

1.

2.

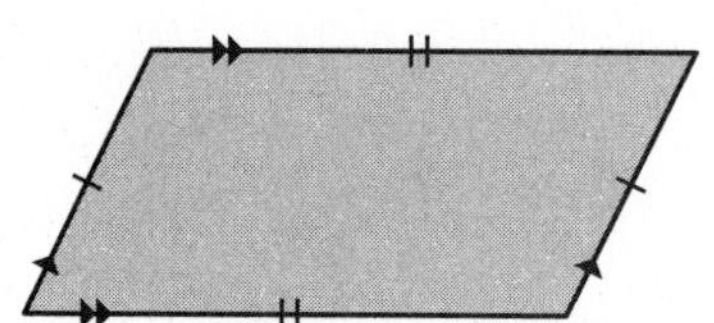

3.

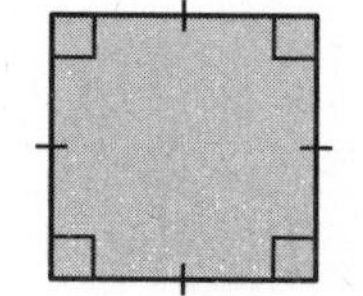

4.

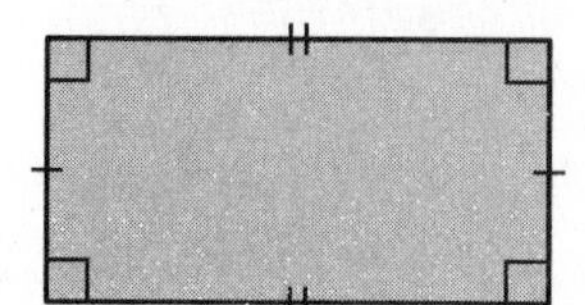

5.

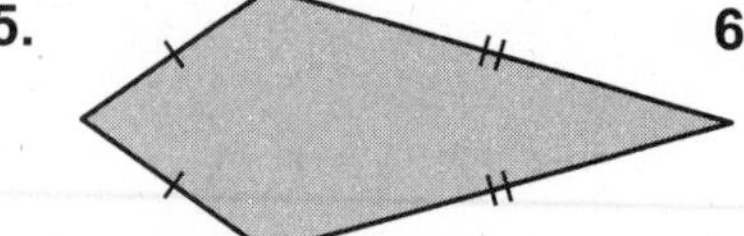

6.

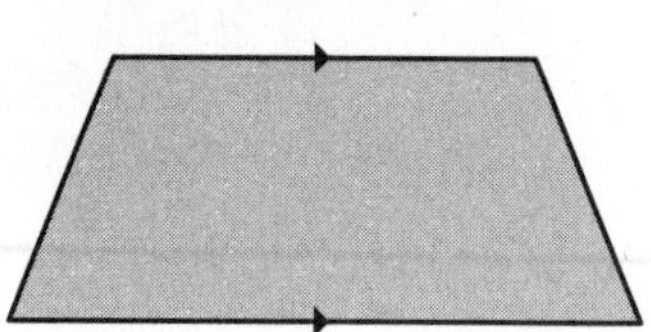

Find the value of *x*.

7.

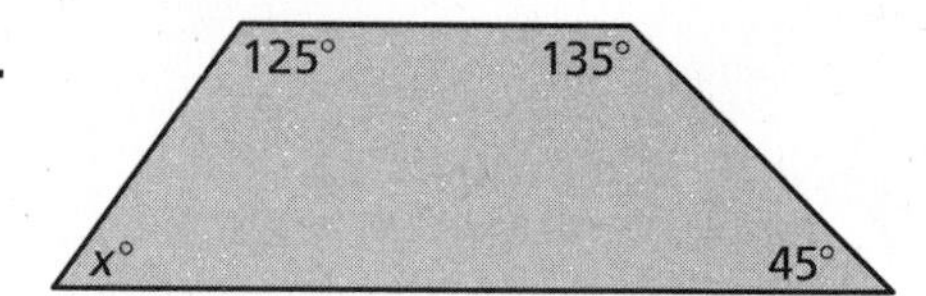

8.

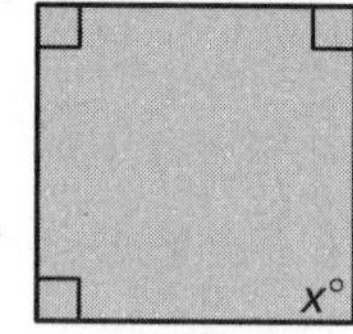

9.

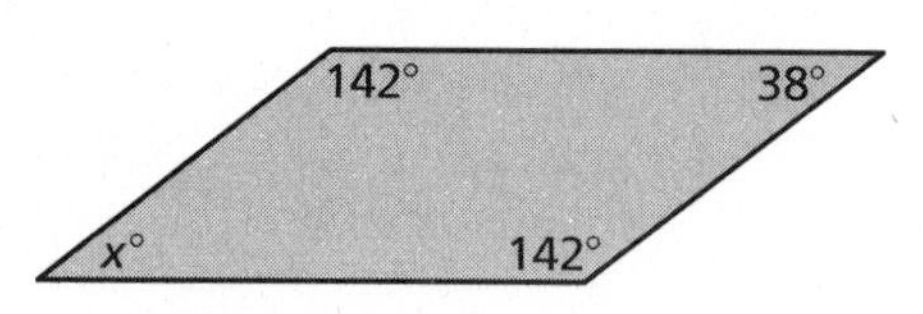

10.

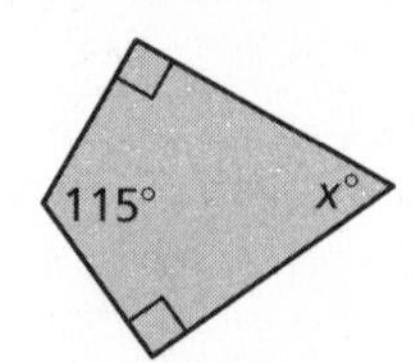

9	7	2	5	6	3	4		10	1	8

Activity 7.5 Start Thinking!

For use before Activity 7.5

Explain why a contractor must know how to read an architect's blueprints.

Activity 7.5 Warm Up

For use before Activity 7.5

Find the missing number.

1. $\frac{8}{9} = \frac{x}{27}$

2. $\frac{4}{14} = \frac{x}{35}$

3. $\frac{8}{36} = \frac{x}{90}$

4. $\frac{12}{3} = \frac{x}{7}$

5. $\frac{20}{14} = \frac{x}{56}$

6. $\frac{18}{4} = \frac{x}{18}$

Lesson 7.5 Start Thinking!

For use before Lesson 7.5

What must you know about the scale of a map before planning a trip?

Lesson 7.5 Warm Up

For use before Lesson 7.5

Find the actual dimension. The scale is 1 cm : 4 ft.

1. model: 7 cm

2. model: 10.5 cm

3. model: 30 cm

4. model: 19 mm

5. model: 0.4 m

6. model: 4.25 m

Name______________________________________ Date__________

7.5 Practice A

1. Use the drawing of the game court and an inch ruler. Each inch in the drawing represents 8 feet.

Server A | Net Area | Receiver A
Server B | | Receiver B

a. What is the actual length of the court?

b. What are the actual dimensions of Receiver A?

c. What are the actual dimensions of the Net Area?

d. The area of Server B is what percent of the area of Server A?

e. What is the ratio of the perimeter of Receiver B to the perimeter of Net Area?

f. What is the ratio of the area of Receiver B to the area of Net Area?

g. Are Receiver B and Net Area similar rectangles?

h. The area of Server A is increased by what percent to get the area of Net Area?

Find the missing dimension. Use the scale factor 1 : 5.

2. Model: 3 ft

Actual: ?

3. Model: 7 m

Actual: ?

4. Model: ?

Actual: 20 yd

5. Model: ?

Actual: 12.5 cm

6. A scale drawing of a rose is 3 inches long. The actual rose is 1.5 feet long.

a. What is the scale of the drawing?

b. What is the scale factor of the drawing?

Name ______________________________ Date __________

7.5 Practice B

1. In the actual blueprint of the bedroom suite, each square has a side length of $\frac{1}{2}$ inch.

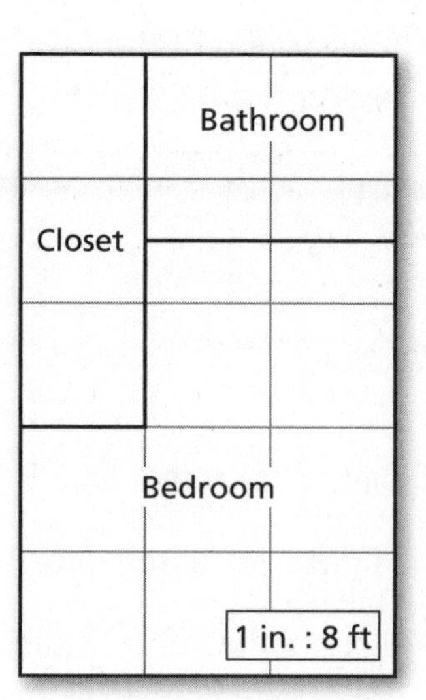

   a. What are the dimensions of the bedroom suite?

   b. What are the dimensions of the bathroom?

   c. What is the length of the longest wall in the bedroom?

   d. What is the ratio of the perimeter of the closet to the perimeter of the bathroom?

   e. What is the ratio of the area of the closet to the area of the bathroom? How can you explain this by looking at the squares in each?

   f. All of the walls in the bedroom suite are covered with drywall. Which will cost the most to drywall—*the closet*, *the bathroom*, or *both are the same*?

   g. All of the floors in the bedroom suite are covered with tile. Which will cost the most to tile—*the closet*, *the bathroom*, or *both are the same*?

   h. What is the area of the bedroom?

Find the missing dimension. Use the scale factor 2 : 5.

2. Model: 10 km

   Actual: ___?___

3. Model: 5 in.

   Actual: ___?___

4. Model: ___?___

   Actual: 24 ft

5. Model: ___?___

   Actual: 32.5 m

6. A scale factor is 1 : 8. Describe and correct the error in finding the model length that corresponds to 48 feet.

   ✗ $\frac{1}{8} = \frac{48 \text{ ft}}{x \text{ ft}}$

   $x = 384$ ft

Name____ Date__________

7.5 Enrichment and Extension

Create a Scale Drawing

Your challenge is to create a scale drawing of a room. It could be your classroom, your bedroom, or another room of your choice. Measure the actual length and width of the room and the dimensions of any furniture in the room. Then decide what your scale will have to be in order to fit your drawing on the grid below. Include furniture and other items in the room drawn to scale. Be sure to label the scale dimensions of the room and the lengths of the items that you include in the room.

Trade papers with another student in your class. Use the scale drawing you are given and the scale to find the actual dimensions of the room and the furniture. Check your answers with the actual dimensions.

Your scale: ______________ = _______________

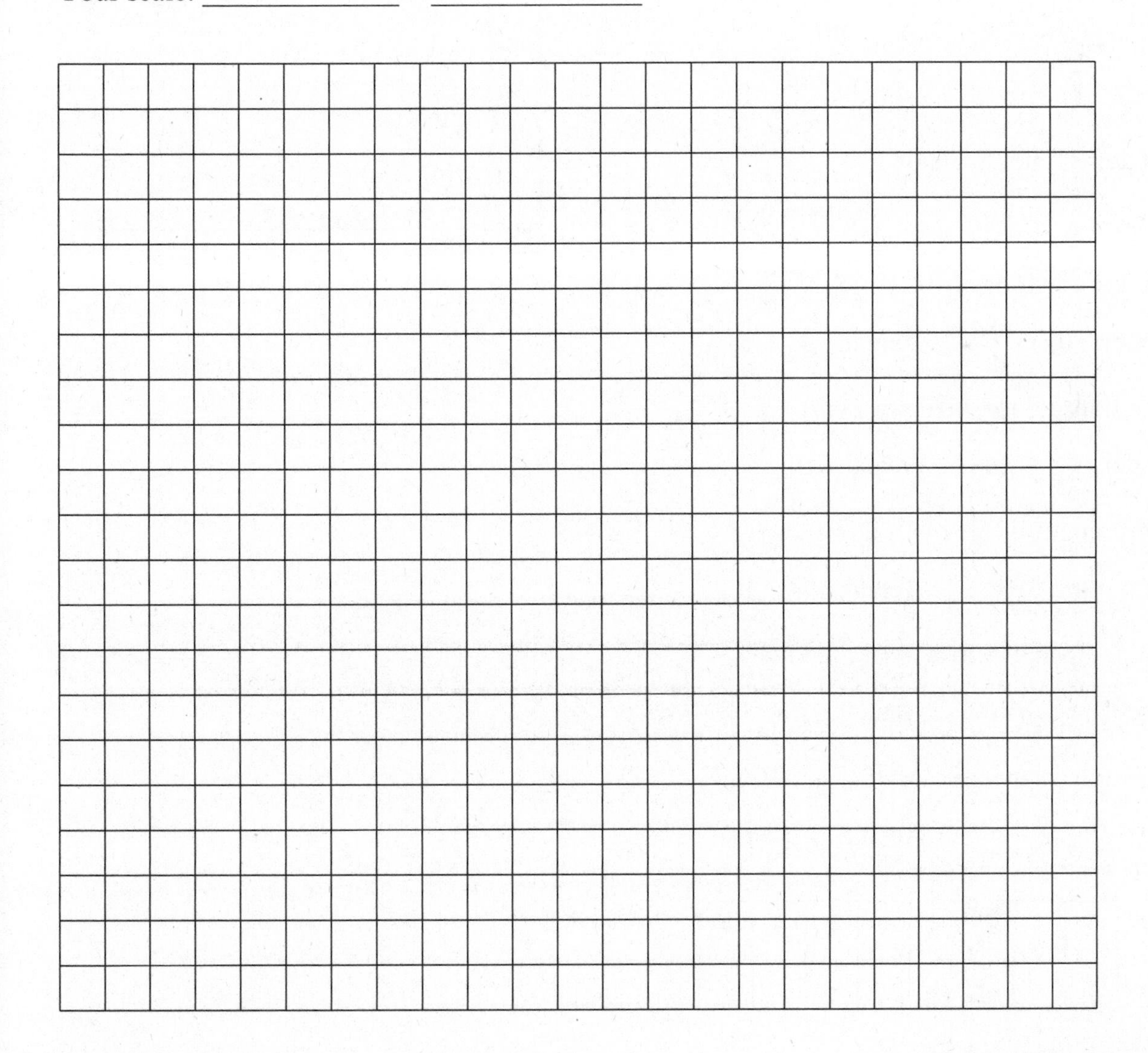

Name ______________________ Date __________

Puzzle Time

What Do Cats Put In Soft Drinks?

Write the letter of each answer in the box containing the exercise number.

Find the missing dimension.

1. Airplane Wingspan, Scale of 1 : 48

   Model: 18 in. Actual: __?__ ft

2. Dinosaur Height, Scale of 1 : 42

   Model: __?__ in. Actual: $12\frac{1}{4}$ ft

3. Railway Train, Scale of 1 : 87

   Model: 11 in. Actual: __?__ ft

4. Rocket Height, Scale of 1 : 15

   Model: __?__ mm Actual: 9.9 m

5. Shark Length, Scale of 1 : 38

   Model: 20 cm Actual: __?__ m

6. Tree house Base, Scale of 1 : 21

   Model: __?__ in. Actual: 7 ft

7. Sofa Length, Scale of 1 : 25

   Map: 3 in. Actual: __?__ ft

8. Bridge Span, Scale of 1 in. : 0.5 mi

   Model: __?__ in. Actual: 7 mi

9. Driving Distance, Scale of 1 in. : 70 mi

   Map: 5.5 in. Actual: __?__ mi

Answers

B. 385

E. 660

C. $79\frac{3}{4}$

M. 14

I. $6\frac{1}{4}$

E. 72

S. $3\frac{1}{2}$

C. 4

U. 7.6

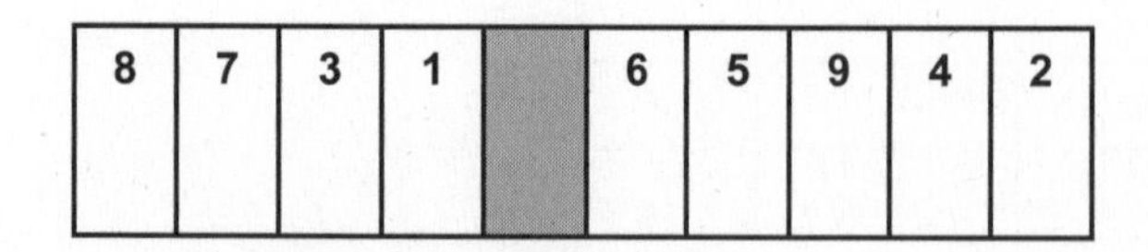

Name ______________________________ Date __________

Technology Connection

For use after Section 7.5

Working with a Scale Model

Scale models can be very useful for planning or rearranging the location of furniture in a room. Although interior decorating professionals use software specifically designed for that purpose, you can use practically any word processing or dynamic geometry software for the same purpose.

EXAMPLE Model the current layout of your classroom with a scale drawing.

SOLUTION

Step 1 Use a tape measure, yardstick, or ruler to find the measurements of the classroom itself and all the furniture in the room. For simplicity, round each measurement to the nearest $\frac{1}{2}$ foot.

Step 2 Decide on a scale that is convenient and convert your measurements to inches. $\left(\text{For example, 1 foot } = \frac{1}{4} \text{ inch.}\right)$

Step 3 Use the Shapes drawing tool to add each piece of furniture to your document. You will need to access the object's properties dialogue box (usually with a right-click of your mouse) to specify the exact size of the object. Remember to use copy and paste for all of the repeated objects like student desks.

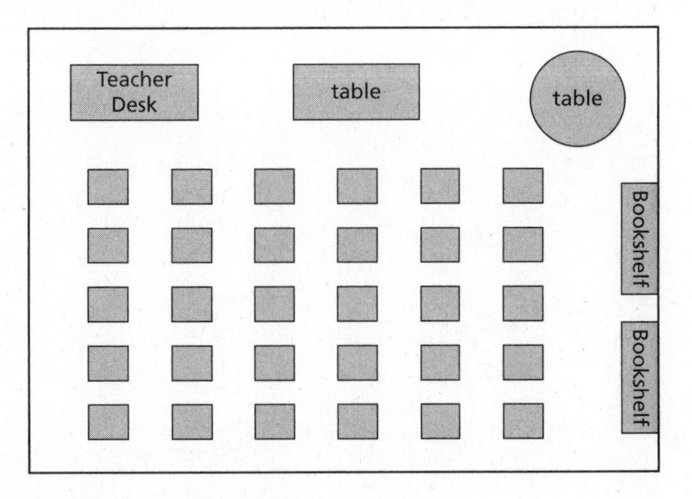

Use the scale model of your classroom in the following exercises.

1. Rearrange the objects in your scale model to find one alternative arrangement for your classroom's furniture.

2. Analyze your new scale model and list some benefits and drawbacks to the functionality of your arrangement. In other words, why would your teacher decide to use your arrangement of furniture over someone else's, and why might someone else's arrangement be better than your own?

Chapter 8

Name______________________________ Date__________

Chapter 8 Circles and Area

Dear Family,

Many families enjoy making crafts. Sewing, scrap-booking, landscaping, gardening, and doing woodwork are just a few activities that use mathematics to form pleasing shapes.

When you lay out a circular garden in a landscape, you cut the edging to match the circumference of the circle. Then you curve the edging around the circular plot to define the area. The area of the circle is used to figure out how much mulch, plants, and flowers you will need.

Similarly, basting the edge of a placemat involves figuring out the length of the edge. For a placemat with semicircles on each end, you use the circumference to find the lengths around the curve. You then add those lengths to the sides of the rectangular portion in the middle.

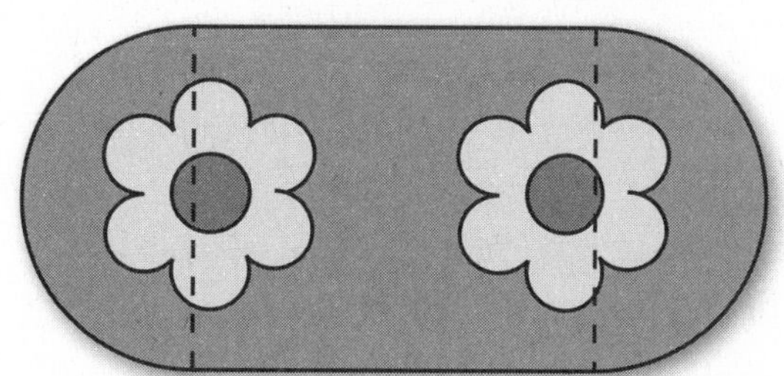

Find a craft to enjoy with your student. Ask your student to help with the following.

- Draw a plan of your project. You can use combinations of geometric shapes (like the rectangle and circle in the placement above) to help find the measurements.
- Find the parts of the project that use measurements around the perimeter and then make a list of the lengths of each edge. Use these lengths to plan for outlining and edging material.
- Find the parts of the project that fill in an area and make a list of the areas for each part. You will use these lengths to plan for interior material, such as the amount of fabric or wood.

As you work with your student, ask him or her whether to cut material exactly to the measurements or leave a little extra—different crafts have different needs. Work out strategies for changing your plans to match the materials you have readily available.

You and your student will have a lasting record of your work together—have fun showing it off to your friends and family!

Nombre ____ Fecha _________

Círeulos y área

Estimada Familia:

Muchas familias disfrutan haciendo manualidades. Costura, álbumes de recortes, diseño de jardines, jardinería y trabajos en madera son solo algunas de las actividades que usan las matemáticas para hacer formas agradables.

Cuando se dispone un jardín circular en un diseño, se cortan los bordes para hacer coincidir la circunferencia del círculo. Luego se curvan los bordes alrededor del plano circular para definir el área. Se usa el área del círculo para averiguar cuánta composta, plantas y flores se necesitarán.

De forma similar, al hacer el hilván de un mantel individual se averigua el largo del borde. Para un individual con semicírculos en cada extremo, se puede usar la circunferencia para hallar los largos alrededor de la curva. Luego se pueden sumar esos largos a los lados de la porción rectangular en el medio.

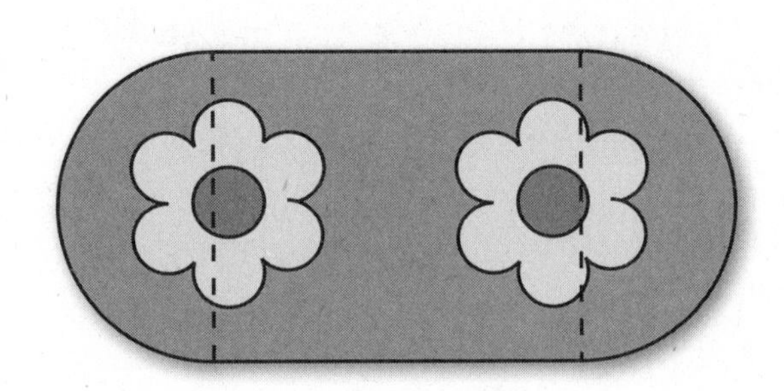

Busque una manualidad que disfrute hacer con su estudiante. Pida a su estudiante ayuda con lo siguiente:

- Dibujen un plano de su proyecto. Pueden usar combinaciones de formas geométricas (como el rectángulo y círculo en la disposición anterior) para ayudar a encontrar las medidas.
- Encuentren las partes del proyecto que usan las medidas alrededor del perímetro y luego hagan una lista de los largos de cada borde. Usen estos largos para planear el delineado y el material del reborde.
- Encuentren las partes del proyecto que encajan en un área y hagan una lista de las áreas para cada parte. Usarán estos largos para planear el material interior, como por ejemplo la cantidad de tela o madera.

A medida que va trabajando con su estudiante, pregúntele si debe cortar el material exacto según las medidas o si debe dejar un poquito extra—las diversas manualidades tienen necesidades diferentes. Diseñen estrategias para cambiar sus planos y hacer que sus materiales se adapten a los materiales que ya tiene disponibles.

Usted y su estudiante tendrán un registro duradero de su trabajo juntos—diviértanse mostrándolo a sus amigos y familiares!

Activity 8.1 Start Thinking!

For use before Activity 8.1

Pi, written π, is the ratio of a circle's circumference to its diameter. The digits of π do not repeat and continue on without end.

"Piems" are poems that can help you remember π to a certain number of digits. The length of each word in a "piem" represents a digit of π. For example:

Piem: **Can I find a trick recalling pi easily?**

$\pi \approx$ **3 . 1 4 1 5 9 2 6**

Write you own "piem" to represent at least the first five digits of π (3.1415).

Activity 8.1 Warm Up

For use before Activity 8.1

Find the perimeter of the polygon.

1.

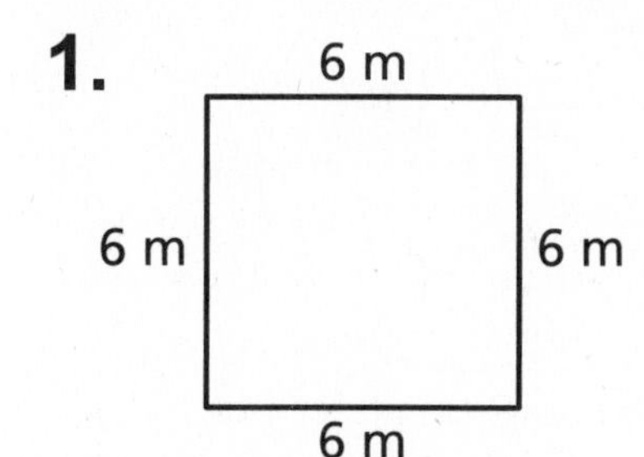

2.

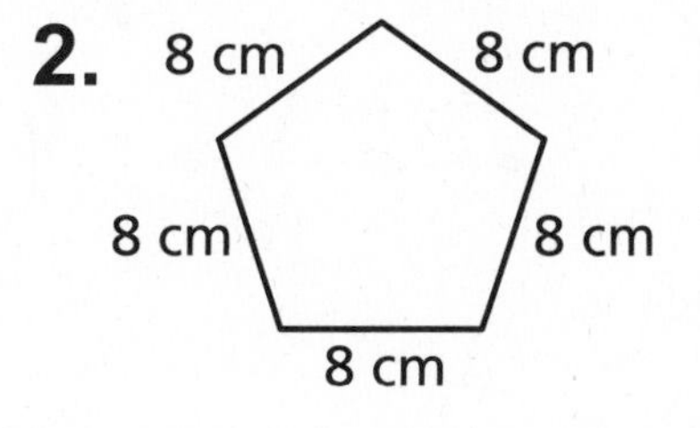

3.

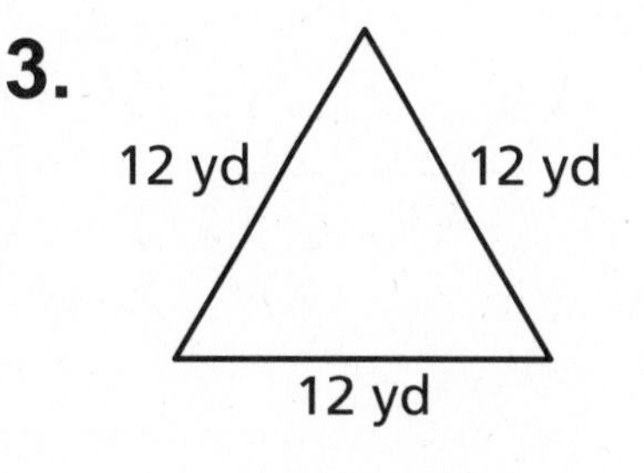

4.

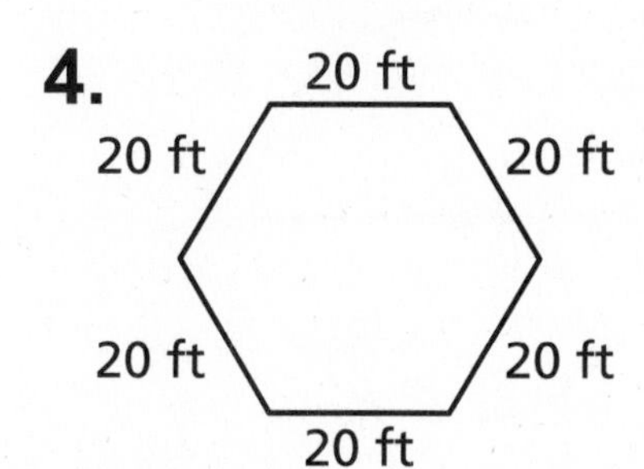

5.

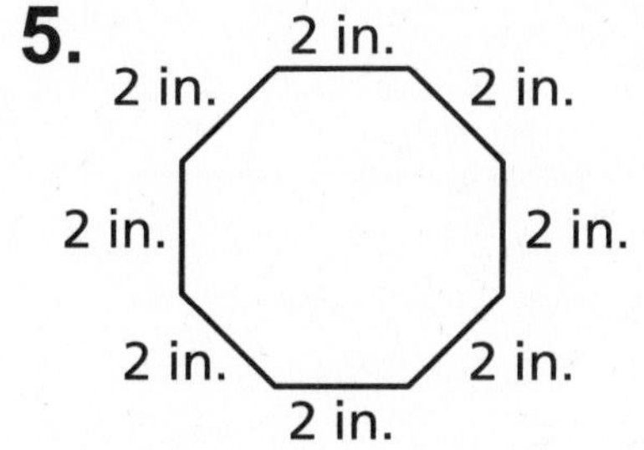

6.

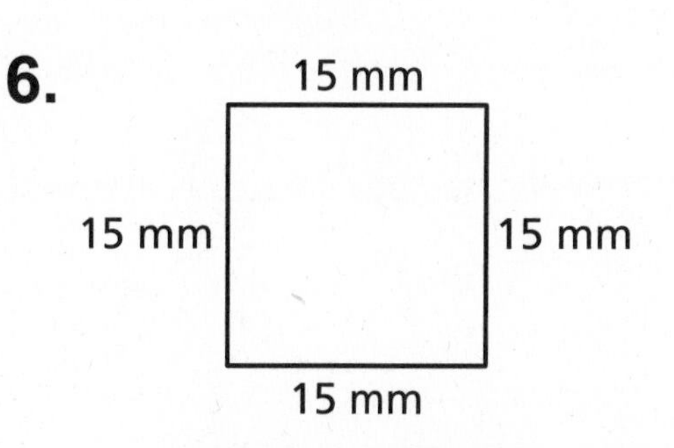

Lesson 8.1 Start Thinking!
For use before Lesson 8.1

Is it easier to measure the diameter or circumference of a tree trunk? Why?

Is it easier to measure the diameter or circumference of a quarter? Why?

In general, when is it easier to measure the diameter of a circular object? When is it easier to measure the circumference?

Lesson 8.1 Warm Up
For use before Lesson 8.1

Find the circumference of the circle. Use 3.14 or $\frac{22}{7}$ for π.

1. 6 in.

2. 2 ft

3. 3.5 cm

4. 20 ft

5. 14 mm

6. 1 in.

Name________________________________ Date__________

8.1 Practice A

Find the diameter of the circle.

1.
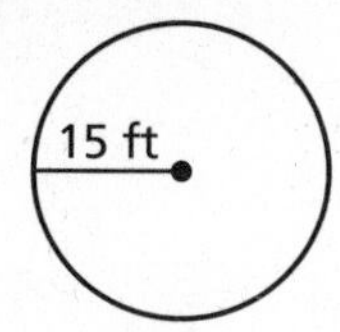

2.
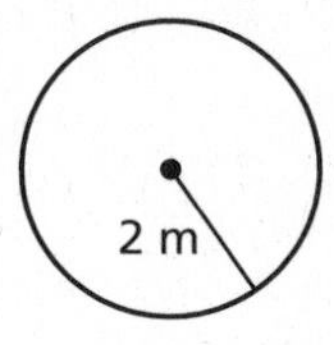

3.
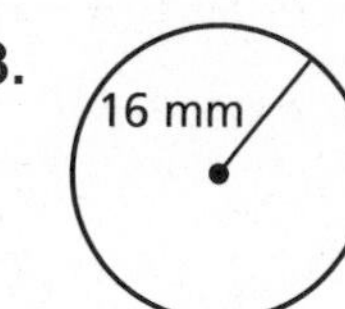

Find the radius of the circle.

4.
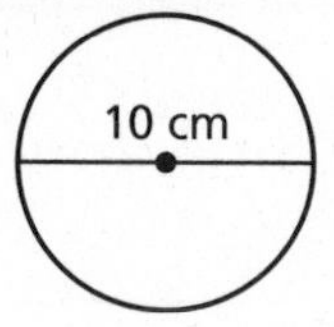

5.
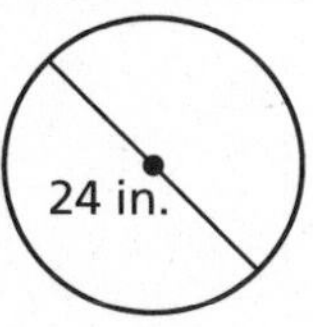

6.
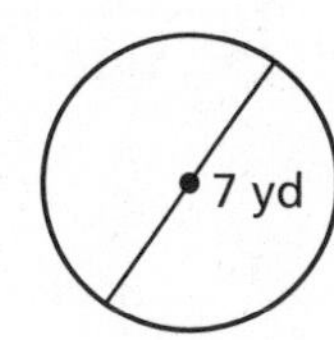

Find the circumference of the circle. Use 3.14 or $\frac{22}{7}$ for π.

7.
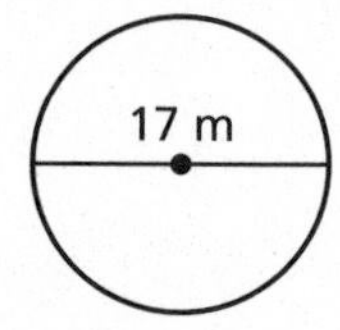

8.
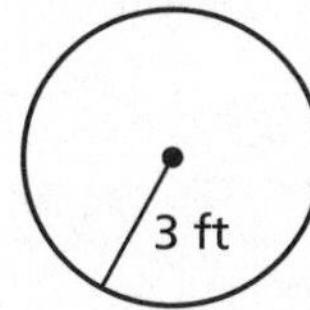

9.
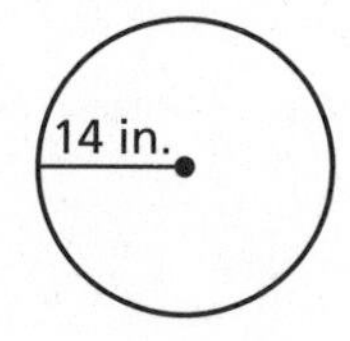

Find the perimeter of the semicircular region.

10.
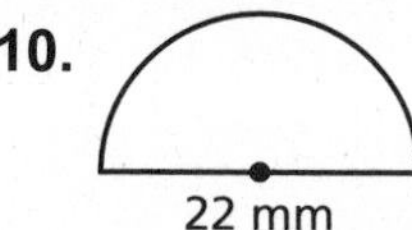

11.
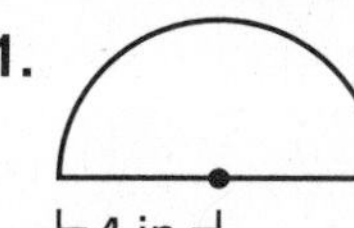

12.
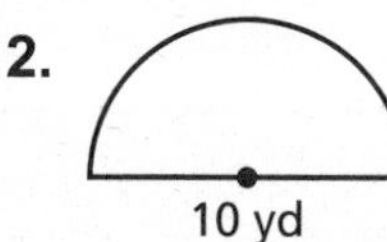

13. A circular ink spot has a circumference of 25.12 millimeters. A minute later, it has a circumference of 75.63 millimeters.

a. Estimate the diameter of the ink spot each minute.

b. How many times greater is the diameter of the ink spot compared to the previous minute?

14. You are enclosing a circular flower garden with a fence that costs \$2.99 per foot. The radius of the garden is 7 feet. How much will it cost to buy the fence? $\left(\text{Use } \frac{22}{7} \text{ for } \pi.\right)$

15. Find the perimeter of the semicircular tabletop shown at the right.

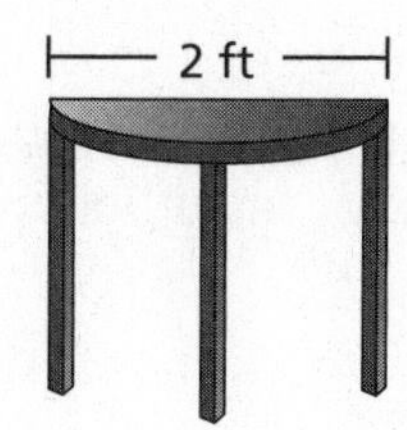

Name ______________________________ Date __________

8.1 Practice B

Find the circumference of the circle. Use 3.14 or $\frac{22}{7}$ for π.

1. 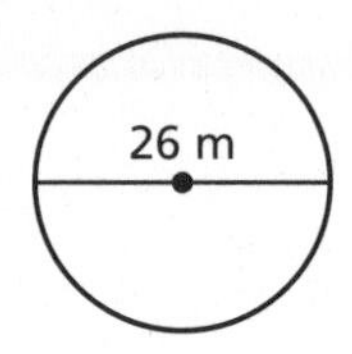

2. 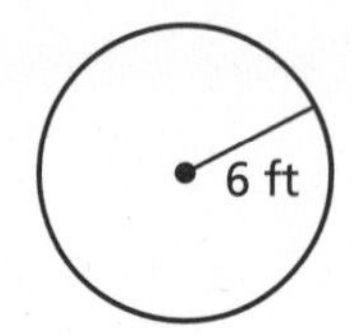

3. 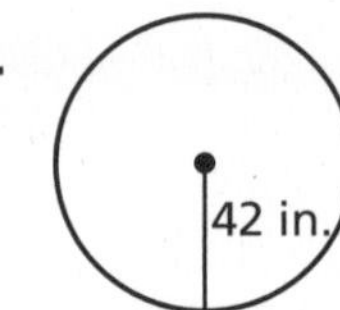

Find the perimeter of the semicircular region.

4. 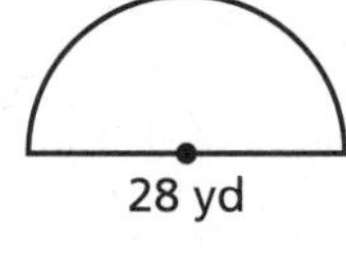

5. 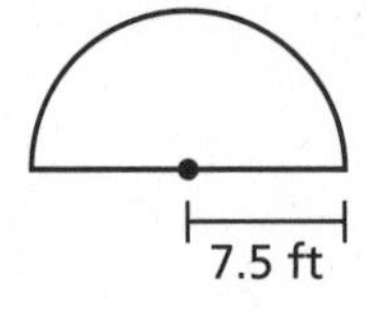

6. 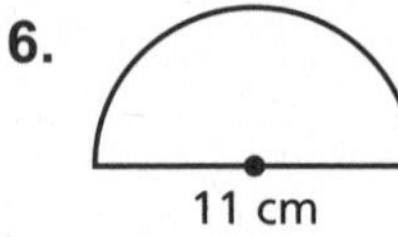

7. Copy and complete the table for Circles A, B, C, and D.

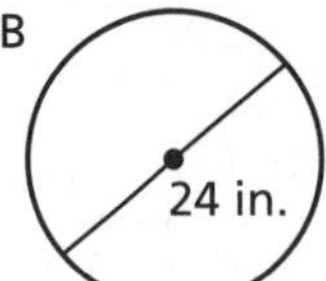

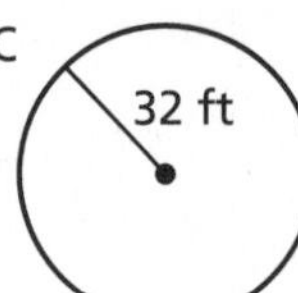

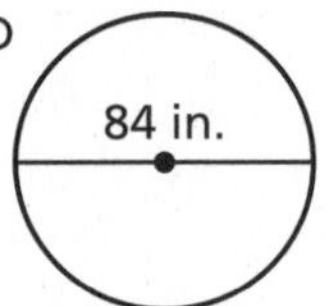

Circle	A	B	C	D
Radius	2.5 ft	? ft	32 ft	? ft
Diameter	? in.	24 in.	? in.	84 in.

8. A coaster has a circumference of 12.56 inches. Suppose the same amount of coaster is visible around the bottom of a glass as shown. What is the circumference of the glass?

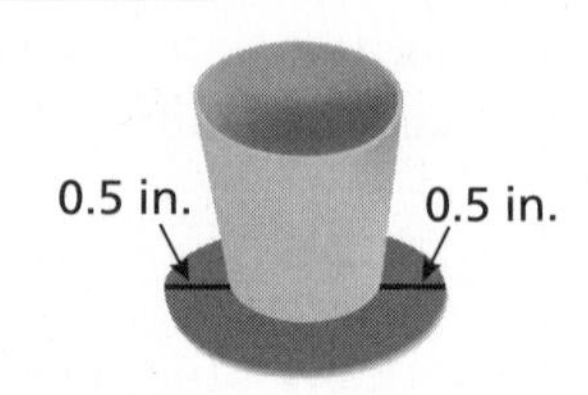

9. Are the side lengths of the squares in Diagram A and Diagram B equivalent? Explain your reasoning?

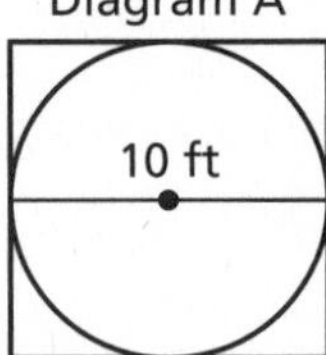

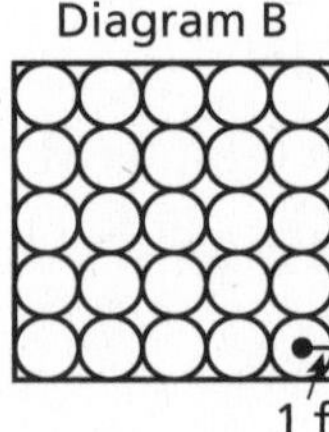

10. You release a ball with a radius of 1 inch into a pipe as shown. How many times will the ball rotate before it falls out of the other end of the pipe?

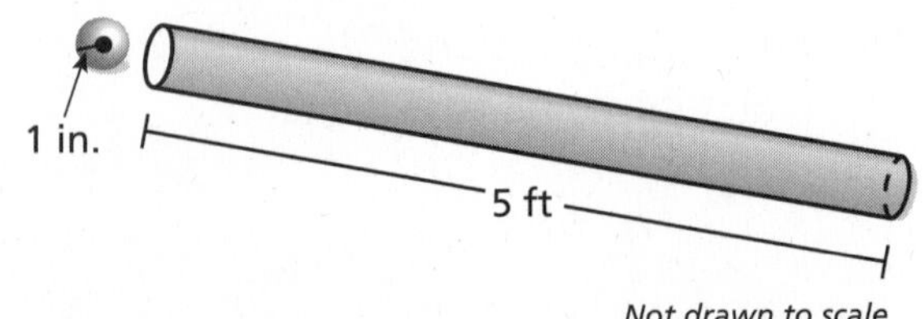

Name______________________________ Date__________

8.1 Enrichment and Extension

Changing Dimensions

Find the circumference of the circle. Then find the circumference if the radius is multiplied by 2. Use 3.14 for π.

1.

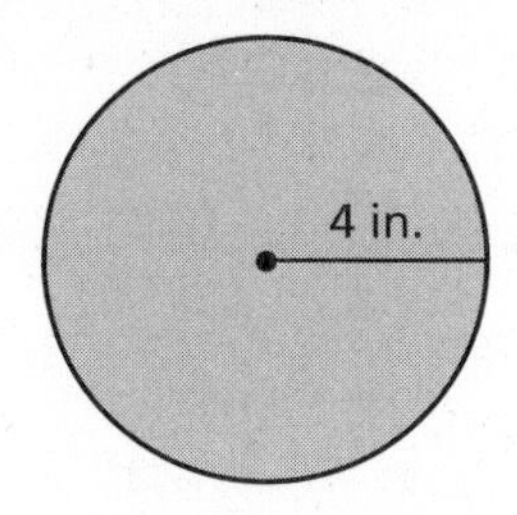

2.

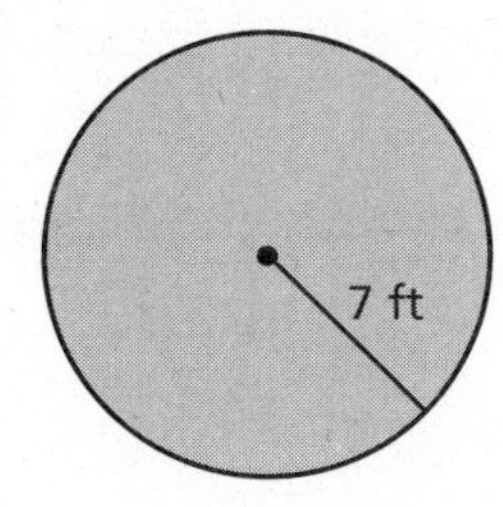

3.

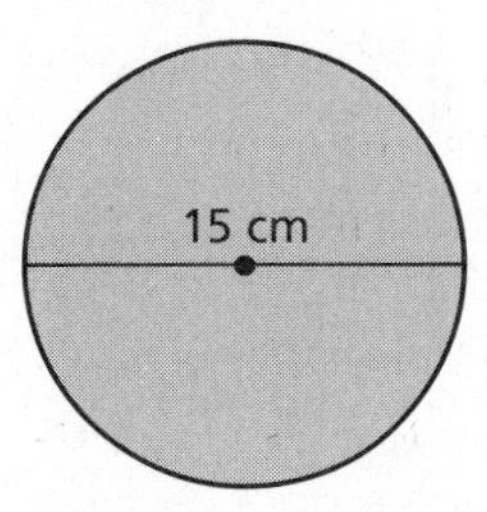

4. What happens to the circumference of a circle when its radius is multiplied by 2?

5. What happens to the circumference of a circle when its radius is multiplied by a positive number n?

Find the perimeter of the semicircle. Then find the perimeter if the radius is multiplied by $\frac{1}{2}$. Use 3.14 for π.

6.

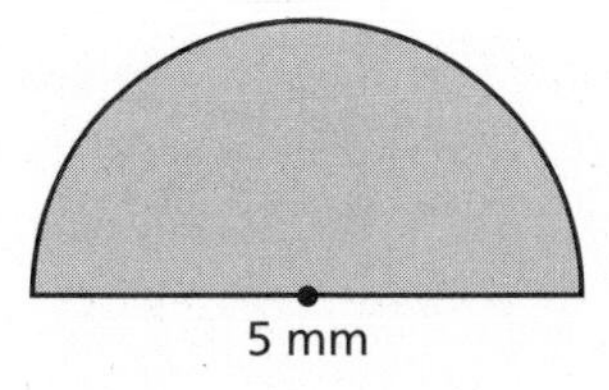

7.

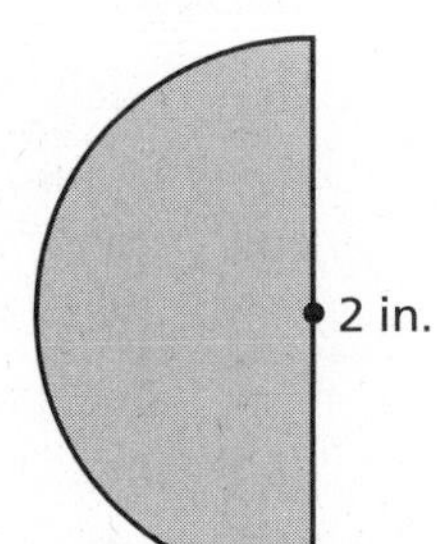

8.

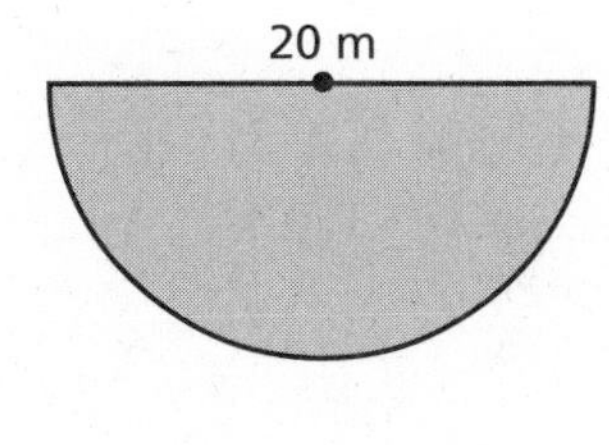

9. What happens to the perimeter of a semicircle when its diameter is multiplied by $\frac{1}{2}$?

10. What happens to the perimeter of a semicircle when its radius is multiplied by a positive number n?

Name ______________________ Date __________

Puzzle Time

Why Was The Gentleman Who Was Selling Watches Unhappy?

A	B	C	D	E	F
G	H				

Complete each exercise. Find the answer in the answer column. Write the word under the answer in the box containing the exercise letter.

62.8 mm MUCH
15 m HE
44 cm TIME
37.68 in. TOO

A. The diameter of a circle is 30 meters. Find the radius.

B. The radius of a circle is 11 millimeters. Find the diameter.

Find the circumference of the circle. Use 3.14 or $\frac{22}{7}$ for π.

C.

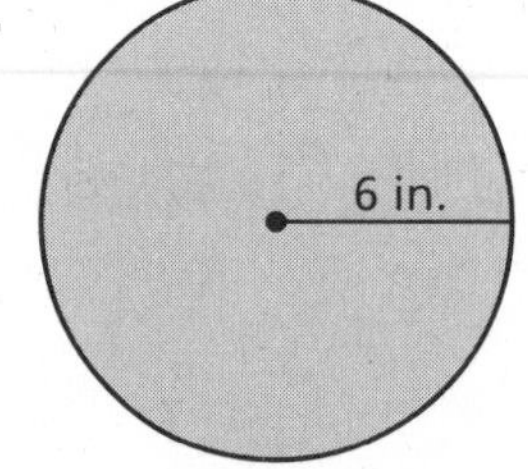

D.

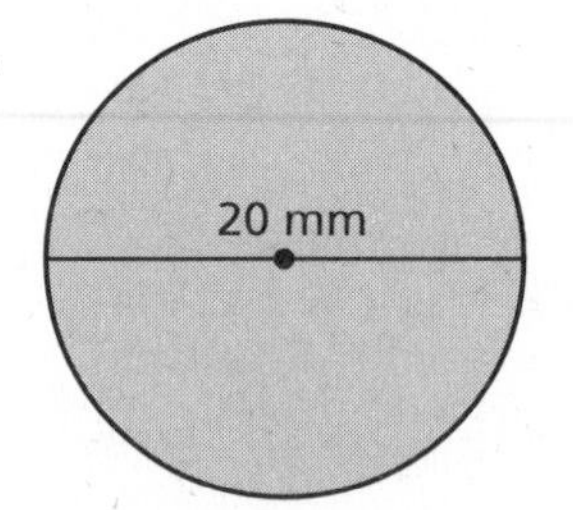

E.

14 cm

F.

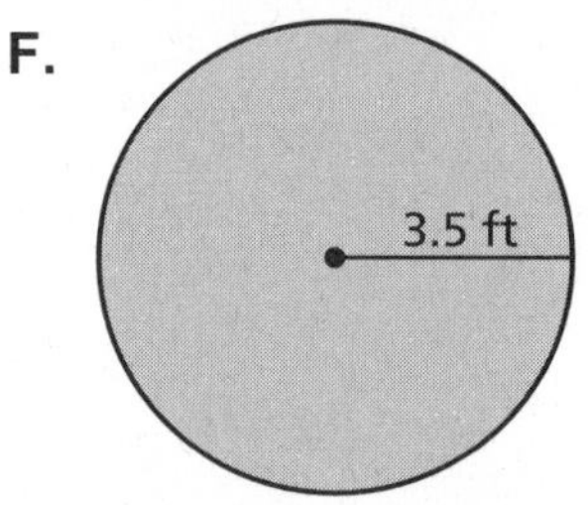

Find the perimeter of the semicircle. Use 3.14 for π.

G.

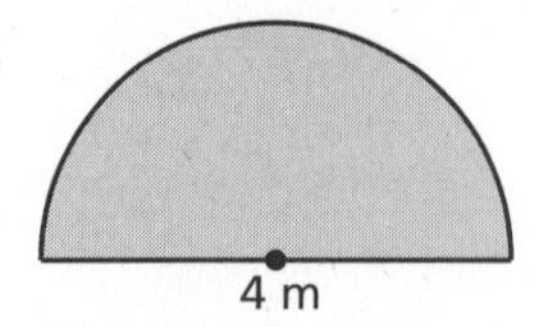

H.

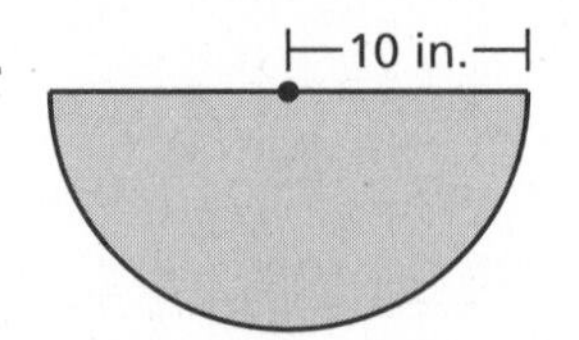

51.4 in. HANDS
22 ft ON
10.28 m HIS
22 mm HAD

Start Thinking!

For use before Activity 8.2

Have you ever used a map to find the distance you need to travel to get from one place to another?

How is using a map similar to finding the perimeter of an irregular shape?

Warm Up

For use before Activity 8.2

Find the perimeter or circumference of the figure described.

1. square with side length 4 cm
2. rectangle with length 5 ft and width 3.5 ft
3. rectangle with length 19 in. and width 7 in.
4. triangle with side lengths 7 m, 8 m, and 10 m
5. circle with radius 9 in.
6. circle with diameter 20 ft

Lesson 8.2 Start Thinking!

For use before Lesson 8.2

Is a race track an example of a composite figure? Why or why not?

What are some other objects that are composite figures?

Lesson 8.2 Warm Up

For use before Lesson 8.2

Estimate the perimeter of the figure.

1.

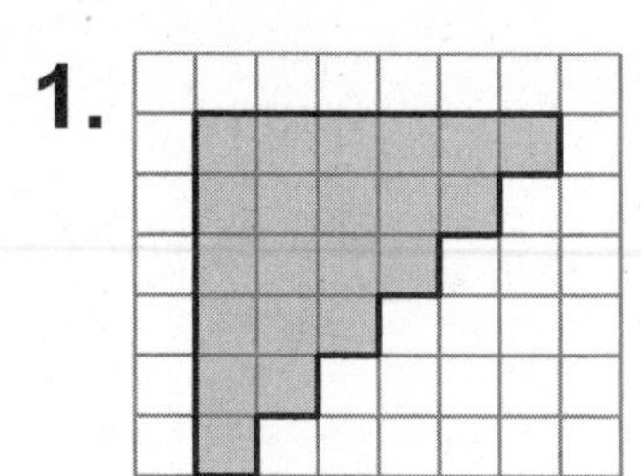

2.

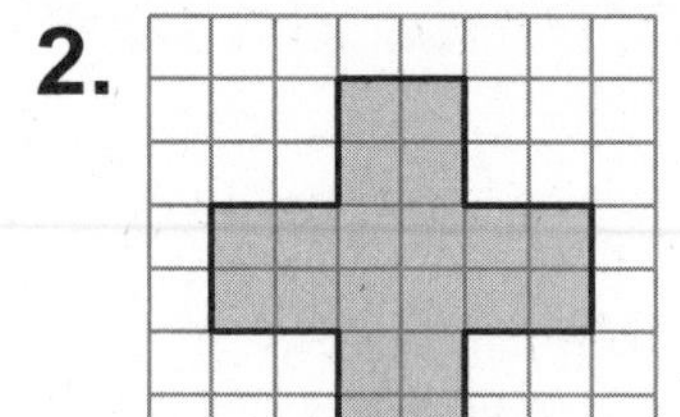

3.

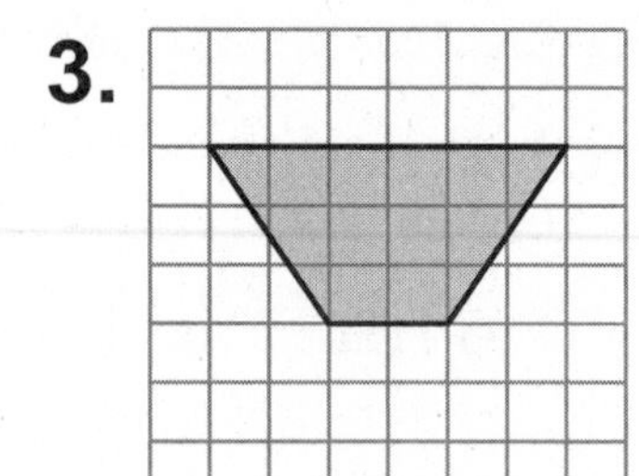

4.

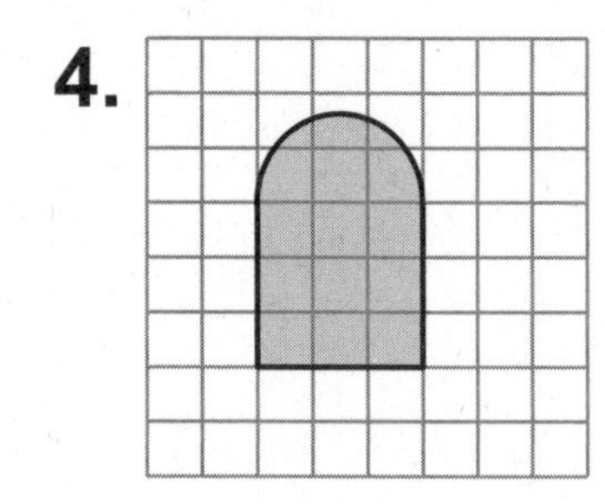

5.

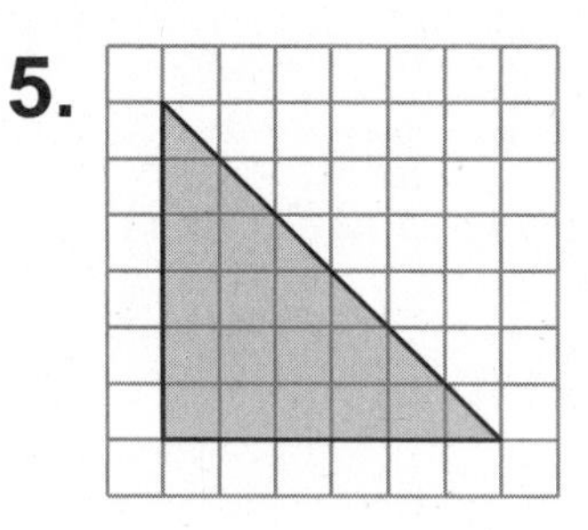

6.

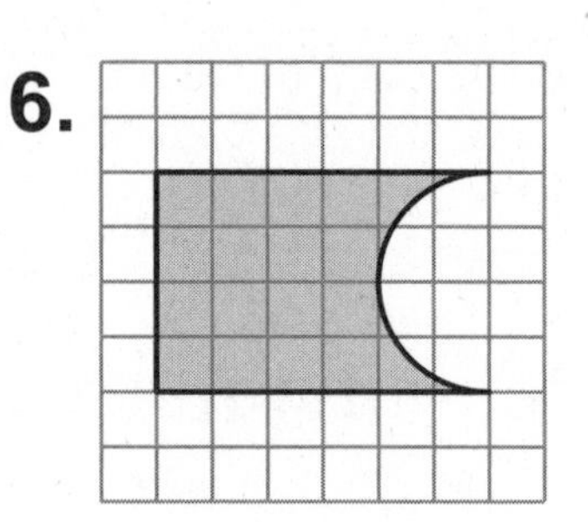

Name____________________ Date__________

8.2 Practice A

Estimate the perimeter of the figure.

1. 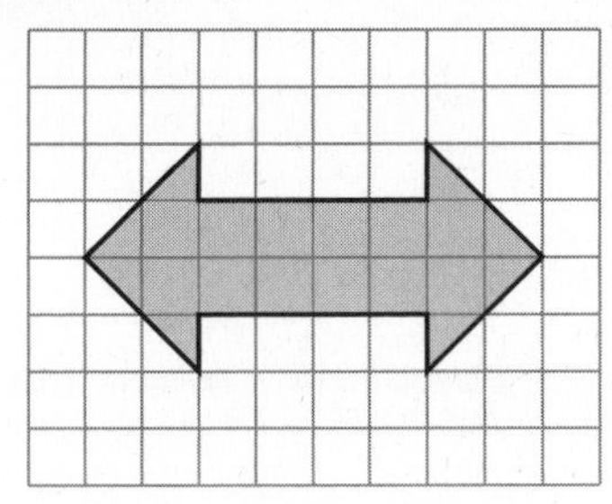

2. 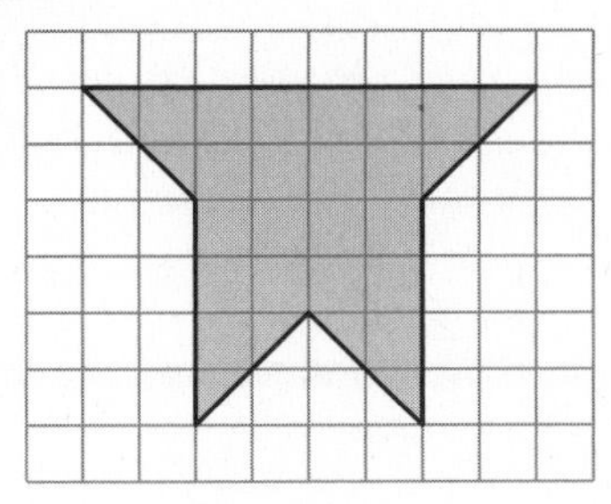

3. 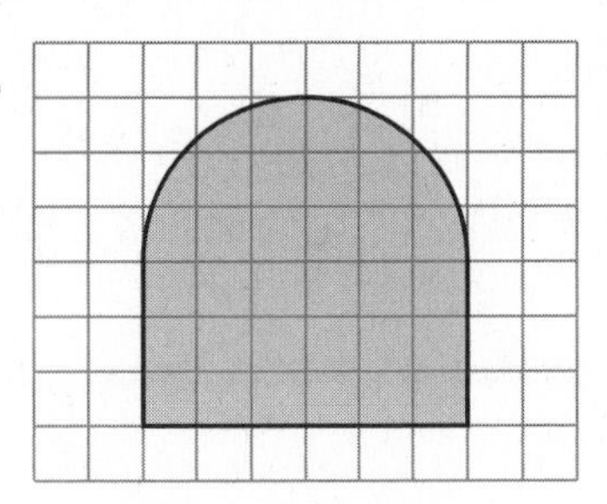

Find the perimeter of the figure.

4. 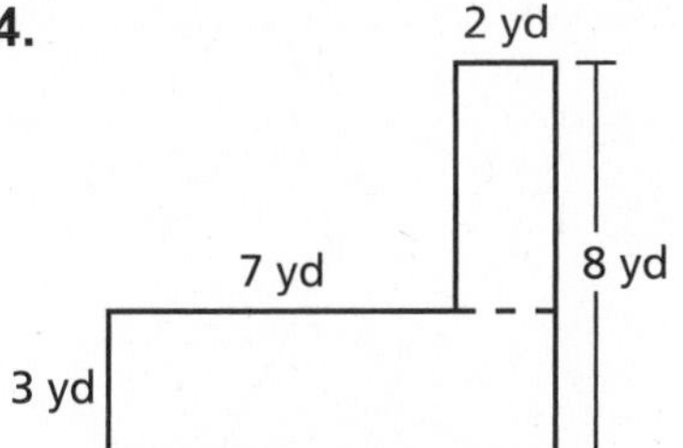

5. 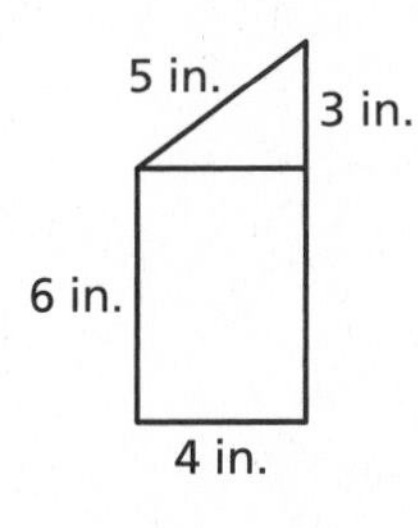

6. 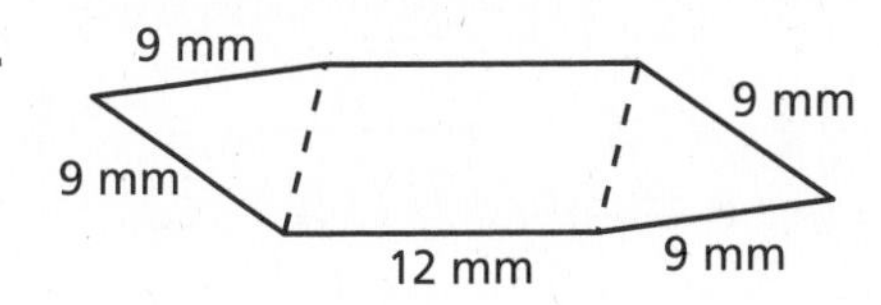

7. 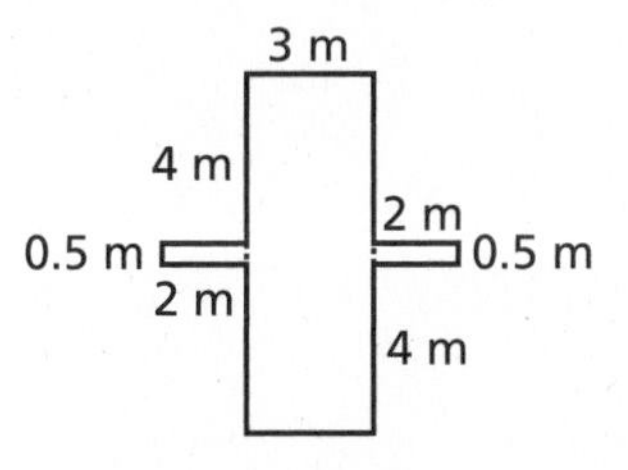

8. 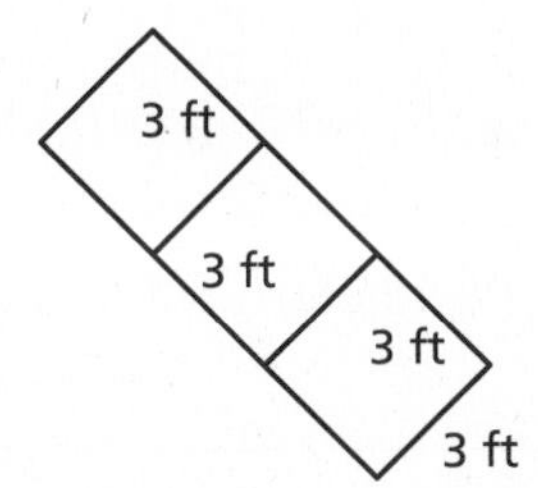

9. 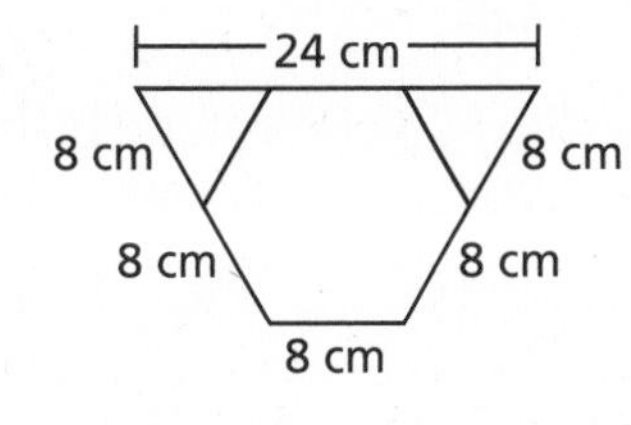

10. A stained glass window has the dimensions shown. What is the perimeter of the hole that should be cut in the wall in order for the window to be installed?

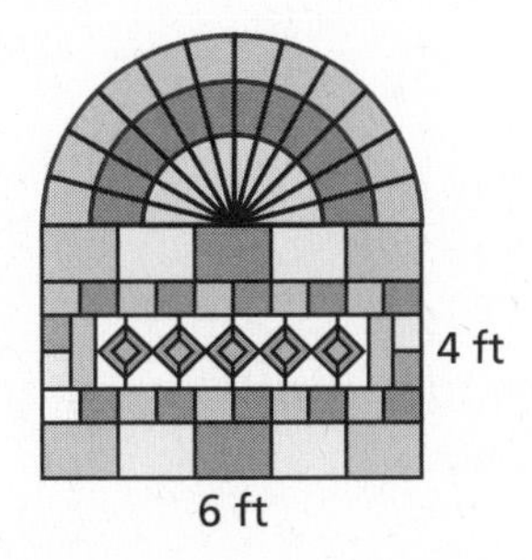

11. The dimensions of a new city park basketball court are shown at the right. A fence is to be built around the court and bleachers. The fence costs \$8.99 per foot. How much will it cost to install the fence?

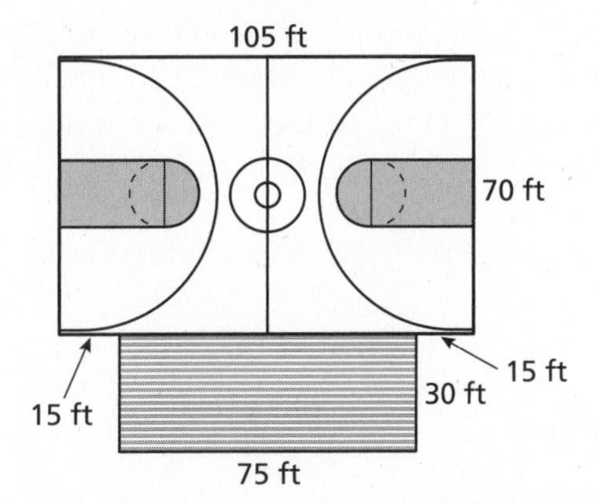

Name ______________________________ Date __________

8.2 Practice B

Estimate the perimeter of the figure.

1.

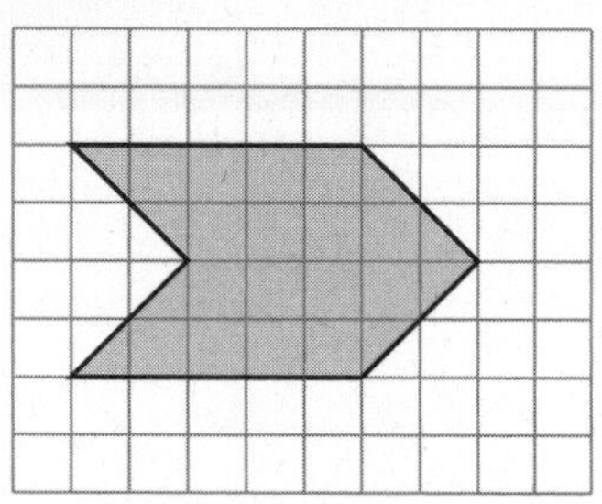

2.

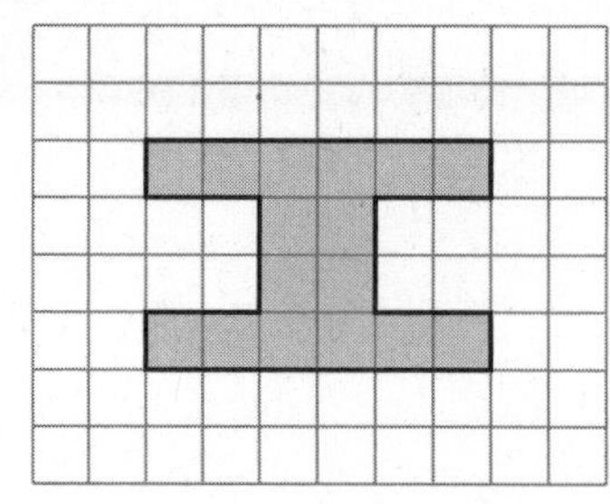

3.

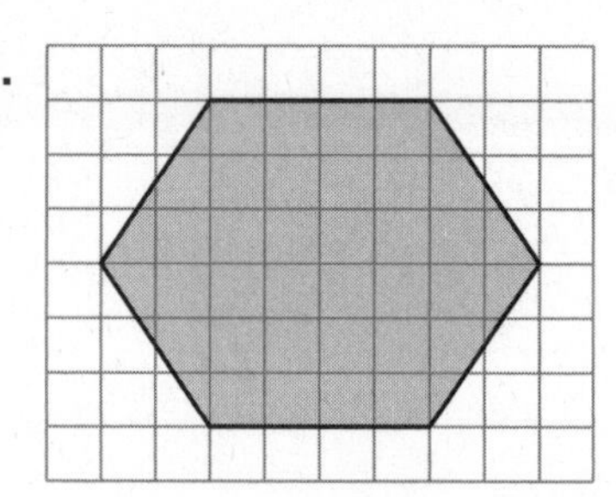

Find the perimeter of the figure.

4.

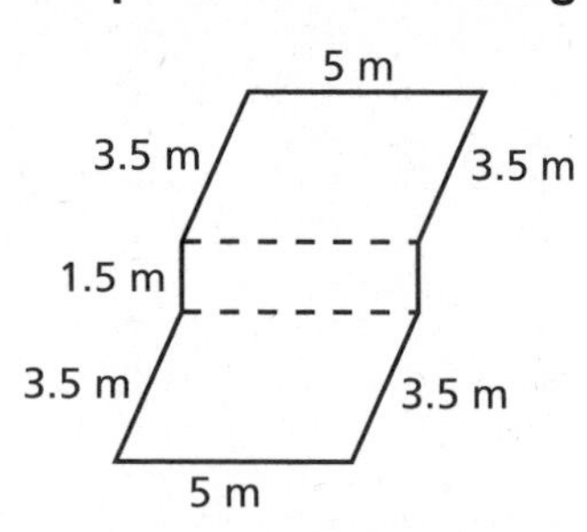

5.

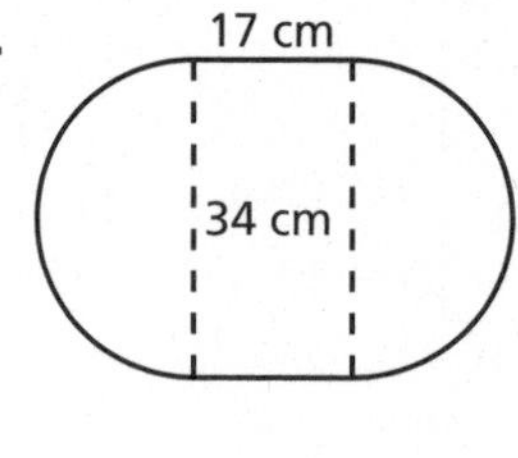

6.

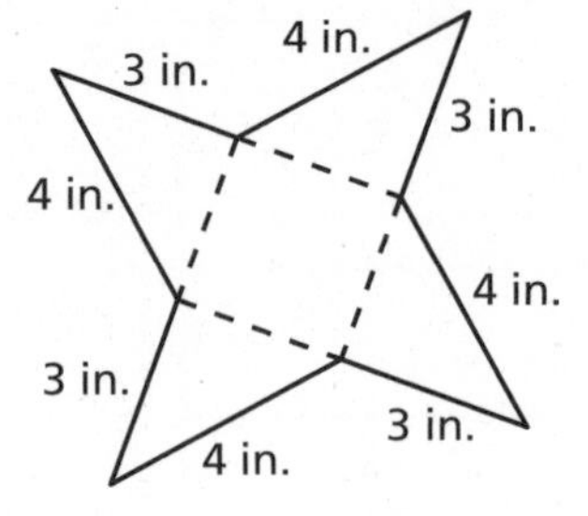

7. Describe and correct the error in finding the perimeter of the figure.

2 ft

7 ft

✗	Perimeter $\approx 2 + 7 + 2 + 21.98$
	$= 32.98$ ft

8. A school has a garden in the shape of a pencil. A fence is to be built around the garden. The fence costs \$2.75 per foot. How much will it cost to install the fence?

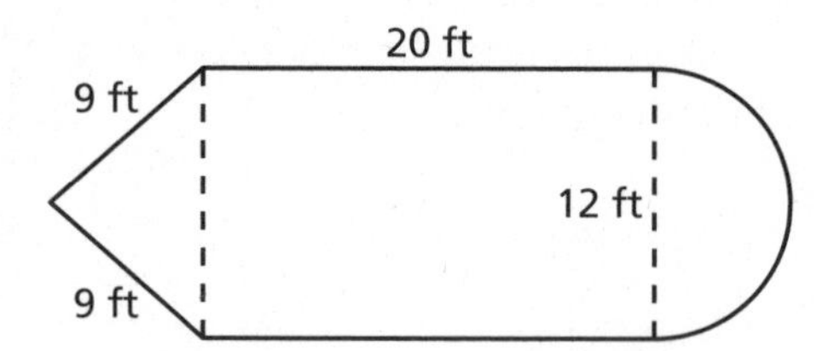

9. A shrub has been cut and trimmed into the shape of an "F." The owner has hired a landscaper to decrease the perimeter of the shrub by 10 feet. Draw a diagram of how the landscaper might do this. Is there more than one way? Explain.

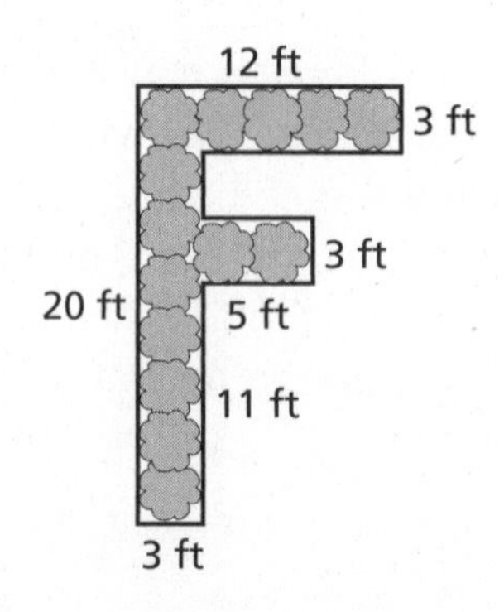

Name____ Date__________

8.2 Enrichment and Extension

Geometry

Find the perimeter of the figure. Use 3.14 for π.

1.

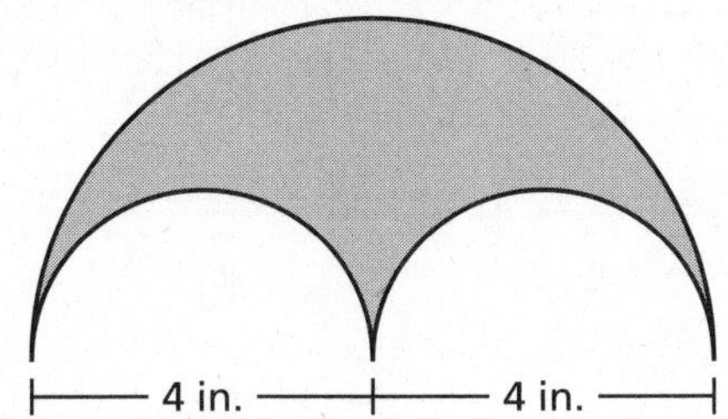

2.

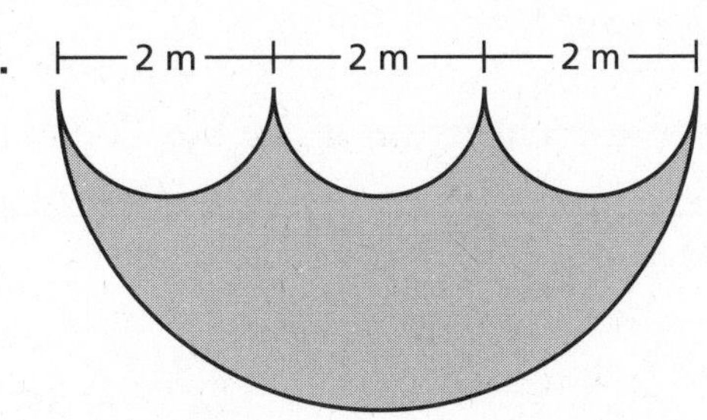

3.

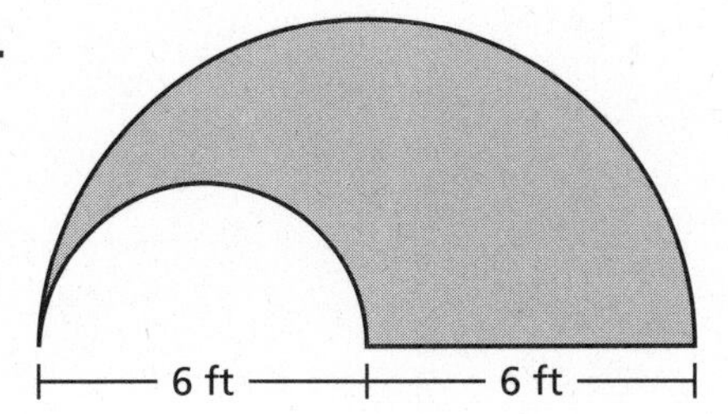

4.

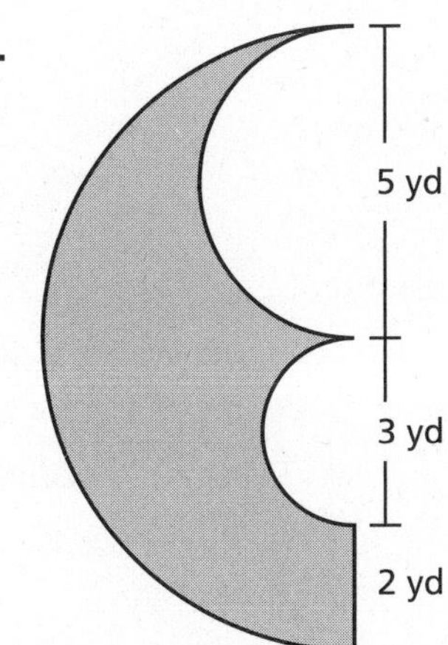

5.

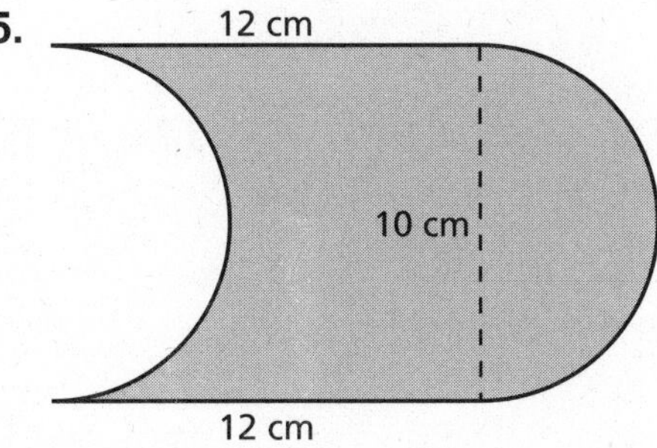

6.

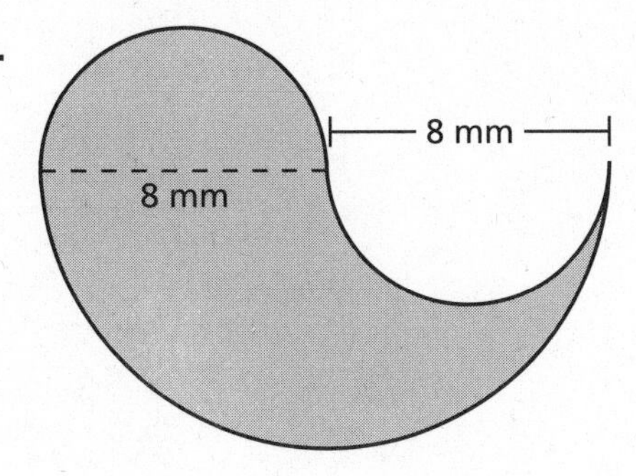

Name ______________________ Date __________

8.2 Puzzle Time

What Is The Building In Your City That Has The Most Stories?

Write the letter of each answer in the box containing the exercise number.

Answers

R. 22.56

B. 40

I. 30.28

Y. 33.42

R. 82.24

L. 22

A. 74.82

Find the perimeter of the figure.

1.

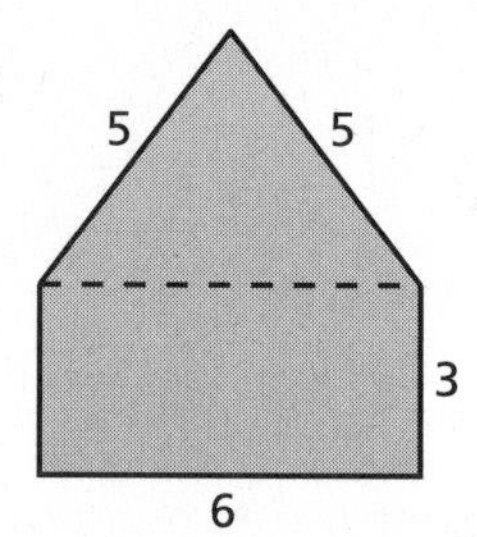

2.

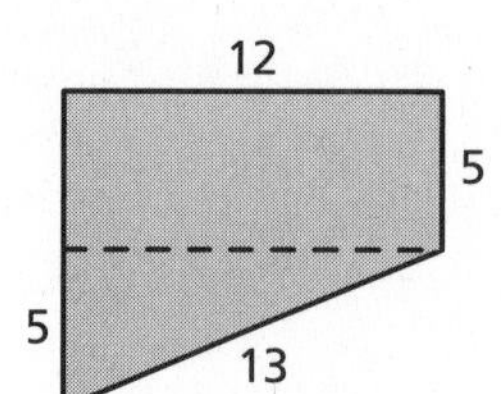

3.

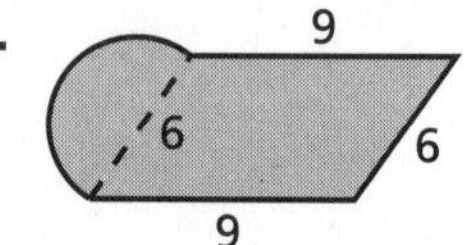

4.

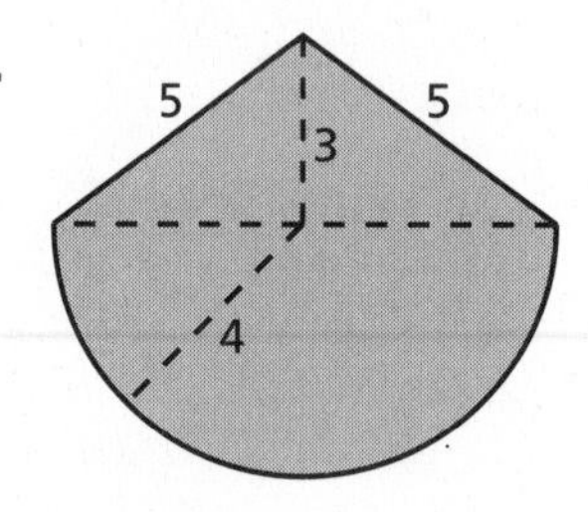

5.

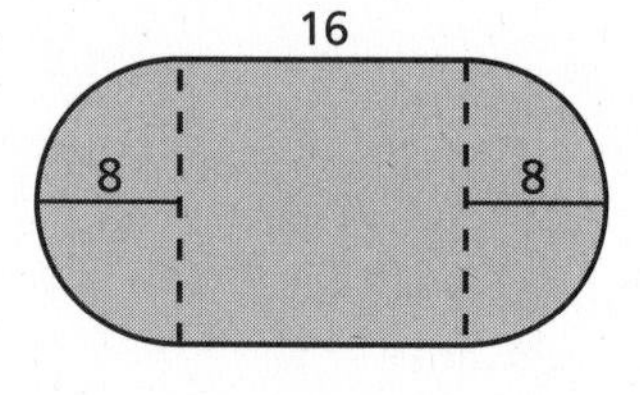

6.

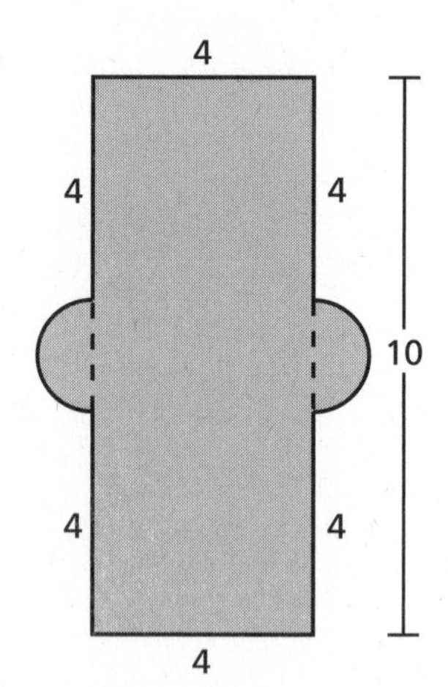

7.

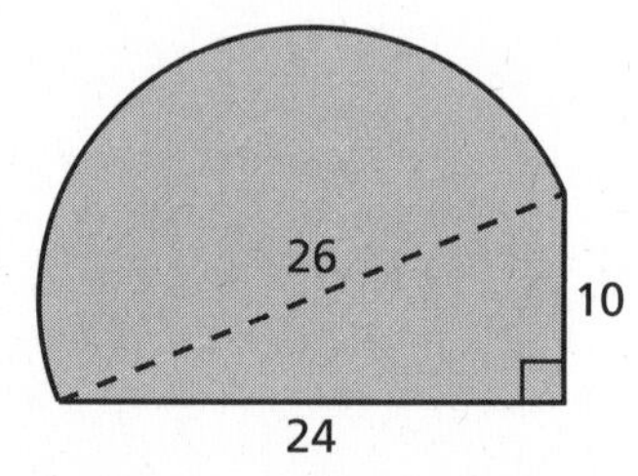

1	6	2	5	7	4	3

Activity 8.3 Start Thinking!

For use before Activity 8.3

You know how to find the area of squares, rectangles, triangles, trapezoids, and parallelograms.

Describe three different methods you could use to estimate the area of a circle.

Activity 8.3 Warm Up

For use before Activity 8.3

Find the area of the triangle.

1. 14 in., 9 in.

2. 4 m, 7 m

3. 12 yd, 11 yd

4. 10 cm, 10 cm

5. 5.6 m, 8 m

6. 1 mm, 2 mm

Lesson 8.3 Start Thinking!

For use before Lesson 8.3

Two approximations for π are $\frac{22}{7}$ and 3.14. In finding the area of a circle, when is it easier to use $\frac{22}{7}$? When is it easier to use 3.14?

Write a word problem involving the area of a circular object. Exchange problems with a classmate and solve your classmate's problem. Is it easier to use $\frac{22}{7}$ or 3.14 for π to solve your classmate's problem?

Lesson 8.3 Warm Up

For use before Lesson 8.3

Find the area of the circle. Use 3.14 or $\frac{22}{7}$ for π.

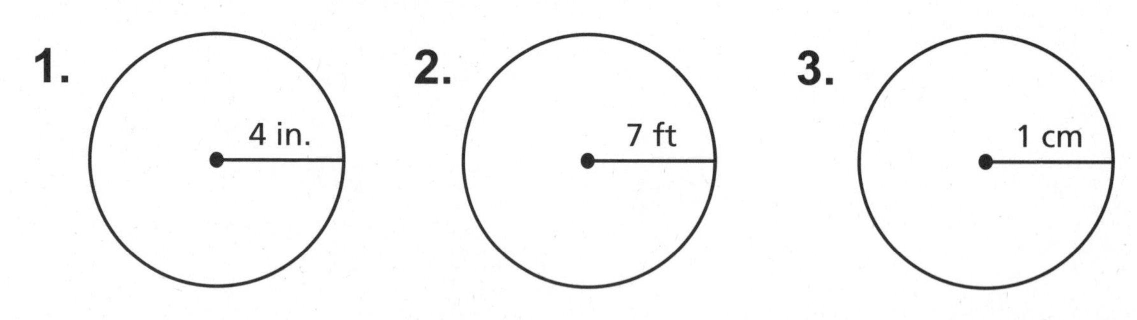

Name________________________________ Date__________

8.3 Practice A

Find the area of the circle. Use 3.14 or $\frac{22}{7}$ for π.

1. 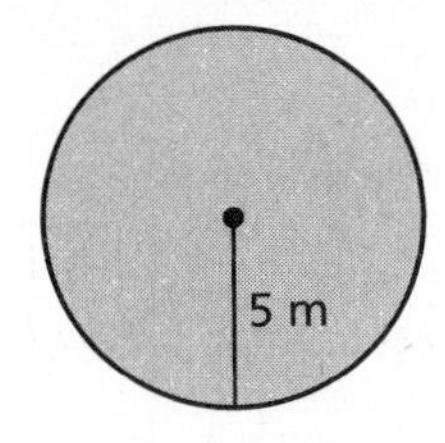

2. 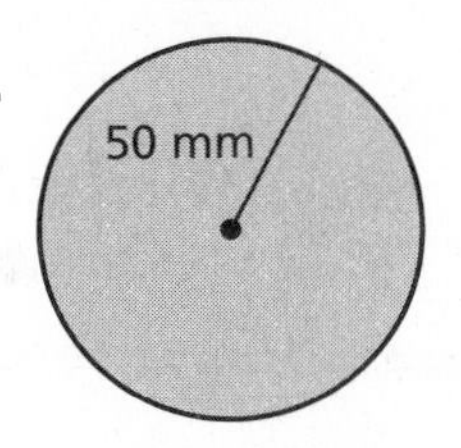

3. 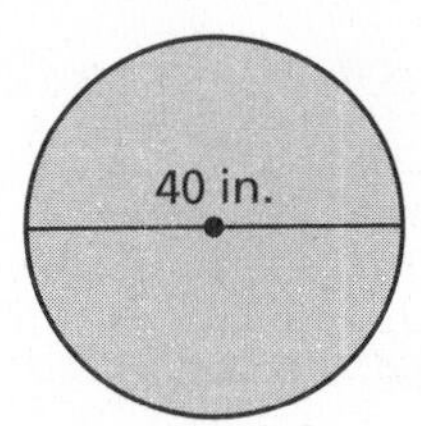

4. 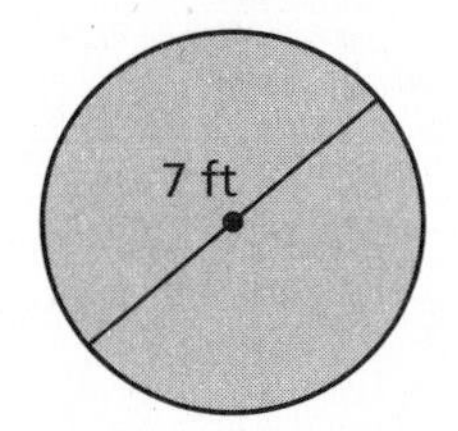

5. 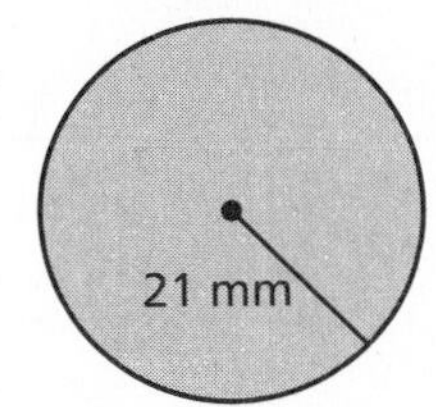

6. 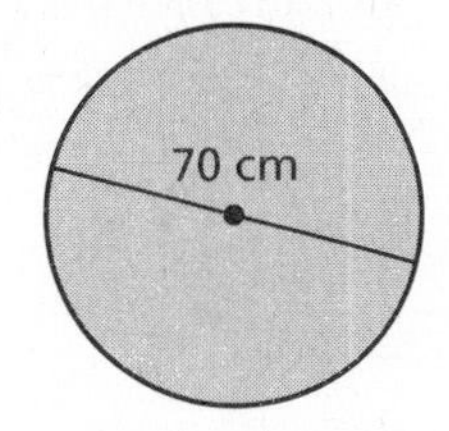

Find the area of the object.

7. 

8. 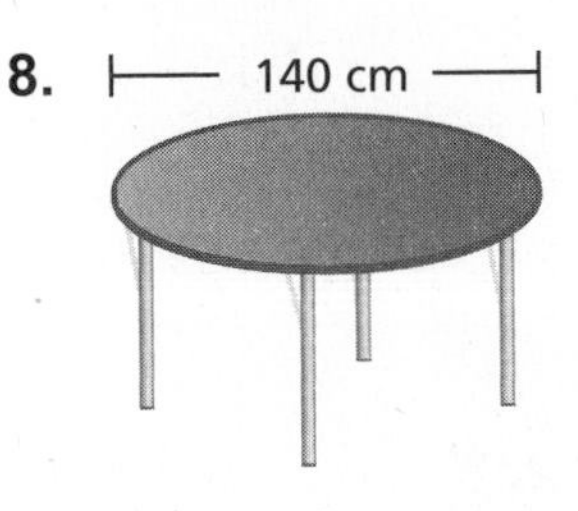

9. 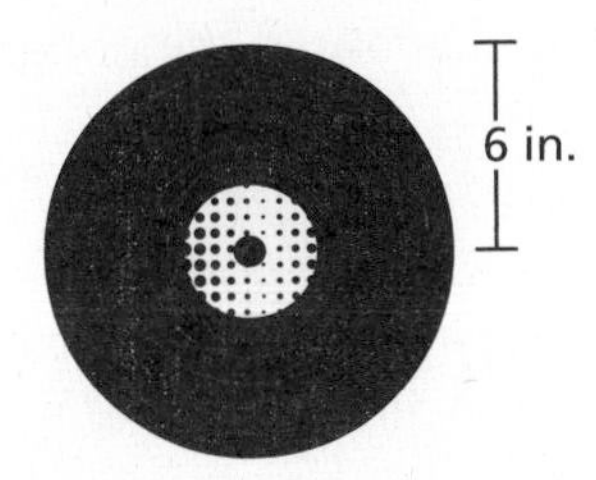

10. You use the compass to draw a circle. What will be the area of the circle?

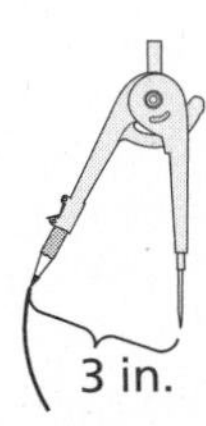

11. What fraction of a square inch is the area of one side of a penny? Use $\frac{22}{7}$ for π. Write your answer in simplest form.

12. To make a pizza, you spread pizza sauce over all but a 1-inch area around the outside edge as shown. What area of the crust is covered with sauce?

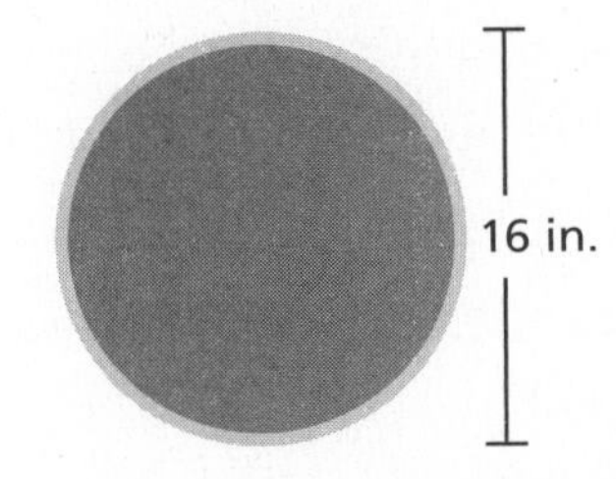

Name ______________________________ Date ________

8.3 Practice B

Find the area of the circle. Use 3.14 or $\frac{22}{7}$ for π.

1.

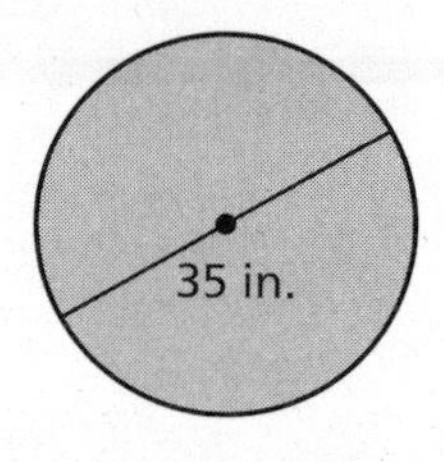

2.

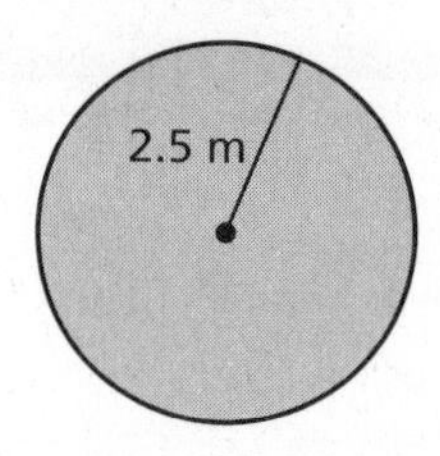

3.

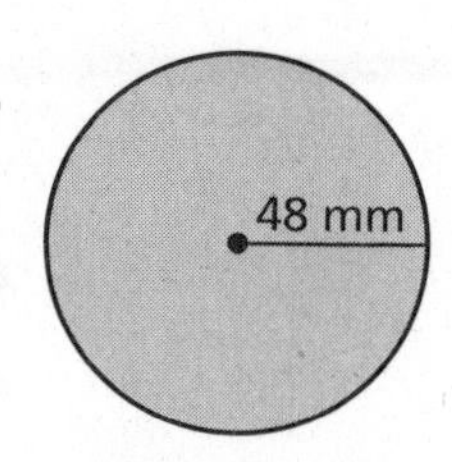

Find the area of the semicircle.

4.

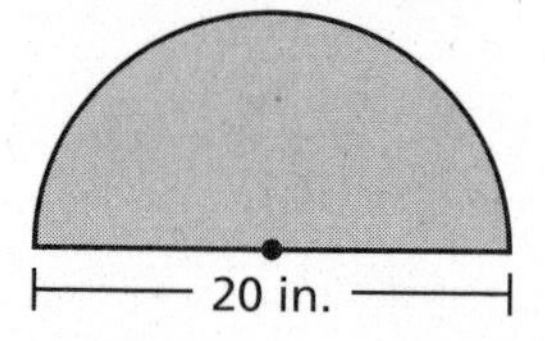

5.

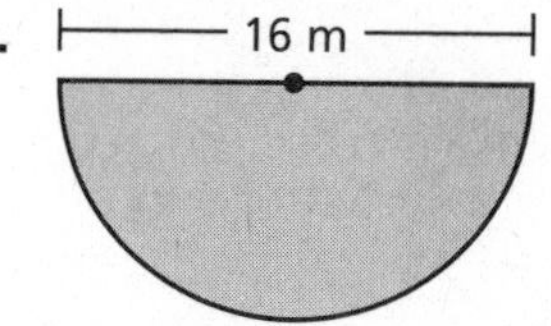

6.

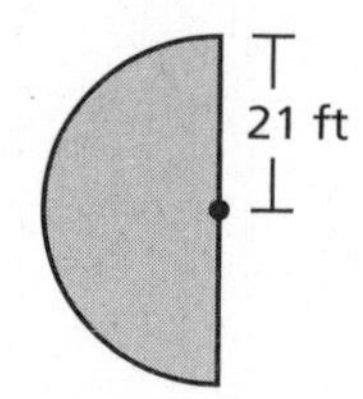

7. The shadow of an object is roughly the same size as the object. What is the area of the circular shadow of the hot air balloon?

8. How many *square feet* of the ground are sprayed by the beach shower?

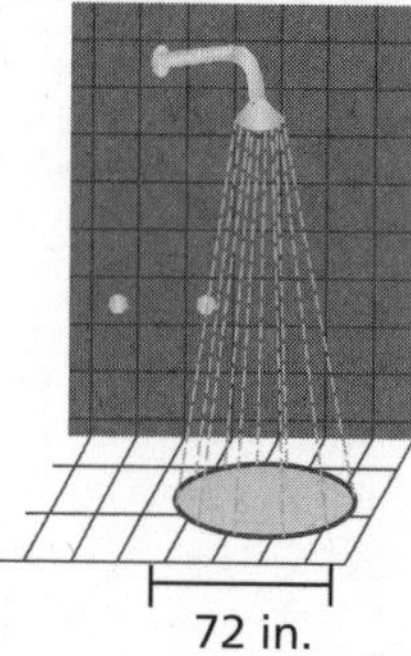

9. The radius of the small circle is half the radius of the large circle.

a. Use the radius r to write a formula for the area of the large circle.

b. Use the radius $\frac{r}{2}$ to write a formula for the area of the small circle.

c. How does the area of a circle compare to the area of another circle whose radius is twice as large? Explain your reasoning.

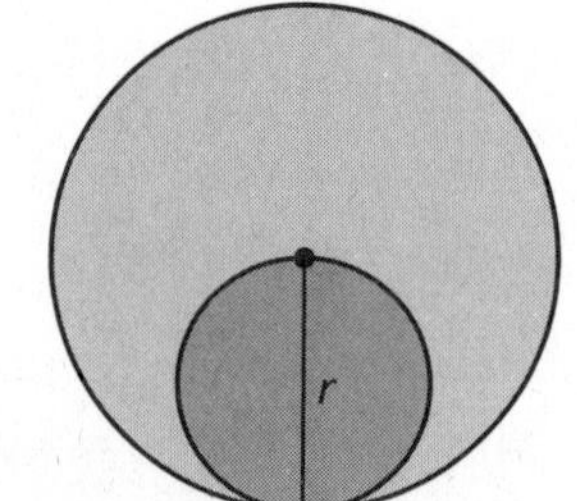

10. The number of square inches of a circle's area is equal to the number of inches of its circumference. What is the radius of the circle? Explain how you found your answer.

Name________________________________ Date__________

8.3 Enrichment and Extension

Where Are You Located?

The circular broadcast areas of two television stations and two radio stations are shown in the diagram. Use 3.14 for π.

	Broadcast radius (mi)
TV Channel 19	150
TV Channel 36	120
Radio 105.5 FM	60
Radio 94.7 FM	40

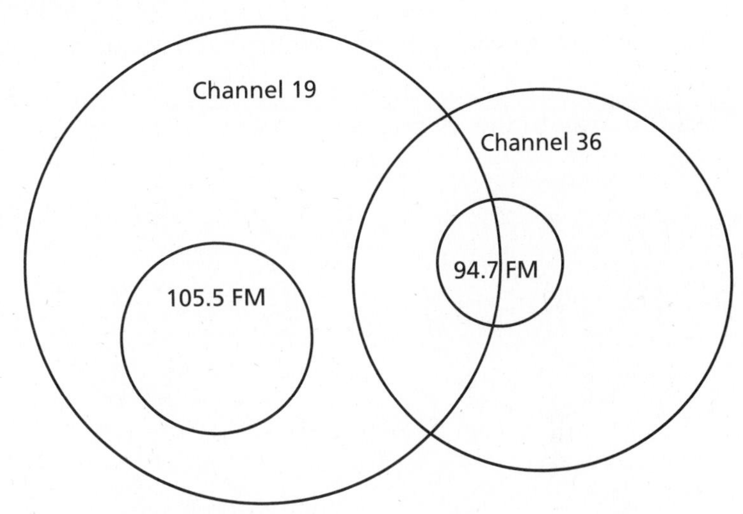

1. How many times larger is the broadcasting area of Channel 19 than the broadcasting area of Channel 36?

2. How many times larger is the broadcasting area of 105.5 FM than the broadcasting area of 94.7 FM?

3. What percent of the broadcast area of Channel 19 is also in the broadcast area of 105.5 FM?

4. What fraction of the broadcast area of Channel 36 is also in the broadcast area of 94.7 FM?

5. Half of the broadcast area of 94.7 FM can watch both television channels.

   a. What percent of the broadcast area of Channel 19 can listen to 94.7 FM?

   b. You can watch one of the television channels and can listen to one of the radio stations. Draw a diagram and shade the area where you may be located. What is the total area of the shaded region?

   c. Your friend can watch Channel 19 but cannot listen to either of the radio stations. You do not know if your friend can watch Channel 36. Draw a diagram and shade the area where your friend may be located. What is the total area of the shaded region?

Name ______________________________ Date __________

8.3 Puzzle Time

What Do Little Piggies Do As Soon As They Get Home From School?

Write the letter of each answer in the box containing the exercise number.

Find the area of the circle. Use 3.14 or $\frac{22}{7}$ for π.

1.

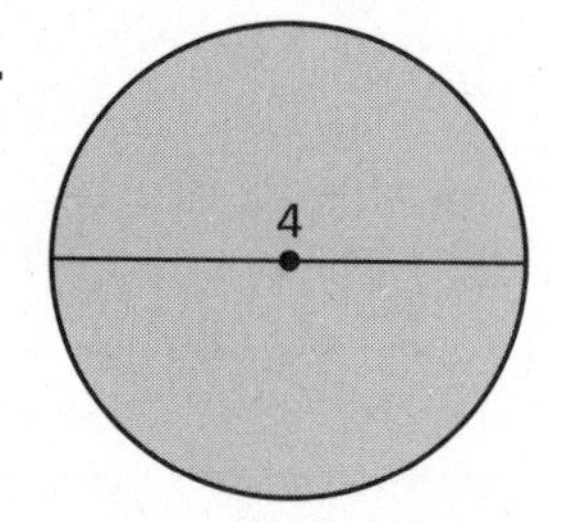

2.

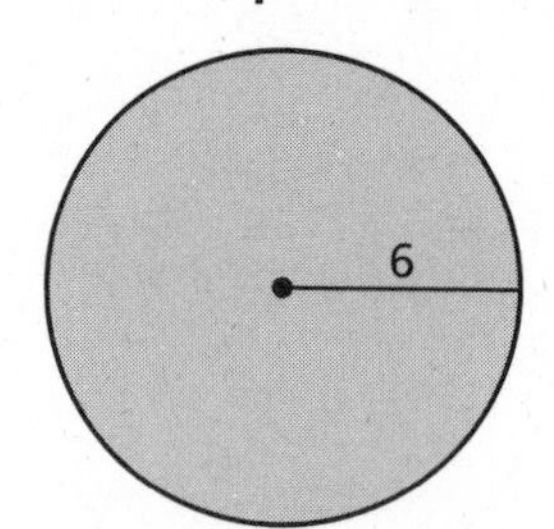

3.

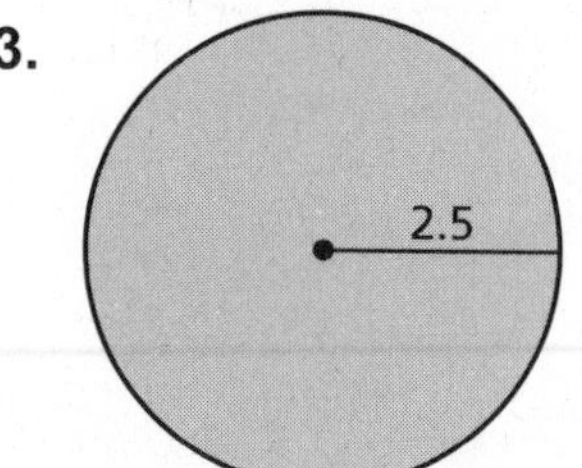

4.

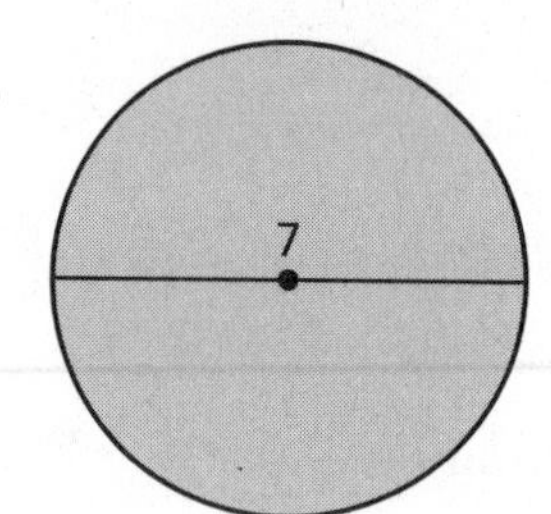

5.

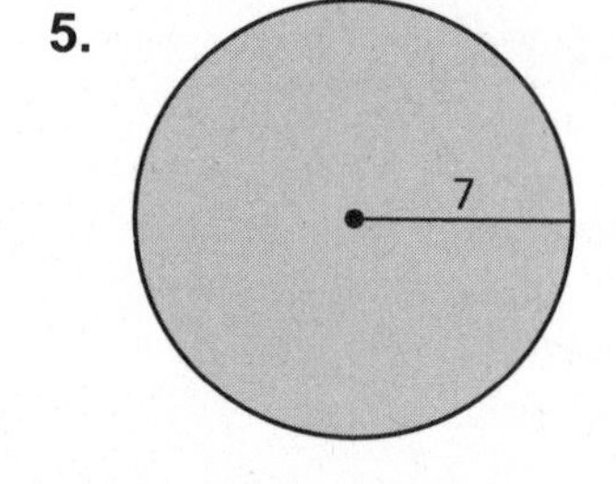

Answers

K. 38.465 units^2

H. 12.56 units^2

M. 19.625 units^2

R. 127.17 units^2

W. 154 units^2

O. 113.04 units^2

A. 157 units^2

Find the area of the semicircle. Use 3.14 for π.

6.

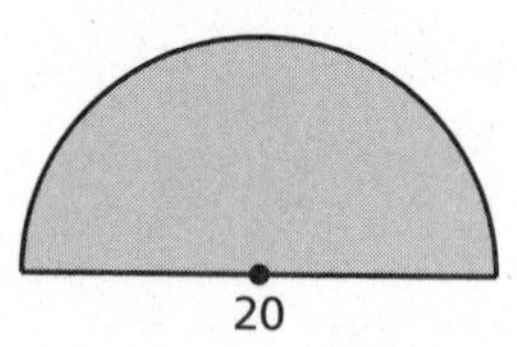

7.

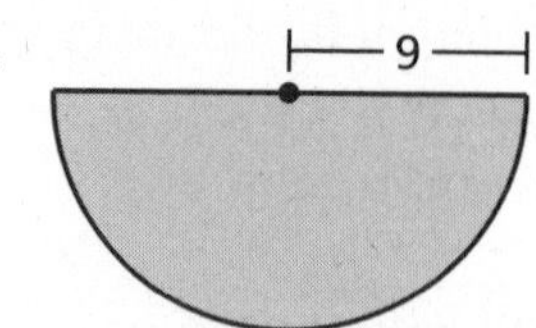

1	6	3	5	2	7	4

Activity 8.4 Start Thinking!

For use before Activity 8.4

When might it be useful to know how to find the area of a real-life composite figure?

Activity 8.4 Warm Up

For use before Activity 8.4

Find the area of the figure described.

1. square with side length 10 ft
2. square with side length 16 in.
3. rectangle with length 15 m and width 10 m
4. triangle with base 5 cm and height 12 cm
5. circle with radius 10 ft
6. circle with diameter 100 yd

Lesson 8.4 Start Thinking!

For use before Lesson 8.4

Draw a picture of a house with a roof and a chimney.

How can you use a ruler and some calculations to find the area covered by your drawing?

Use a ruler and the method you described to estimate the area of the house.

Lesson 8.4 Warm Up

For use before Lesson 8.4

Find the area of the figure.

1. 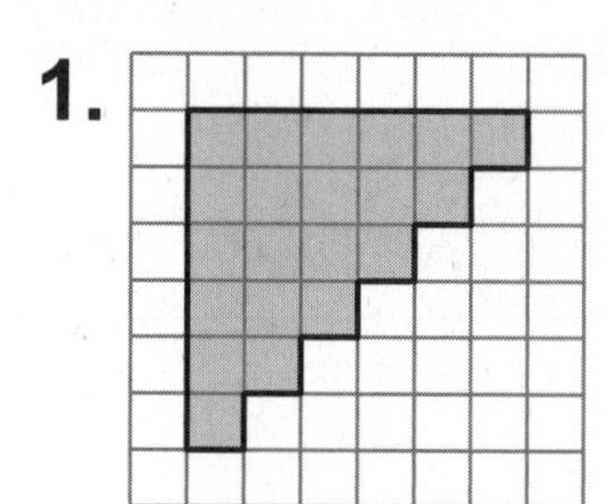

2. 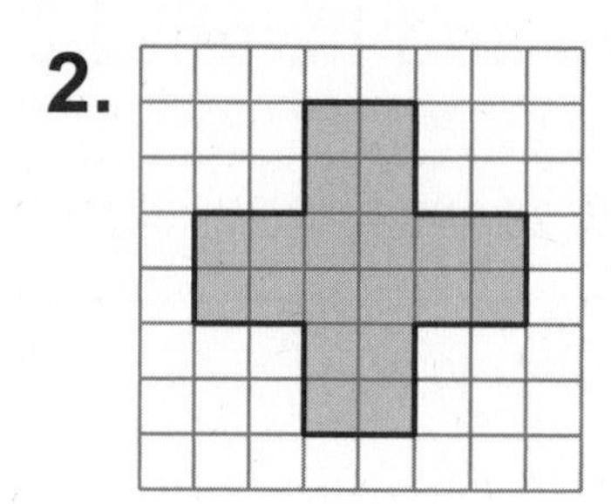

3. 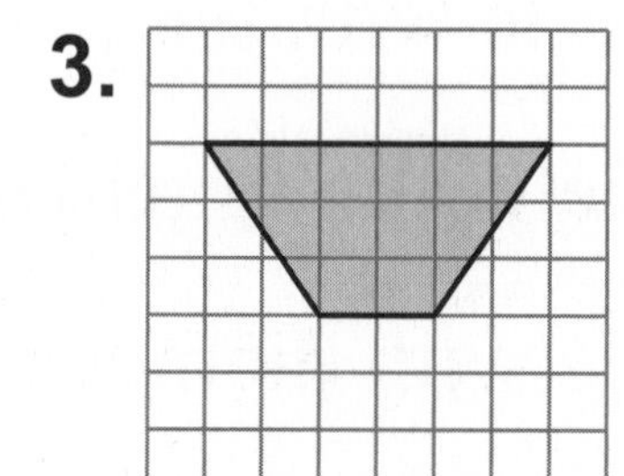

4. 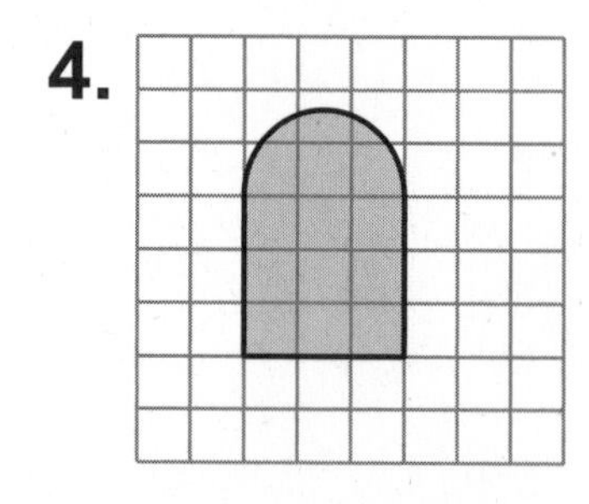

5. 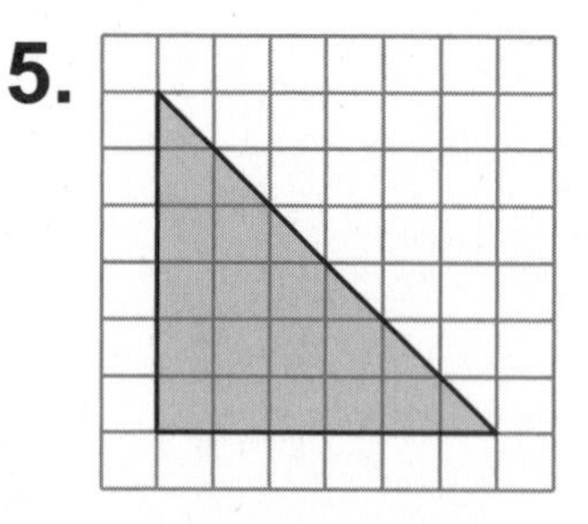

6. 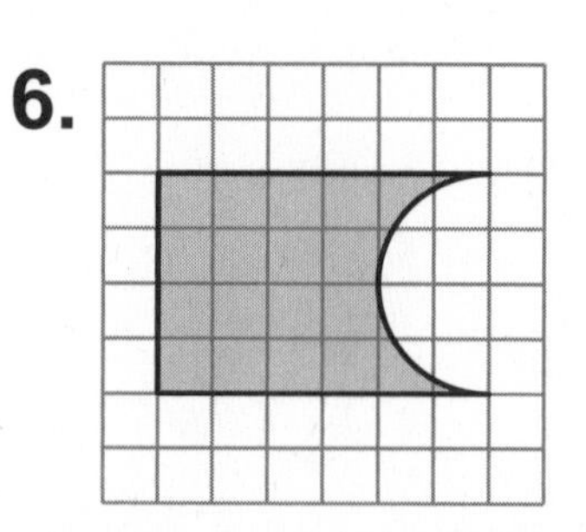

Name______________________________ Date__________

8.4 Practice A

Find the area of the figure.

1.

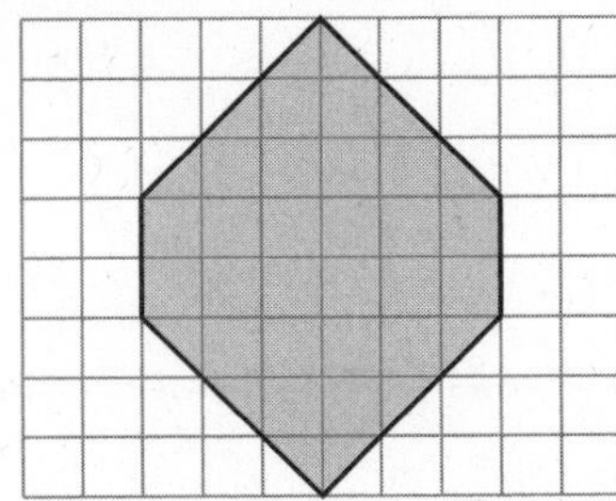

2.

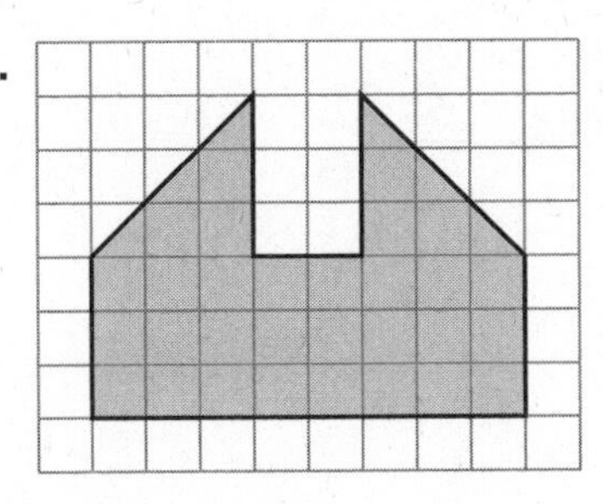

3.

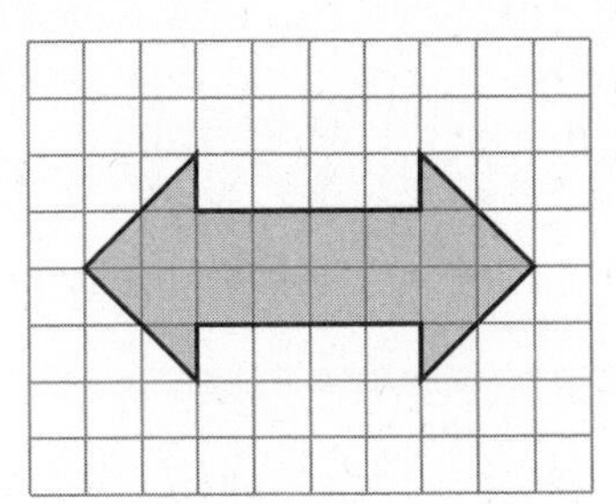

4.

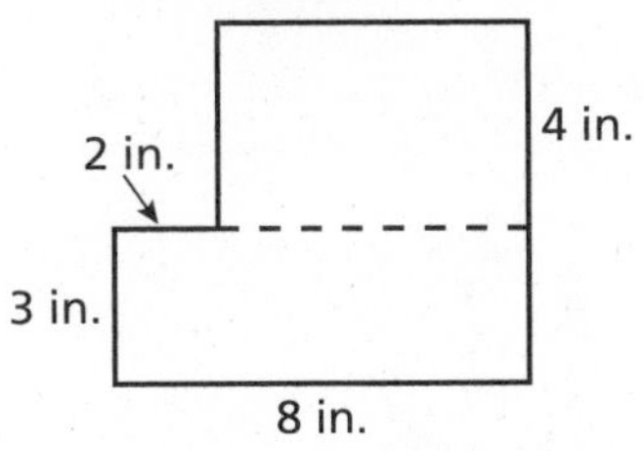

5.

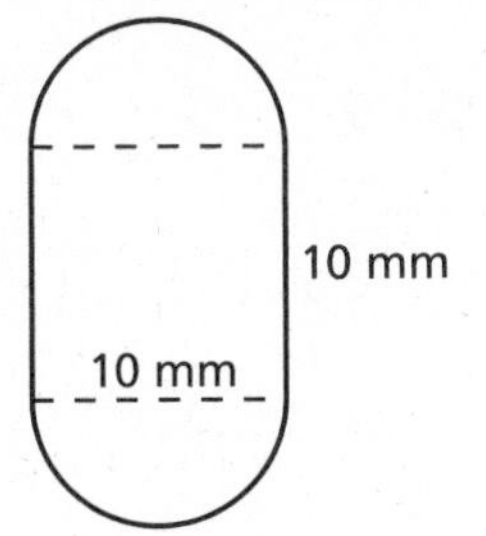

6.

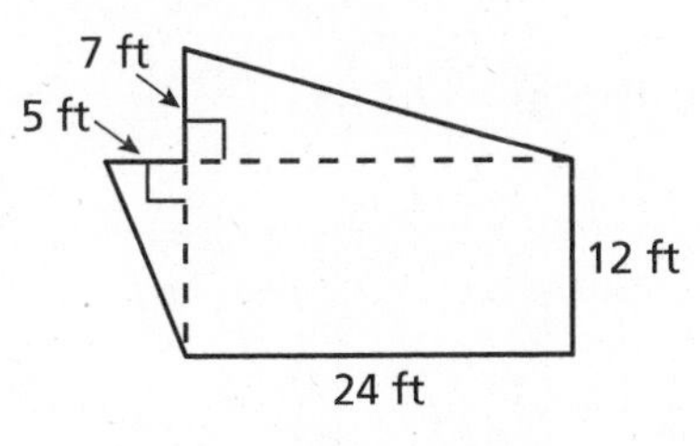

7.

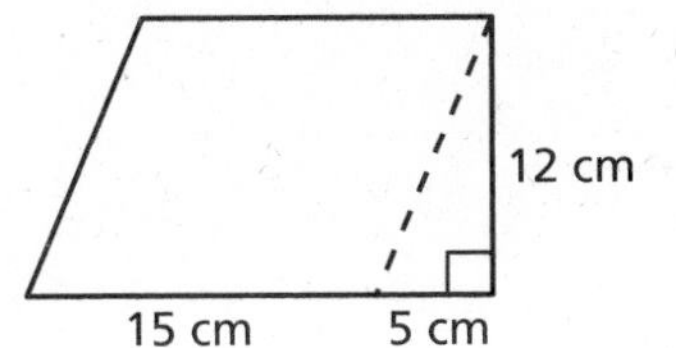

8.

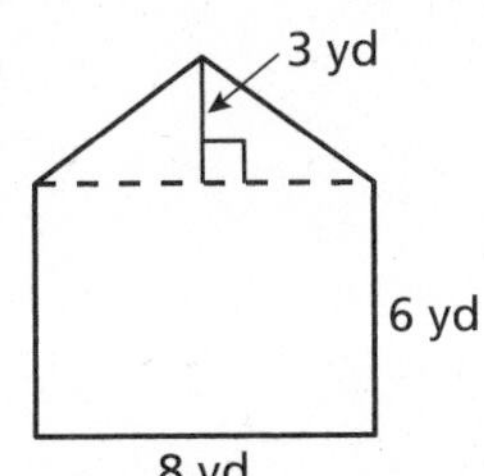

9.

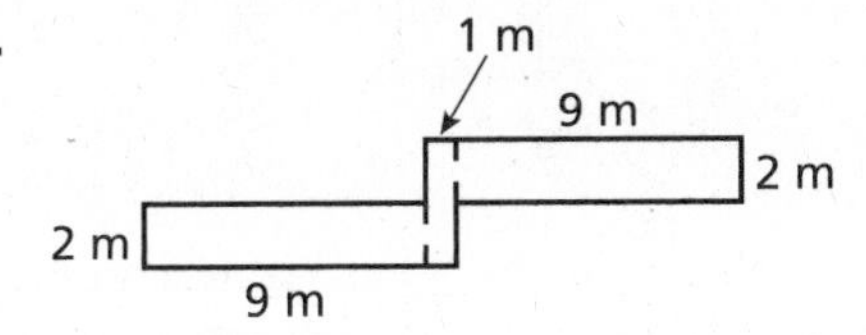

10. A garden is made up of two squares and a quarter circle. What are the perimeter and area of the garden?

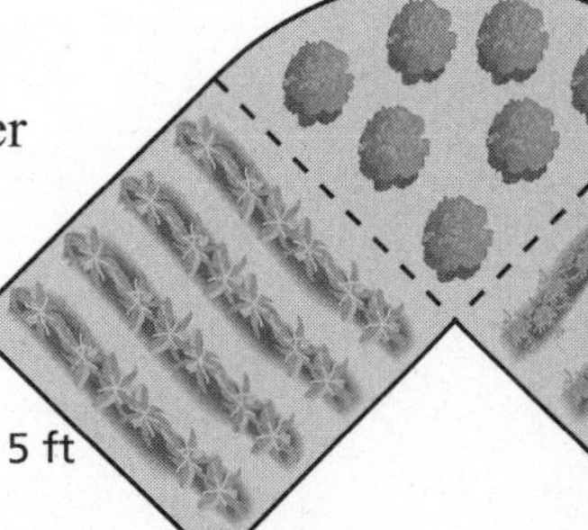

11. A pinwheel design for a quilt is shown at the right. Each square of the design has a side length of 3 inches.

a. Find the total area of the design.

b. Find the total area of the four large shaded triangles.

c. The quilt will be made of 30 pinwheel designs. The material to make the large triangles costs \$0.08 per square inch. How much will it cost to purchase the material to make the triangles for the quilt?

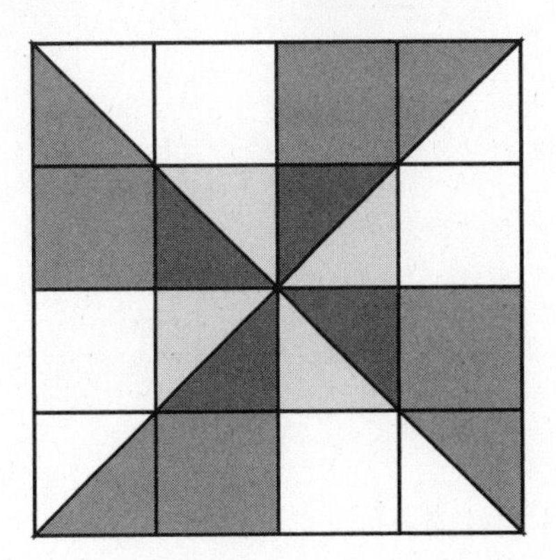

Name ________________________________ Date __________

8.4 Practice B

Find the perimeter and area of the figure.

1\.

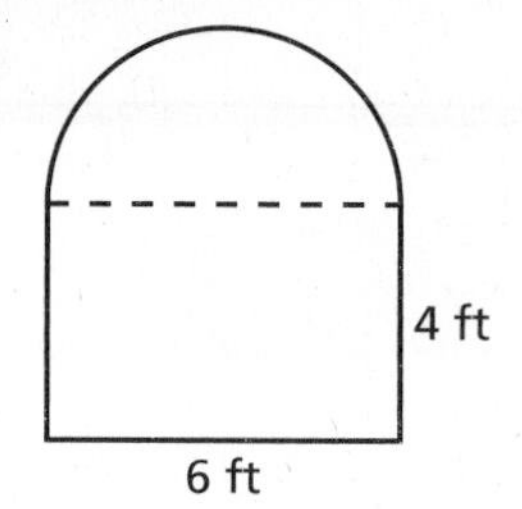

2\.

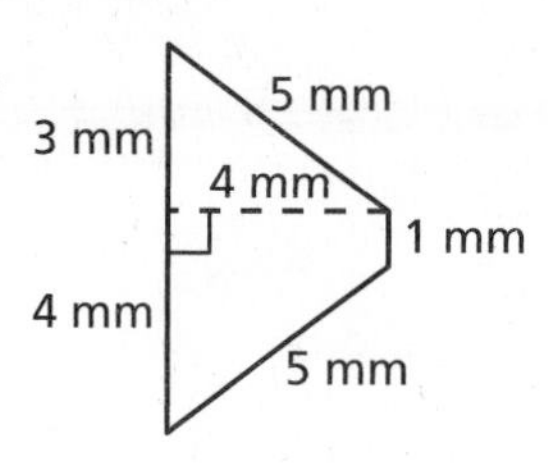

3\.

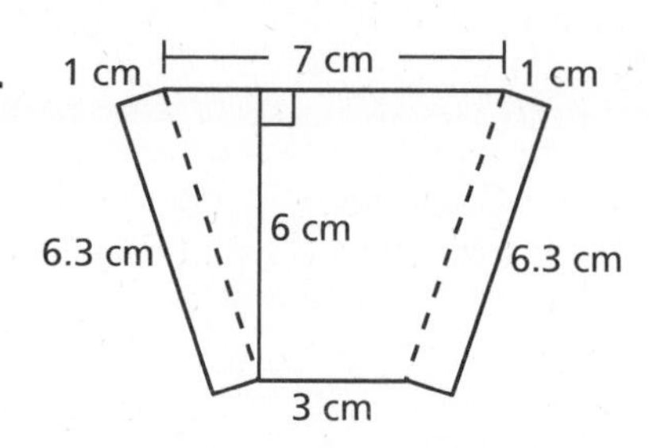

Find the area of the shaded region of the figure.

4\.

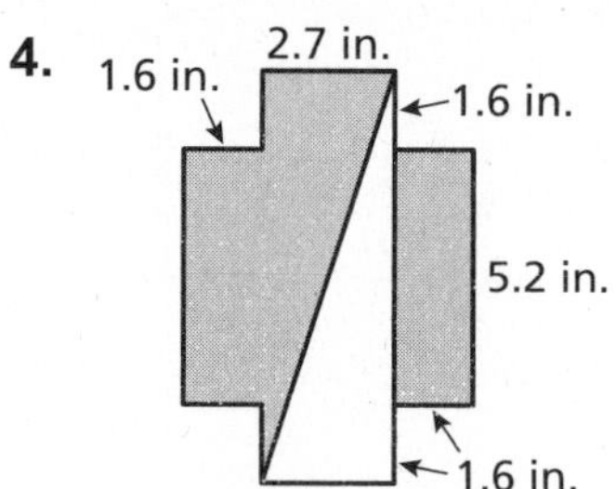

5\.

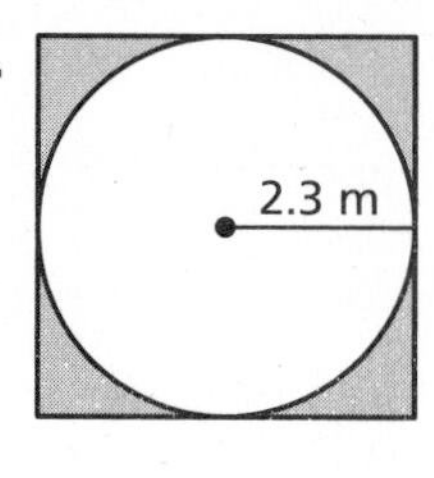

6\.

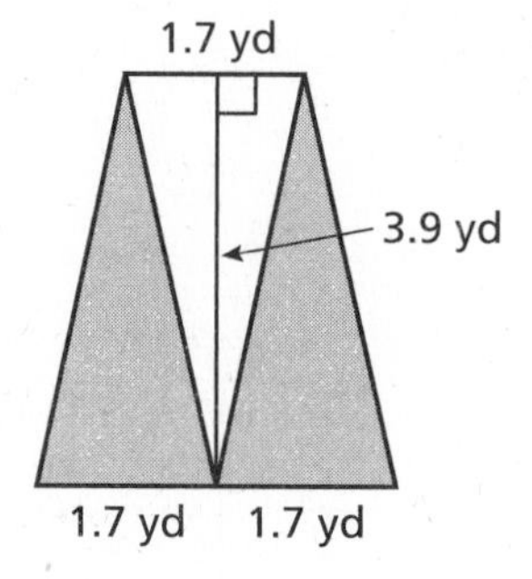

7\. Describe and correct the error in finding the area of the shaded region of the figure.

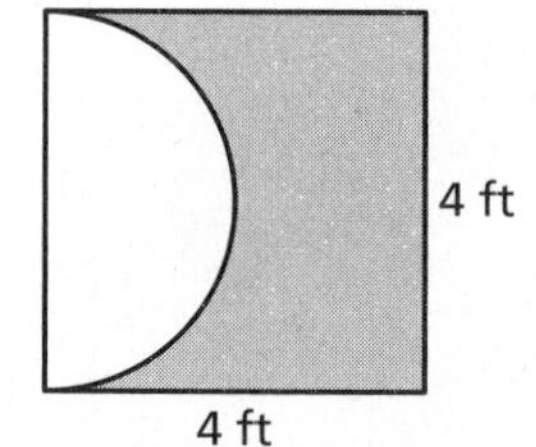

$$\text{Shaded area} \approx (4 \bullet 4) - (3.14 \bullet 2^2)$$
$$= 16 - 12.56$$
$$= 3.44 \text{ ft}^2$$

8\. Jackson Well Middle School's logo is shown at the right.

a. Find the perimeter of the logo.

b. Find the area of the logo.

c. Will the logo fit on a notebook cover that is 11 inches long and 8.5 inches wide? Explain.

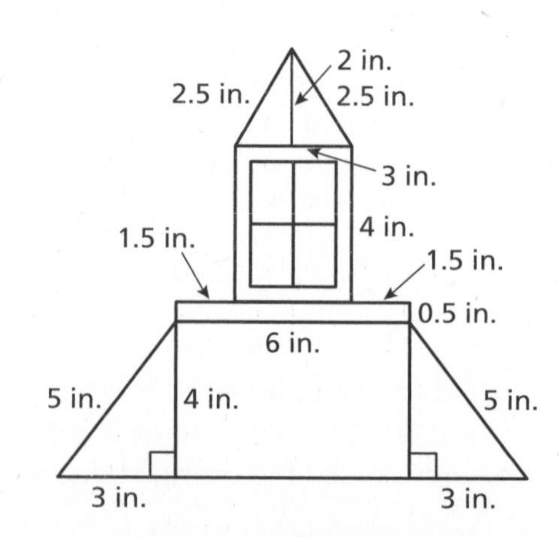

Jackson Well Middle School

Name_______________________________ Date__________

8.4 Enrichment and Extension

Composite Letters

The letters consist of rectangles, semicircles, and circles. Find the area of the letter.

1.

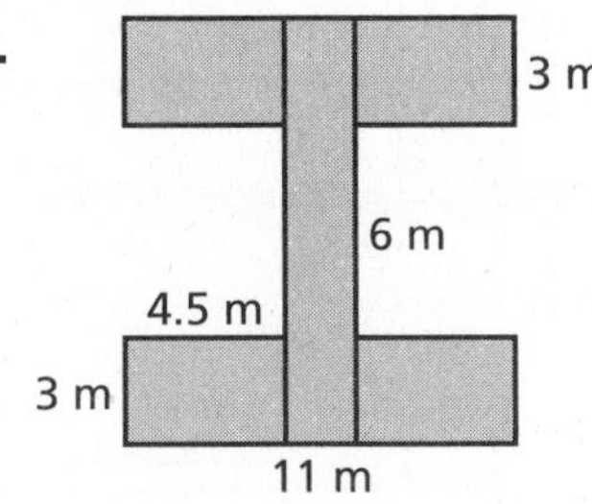

2.

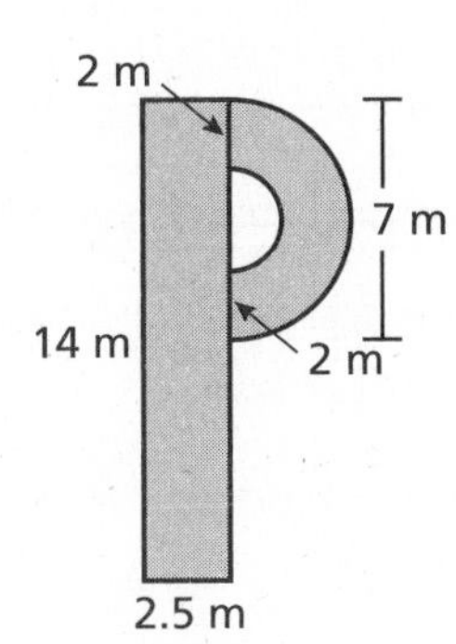

3.

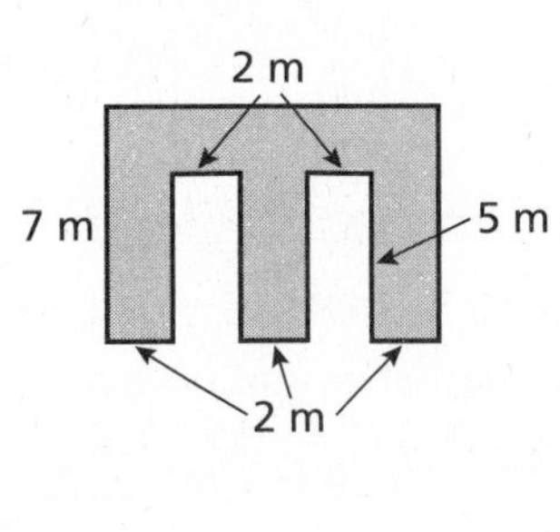

4.

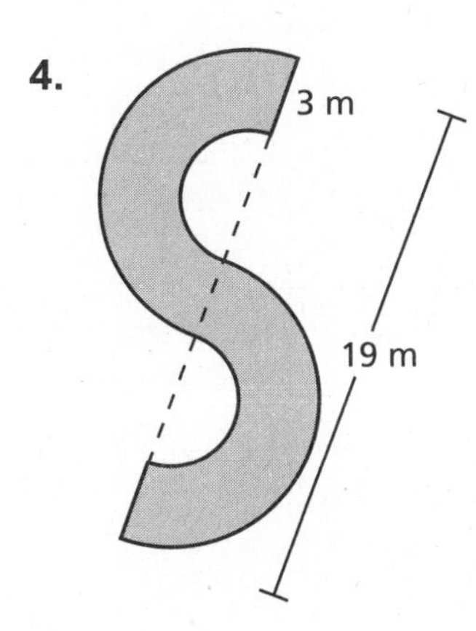

5.

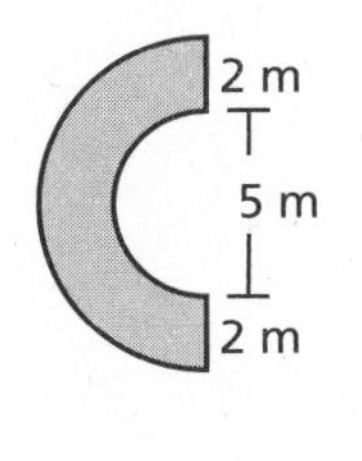

6.

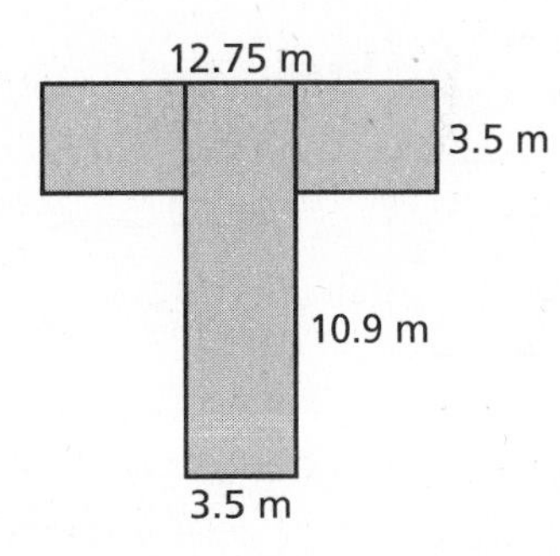

7.

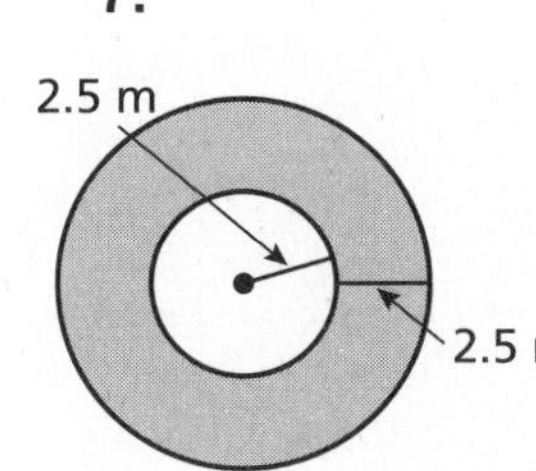

8.

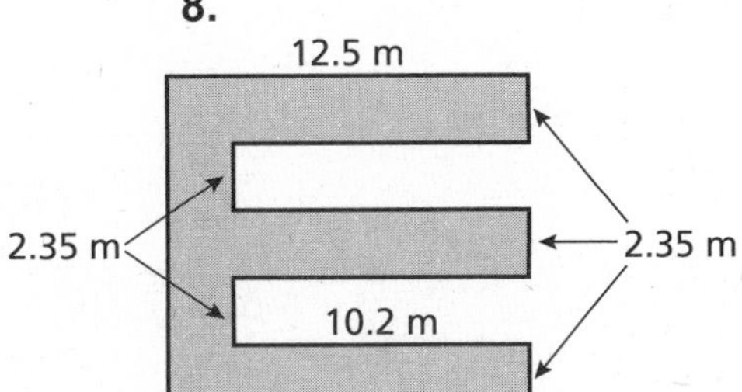

9.

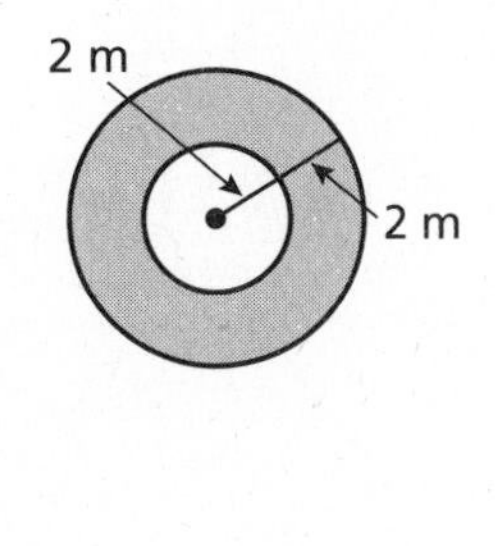

10. Order the letters from least to greatest area. What do they spell?

Name ______________________________ Date __________

Puzzle Time

What Did One Flea Say To Another Flea?

A	B	C	D	E	F
G	H				

Complete each exercise. Find the answer in the answer column. Write the word under the answer in the box containing the exercise letter.

61.12 cm^2 **HOP**
126 cm^2 **A**
46 cm^2 **DO**
100.48 cm^2 **DOG**

Find the area of the shaded region of the figure.

A.

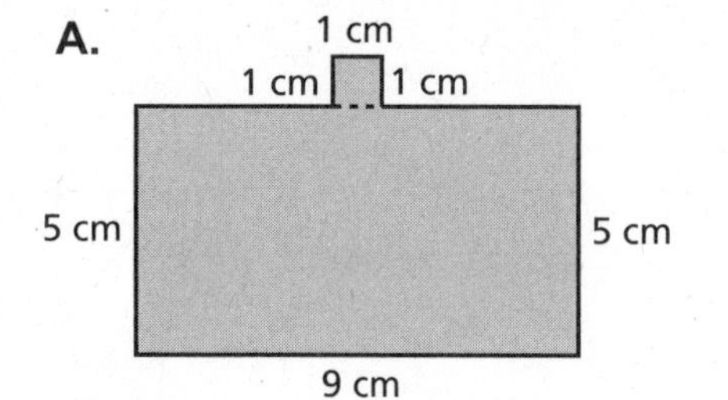

B.

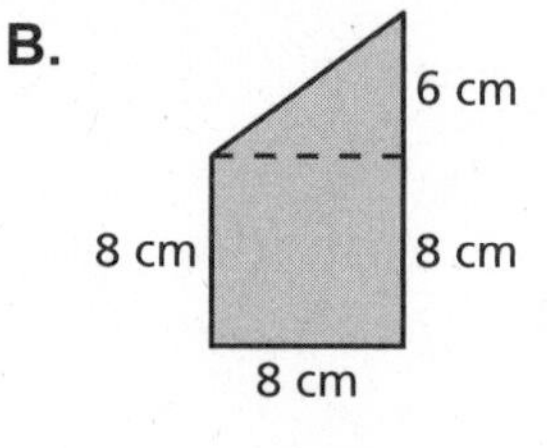

C.

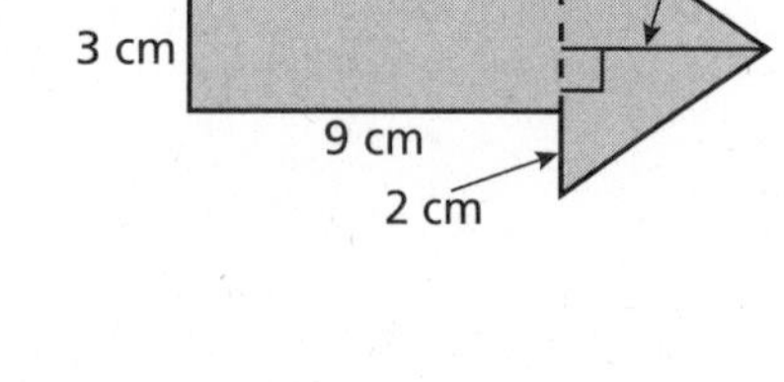

D.

3 cm

6 cm

E.

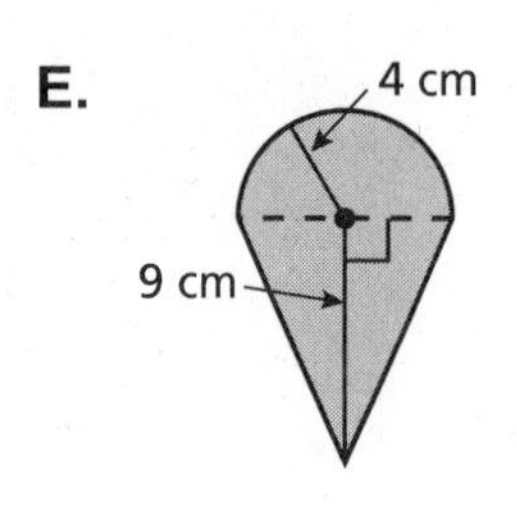

F.

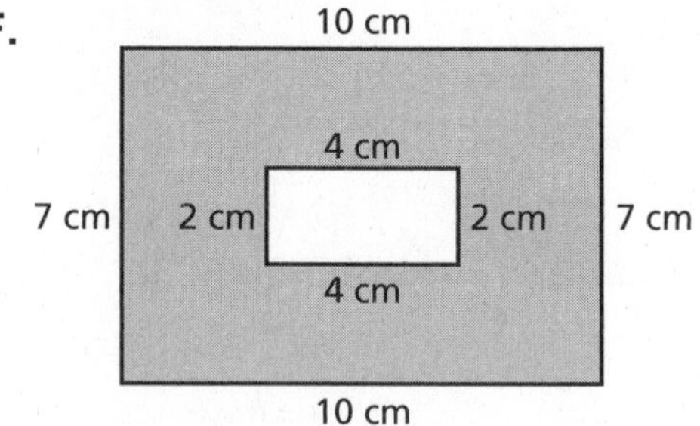

G.

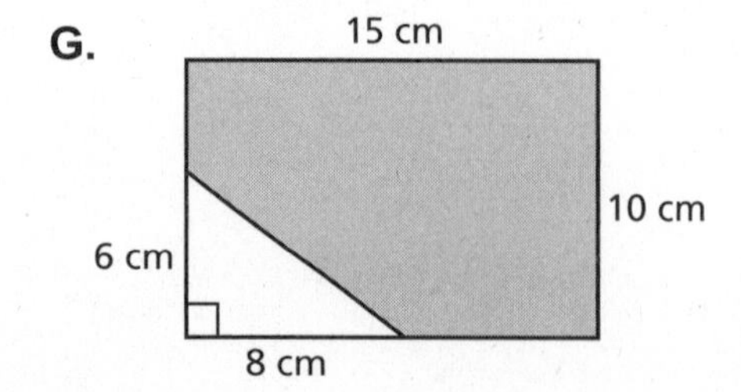

H.

2 cm

4 cm

44.5 cm^2 **WALK**
141.3 cm^2 **OR**
88 cm^2 **WE**
62 cm^2 **ON**

Name____ Date__________

Technology Connection

For use after Section 8.3

Choosing and Using Estimates for π

Pi—the ratio of a circle's circumference to its diameter—is a nonrepeating infinite decimal. To perform calculations using pi (π) you need to choose an approximation. A common decimal approximation of pi is $\pi \approx 3.14$.

A common fraction approximation of pi is $\pi \approx \frac{22}{7}$.

Some calculators have a button for π which will enter a decimal value for π that is more precise than 3.14. If your calculator has such a button you can find the value it is using by pressing the π button ([π] or [2nd] [π], depending on whether π has its own button or is a secondary function) and [=] or [ENTER].

A circle has a diameter of 83 centimeters.

1. Find the circumference using $\pi \approx 3.14$. Write down all the digits displayed on your calculator.

2. Find the circumference using $\pi \approx \frac{22}{7}$. Write down all the digits displayed on your calculator.

3. If your calculator has a button for π, find the circumference again using the button. Write down all the digits displayed on your calculator.

4. Compare your answers to Exercises 1–3. If you rounded your answers to the nearest hundredth, would the answers be the same?

5. To what place do you think you should round your answers when working with π? Explain your reasoning.

6. Find the circumference of a circle with a diameter of 138 miles.

Chapter 9

Name____ Date__________

Surface Area and Volume

Dear Family,

Does your family have an aquarium? Aquatic creatures make beautiful and interesting pets, and allergies are rarely an issue.

Fish can live in simple aquariums or more complex ones. The type of aquarium you choose will determine the supplies you will need, but all aquariums need a clean base of gravel. The gravel provides a place for beneficial microorganisms to grow. These organisms will help keep your aquarium clean and your animals and plants healthy. Plants help provide oxygen in the water and use some of the animal waste to keep the aquarium cleaner. You will also need clean water that is free of chlorine.

To set up your aquarium, ask your student to help make a plan. Here are some things you need to consider.

- Multiply the area of the base of your tank by the height to find the volume of the tank. This will give you an estimate of how much de-chlorinated water you need to have on hand. Tank sizes are often calculated in terms of gallons—work with your student to convert the volume of water to gallons.
- Find the amount of gravel you need by multiplying the height of the gravel in the tank by the area of the base of the aquarium. Make sure your gravel is rinsed and free of chemicals before putting it in the aquarium.
- Use the volume of water in the tank and the surface area of the tank to find out how many plants and animals your aquarium can safely hold. Local hobbyists and aquarium shops can help you figure out how to stock your aquarium.

You will need to change the water in your aquarium regularly (about a third of the volume every week or so). How much de-chlorinated water will you need to have on hand? Depending on the animals and plants you choose, you may also have to use a filter system and a heater.

In no time at all you'll be enjoying your new pets!

Nombre ______________________________ Fecha ________

Area de la superficie y volumen

Estimada Familia:

¿Hay un acuario en su familia? Las criaturas acuáticas son mascotas hermosas e interesantes, y casi no generan problemas de alergias.

Los peces pueden vivir en acuarios simples o complejos. El tipo de acuario que elija determinará los suministros que necesita, pero todos los acuarios necesitan una base limpia de grava. La grava proporciona un lugar para el crecimiento de microorganismos beneficiosos. Estos organismos ayudarán a mantener limpio su acuario y saludables a sus plantas y animales. Las plantas pueden proporcionar oxígeno en el agua y usar algo de los desechos animales para mantener el acuario más limpio. También necesitará agua limpia que no contenga cloro.

Para armar un acuario, pida a su estudiante que le ayude a hacer un plan. He aquí algunas cosas que tendrá que tomar en cuenta:

- Multipliquen el área de su tanque por el alto para encontrar el volumen del tanque. Esto les dará un cálculo de cuánta agua sin cloro se necesita tener a la mano. Los tamaños de los tanques a menudo se calculan en términos de galones—trabaje con su estudiante para convertir el volumen de agua en galones.
- Encuentren la cantidad de grava que se necesita multiplicando el alto de la grava en el tanque por el área de la base del acuario. Asegúrese de enjuagar su grava y que no contenga sustancias químicas antes de colocarlas en el acuario.
- Usen el volumen de agua en el tanque y al área superficial del tanque para averiguar cuántas plantas y animales pueden caber en su acuario. Quienes tienen por pasatiempo criar peces y las tiendas de acuarios, pueden ayudarlos a averiguar cómo implementar su acuario.

Necesitará cambiar el agua de su acuario de forma regular (alrededor de un tercio del volumen cada semana). ¿Cuánta agua sin cloro necesitará tener a mano? Según los animales y plantas que escoja, también querrá usar un sistema de filtro y un calentador de agua.

¡Muy pronto estarán disfrutando a sus nuevas mascotas!

Start Thinking!

For use before Activity 9.1

How can you determine the amount of cardboard used to make a cereal box? List at least two different methods.

Warm Up

For use before Activity 9.1

Evaluate the expression.

1. $2(2)(5) + 2(2)(3) + 2(5)(3)$

2. $2(1)(4) + 2(1)(2) + 2(4)(2)$

3. $2(6)(3) + 2(6)(1) + 2(3)(1)$

4. $2(3)(7) + 2(3)(5) + 2(7)(5)$

5. $2(2)(2) + 2(2)(4) + 2(2)(4)$

6. $2(4)(8) + 2(4)(10) + 2(8)(10)$

Lesson 9.1 Start Thinking!

For use before Lesson 9.1

How are the concepts of *area* and *surface area* similar? How are they different?

What kind of units are used to measure surface area?

Lesson 9.1 Warm Up

For use before Lesson 9.1

Use one-inch cubes to form a rectangular prism that has the given dimensions. Then find the surface area of the prism.

1. $2 \times 2 \times 3$

2. $1 \times 1 \times 5$

3. $3 \times 2 \times 4$

4. $1 \times 3 \times 5$

Name____ Date__________

9.1 Practice A

Find the surface area of the prism.

1.

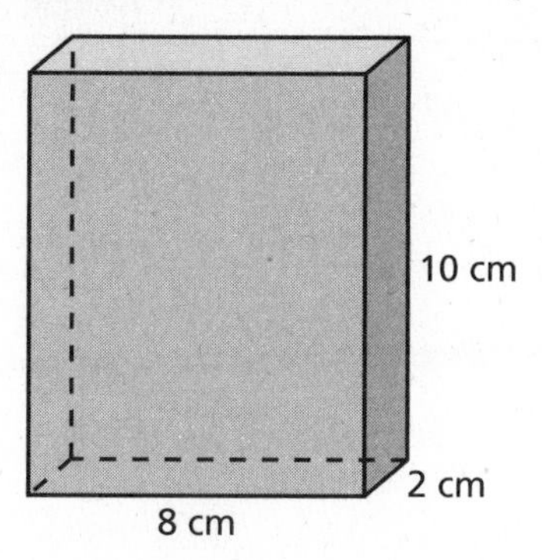

2.

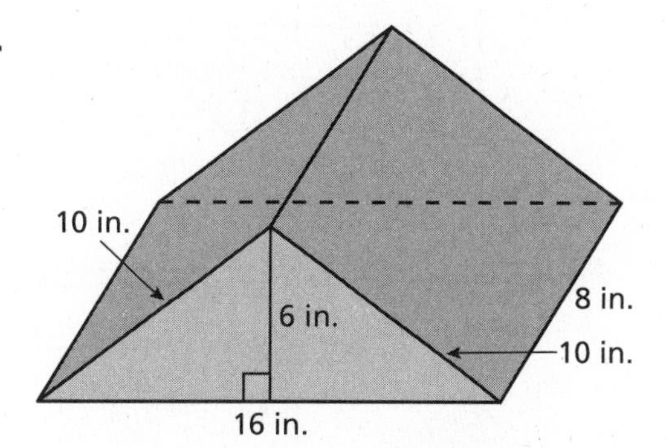

3.

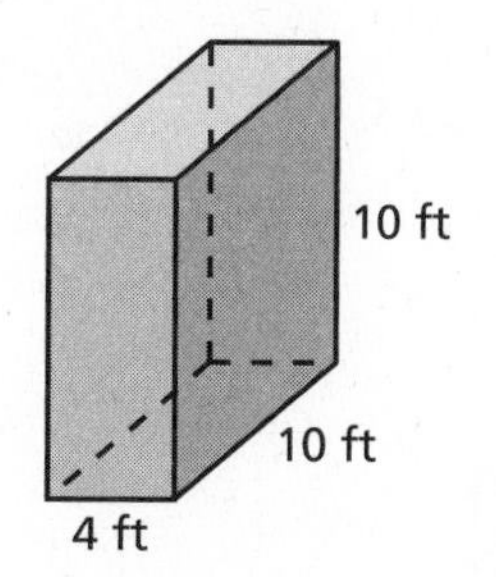

4.

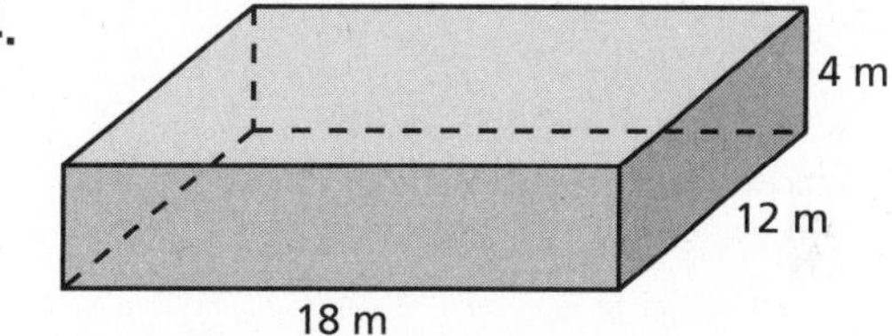

5.

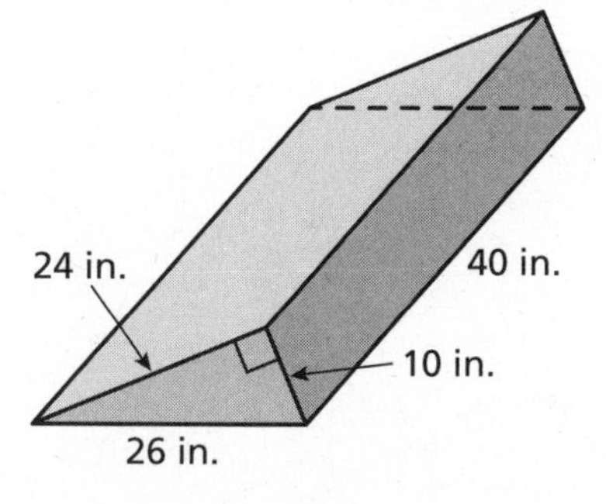

6.

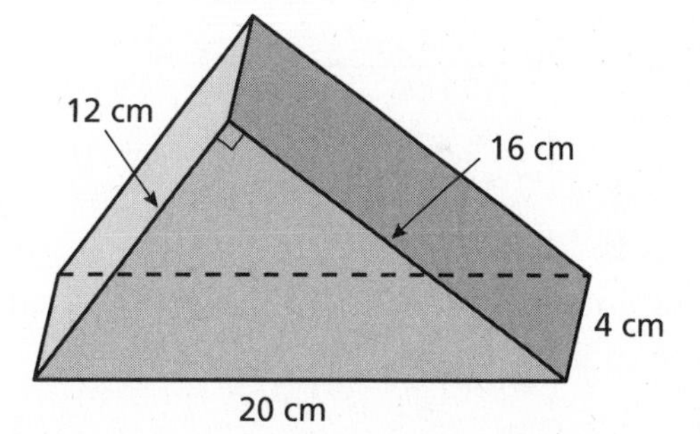

7. The inside of a baking pan is to be lined with tinfoil. The pan is 12 inches long, 9 inches wide, and 1.5 inches tall. How many square inches of tinfoil are needed?

8. Draw and label a rectangular prism that has a surface area of 96 square meters.

Name ______________________________ Date __________

9.1 Practice B

Find the surface area of the prism.

1.

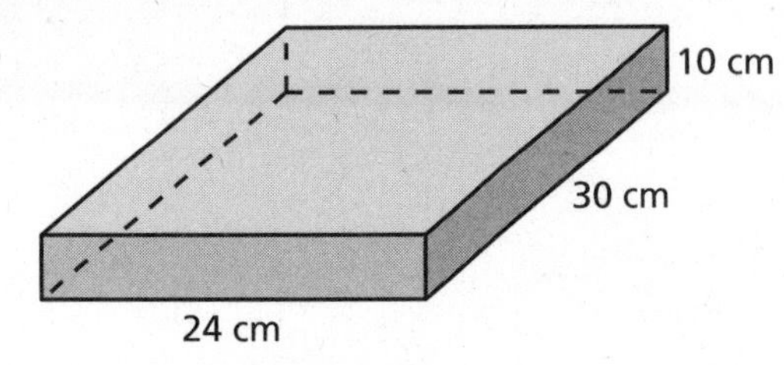

2.

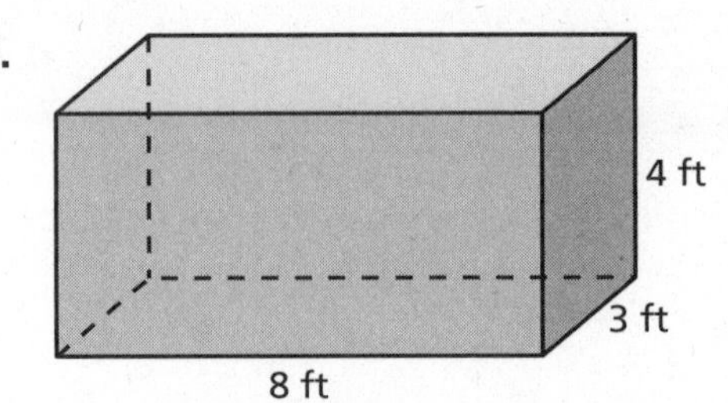

3.

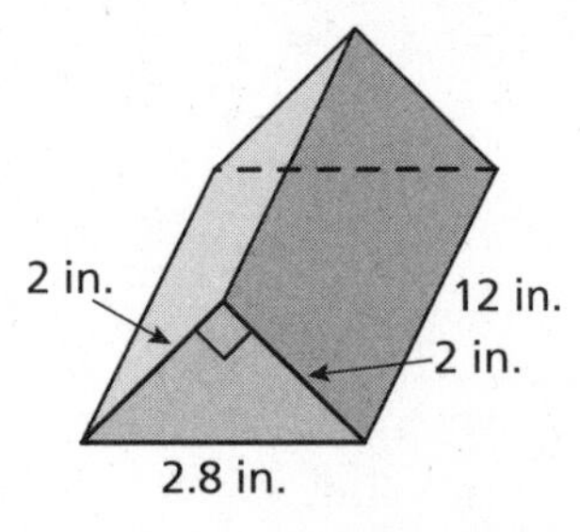

4.

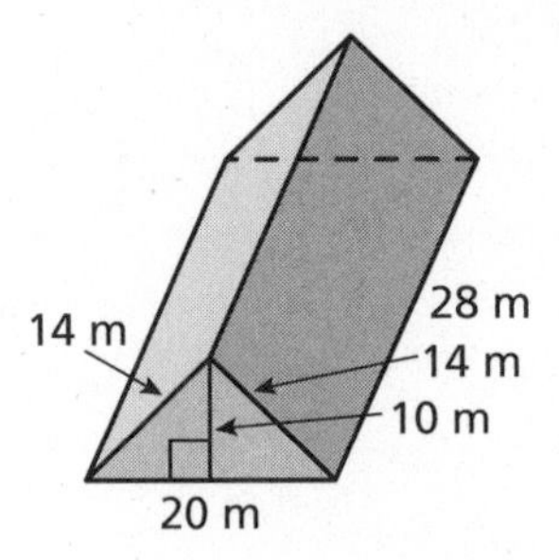

5.

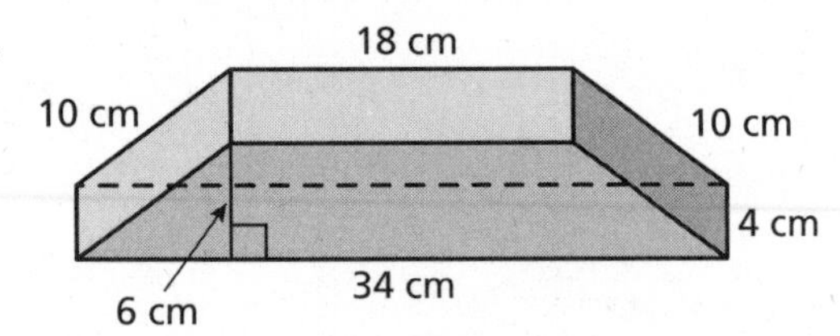

6.

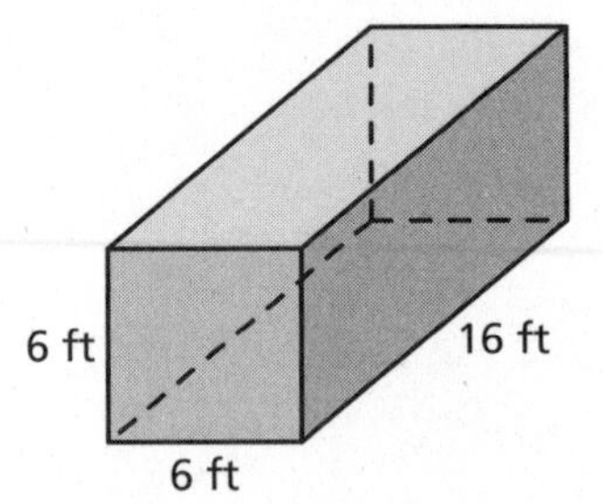

7. A graphing calculator is in the approximate shape of a rectangular prism.

a. Estimate the total surface area of the calculator.

b. The window of the calculator is 6.5 centimeters long and 4.5 centimeters wide. Estimate the surface of the graphing calculator without the window.

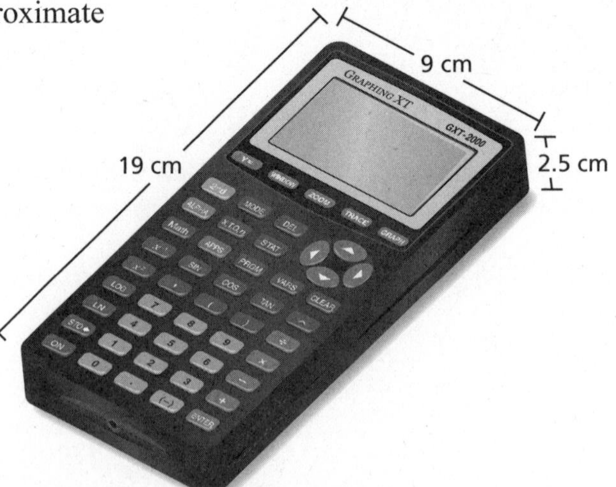

8. The least amount of wrapping paper needed to wrap a cube-shaped gift is 150 square inches. How long is one side of the gift?

Name____ Date__________

9.1 Enrichment and Extension

Boxing Up Basketballs and Cereal

Olivia works in the design department of a packaging company. Help her by answering the following questions.

1. Olivia has to design a plastic shipping container that will hold 12 basketballs in individual boxes. The basketballs have a radius of 4.5 inches and fit exactly in their individual boxes that are cubes.

   a. Give the dimensions (in inches) of 4 different plastic shipping containers that would fit the boxes exactly. Two containers with the same dimensions in a different order do not count as different containers. Find the surface area of each of your designs.

   b. Divide each surface area from part (a) by 144 to convert it to square feet. Explain why you divide by 144.

   c. Olivia's company made 100 containers one month with the design that uses the most plastic. The next month, they made 100 containers with the design that uses the least plastic. How much plastic (in square feet) did the company save in the second month?

2. Next, Olivia was asked to consider some new designs for a cereal box that was originally 7.7 inches by 2.6 inches by 11.8 inches. Each of the new designs will hold roughly the same amount of cereal as the original.

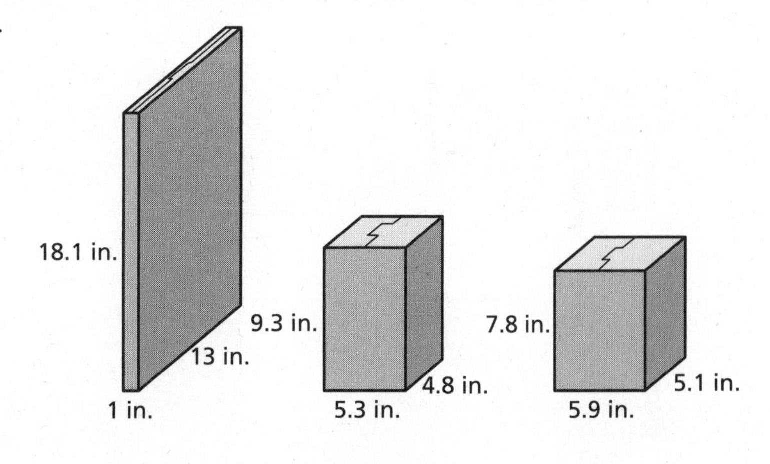

   a. Find how much cardboard (in square inches) it would take to make the original cereal box as well as each of the new designs.

   b. Olivia's company made 1000 cereal boxes with the design that uses the least cardboard. How many square feet of cardboard would they save compared to making 1000 of the original boxes?

   c. What are some advantages to the design with the least surface area? disadvantages? What design do you think Olivia should recommend? Explain your reasoning.

3. Look for a pattern in Exercises 1 and 2. Predict what kind of rectangular prism has the least surface area.

Name ______________________________ Date __________

Puzzle Time

What Did The Little Tire Want To Be When He Grew Up?

Write the letter of each answer in the box containing the exercise number.

Find the surface area of the prism.

1.

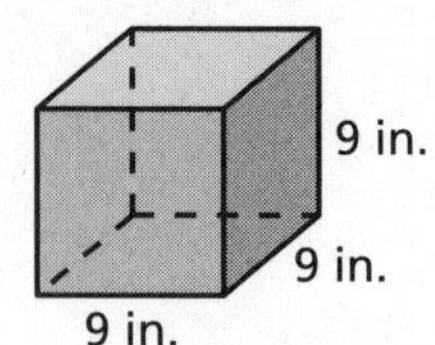

2.

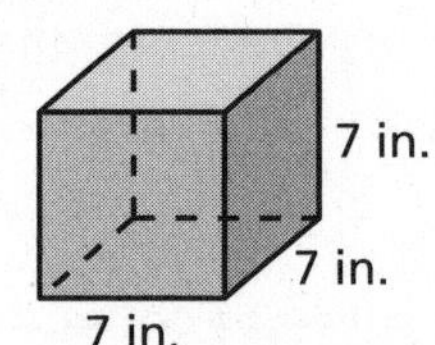

3.

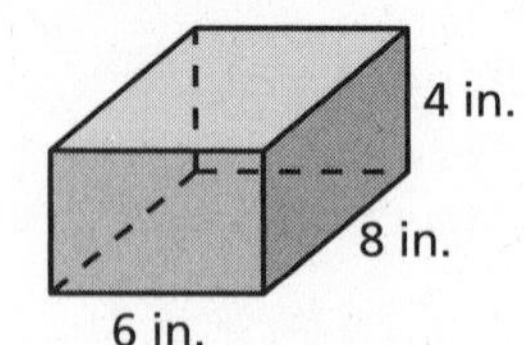

4.

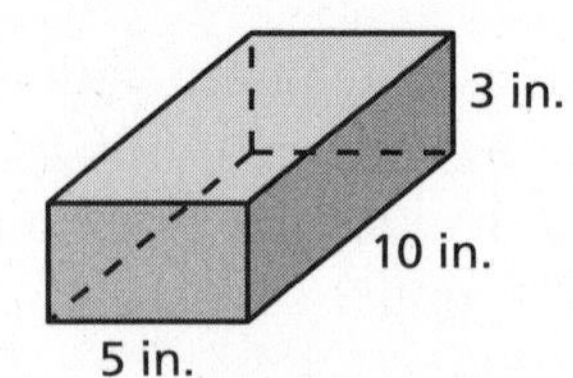

5.

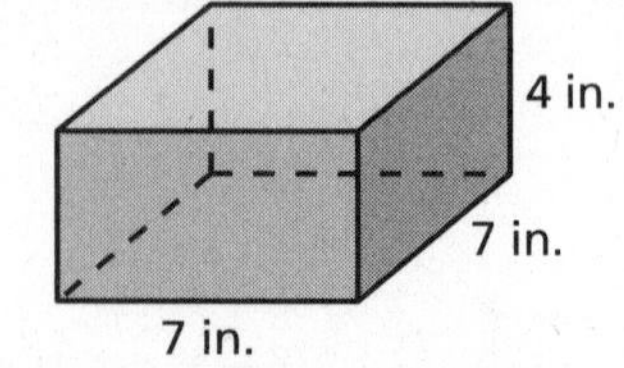

6.

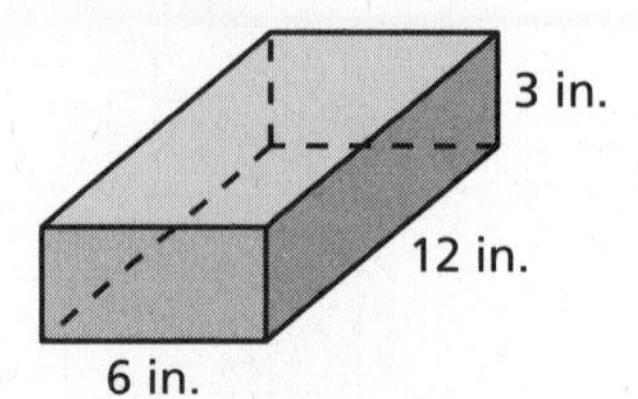

7.

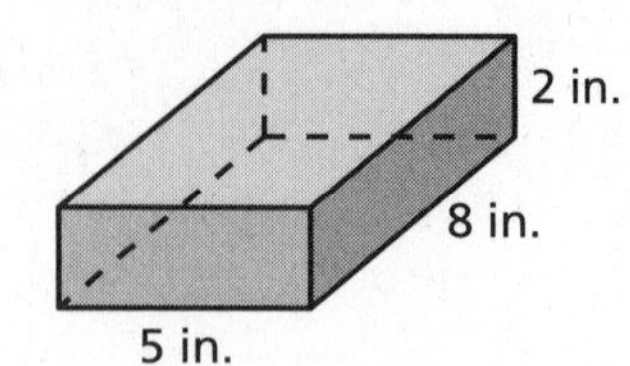

8.

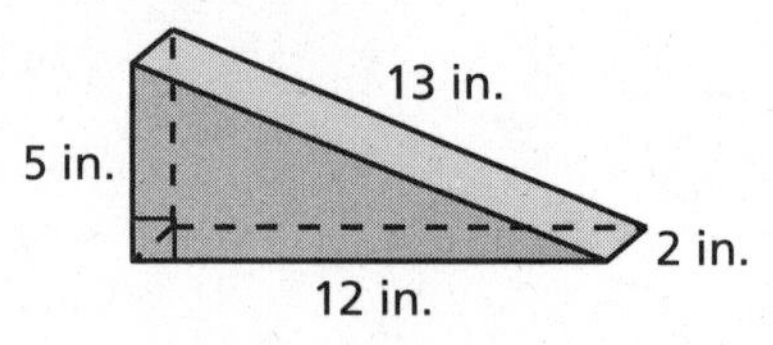

9.

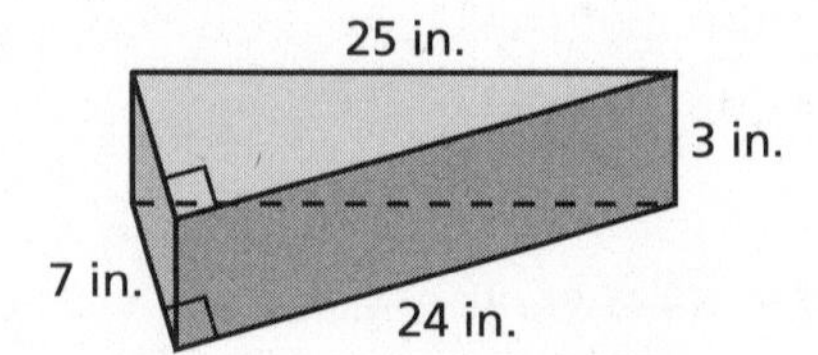

Answers

H. 190 in.^{2}

L. 208 in.^{2}

I. 210 in.^{2}

B. 120 in.^{2}

W. 132 in.^{2}

G. 294 in.^{2}

A. 486 in.^{2}

E. 336 in.^{2}

E. 252 in.^{2}

1		8	5	2		7	4	9	6	3

Are the sides of a pyramid always triangles? Explain.

Is the base of a pyramid always a triangle? Explain.

Find the area.

1. 12 cm, 4 cm

2. 6 in., 7 in.

3. 19 ft, 18 ft

4. 17 cm, 23 cm

Start Thinking!

For use before Lesson 9.2

Your neighbor needs to put a new roof on his gazebo. The roof is an octagonal pyramid. Why would knowing the surface area of the roof be useful information?

Warm Up

For use before Lesson 9.2

Use the net to find the surface area of the regular pyramid.

1.

12 in.
9 in.

2.

7 ft
Base area: 43.3 ft^2
10 ft

Name____ Date__________

9.2 Practice A

Use the net to find the surface area of the regular pyramid.

1.

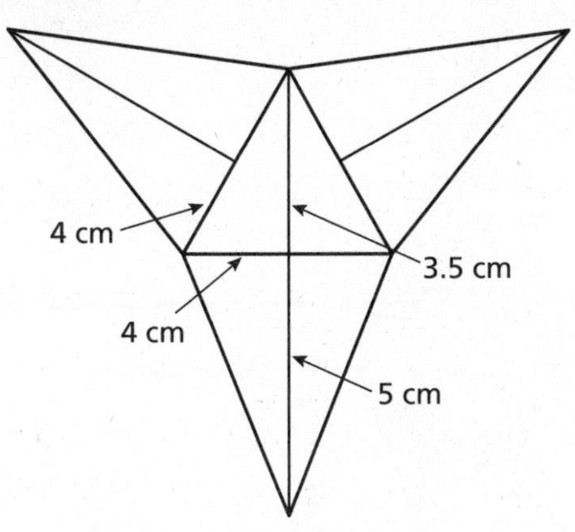

2.

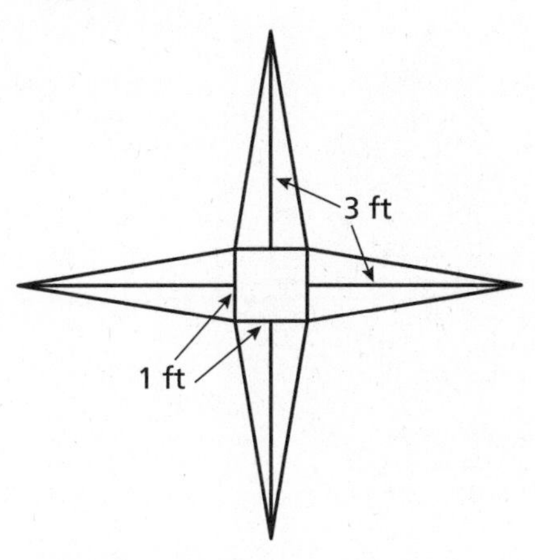

Find the surface area of the regular pyramid.

3.

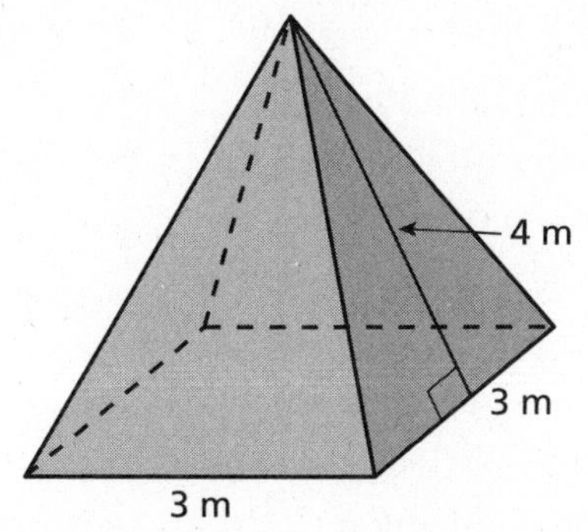

4.

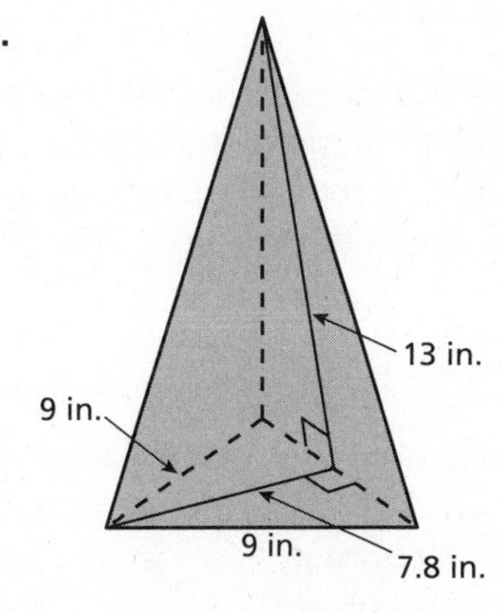

5. Your friend is purchasing an umbrella with a slant height of 4 feet. There are a variety of such umbrellas to choose from.

a. A red umbrella is shaped like a regular pentagonal pyramid with a side length of 3 feet. Find the lateral surface area of the red umbrella.

b. A yellow umbrella is shaped like a regular hexagonal pyramid with a side length of 2.5 feet. Find the lateral surface area of the yellow umbrella.

c. A blue umbrella is shaped like a regular octagonal pyramid with a side length of 1.9 feet. Find the lateral surface area of the blue umbrella.

d. Based on lateral surface areas, would you suggest that your friend pick the umbrella that is her favorite color? Explain.

Name ______________________________ Date __________

9.2 Practice B

Find the surface area of the regular pyramid.

1.

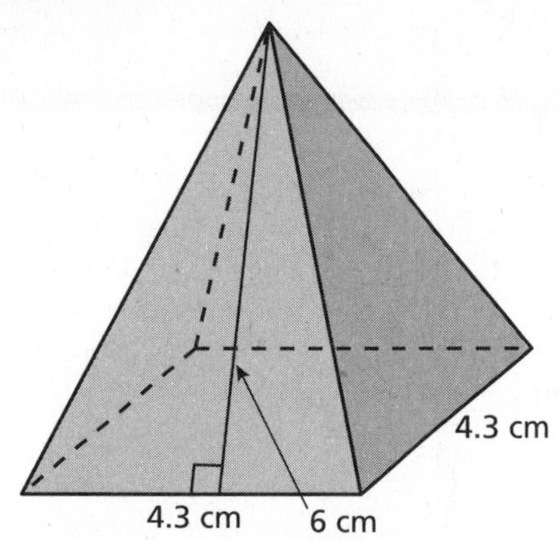

2.

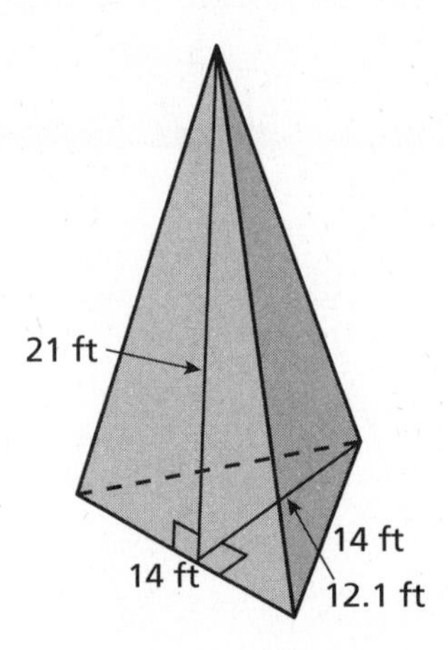

3. Researchers have determined that a hip roof offers the most protection to a house during a hurricane.

a. The house has a square base with a side length of 50 feet. The house has a variation of a hip roof in the shape of a regular pyramid with a square base. The roof extends 1 foot beyond the walls of the house on all sides. What is the length of each side of the base of the roof?

b. The slant height of the roof is 35 feet. Find the sum of the areas of the lateral faces of the pyramid.

c. A metal roof covering offers the most protection to a house during a hurricane. The cost of installing metal roof covering is $350 for every 100 square feet of roof. What is the cost of installing a metal roof covering on the house?

4. The surface area of a regular triangular pyramid is 197.1 square meters. The slant height is 12 meters. The area of the base is 35.1 square meters. The base length is 9 meters. What is the height of the triangular base?

5. The surface area of a regular pentagonal pyramid is 125 square yards. The base length is 5 yards. The area of the base is 37.5 square yards. What is the slant height of the pyramid?

Name______________________________ Date__________

9.2 Enrichment and Extension

Scaling Down the Pyramids

Imagine that you are planning to make scale models of the square pyramids described below. (See Activity 1 in Section 9.2 in your textbook for pictures of the pyramids.) You will be making them out of plywood. Plywood is sold in sheets that are 4 feet by 8 feet.

Pyramid	Actual side length (m)	Actual slant height (m)
Cheops Pyramid in Egypt	230	186
Muttart Conservatory in Edmonton	26	27
Louvre Pyramid in Paris	35	28
Pyramid of Caius Cestius in Rome	22	29

1. You have decided that your scale should be 1 m = 0.5 cm. Why should you use this as your scale instead of 1 m = 1 cm?

2. Find the side length and slant height of the models, and complete the table.

Pyramid	Model side length (cm)	Model slant height (cm)
Cheops Pyramid in Egypt		
Muttart Conservatory in Edmonton		
Louvre Pyramid in Paris		
Pyramid of Caius Cestius in Rome		

3. What is the least amount of plywood (in square centimeters) you would need to make all the models? How many sheets of plywood is this?

4. When purchasing the plywood, the salesman offers half sheets of plywood that are 4 feet by 4 feet.

a. Using your answer to Exercise 3, can you replace one of the full sheets of plywood with a half sheet of plywood? Explain.

b. Check to see if your answer to part (a) is correct by showing how the models can be cut from the plywood. Explain.

Name ______________________________ Date __________

9.2 Puzzle Time

Where Do You Find Baby Soldiers?

Write the letter of each answer in the box containing the exercise number.

Find the surface area of the regular pyramid.

1. Square base: side length = 5 cm; slant height = 12 cm
2. Square base: side length = 8 cm; slant height = 15 cm
3. Square base: side length = 9 cm; slant height = 14 cm
4. Triangular base: side length = 6 cm; slant height = 8 cm; height of base triangle = 5.2 cm
5. Triangular base: side length = 14 cm; slant height = 18 cm; height of base triangle = 12.1 cm
6. Triangular base: side length = 12 cm; slant height = 15 cm; height of base triangle = 10.4 cm

Find the surface area of the *lateral faces* of the regular pyramid.

7. Pentagonal base: side length = 7 cm; slant height = 12 cm
8. Hexagonal base: side length = 10 cm; slant height = 13 cm
9. Octagonal base: side length = 12 cm; slant height = 16 cm
10. The top of a play canopy tent forms a pyramid with a square base. The sides of the base are 8.5 centimeters. The slant height is 5.4 centimeters. How much canvas is needed to make the canopy?
11. The base of a glass paperweight is a regular hexagon with a side length of 6 centimeters. The area of the base is 93.6 square centimeters. The slant height is 12 centimeters. What is the surface area of the paperweight?

Answers

- **N.** 332.4 cm^2
- **Y.** 309.6 cm^2
- **A.** 304 cm^2
- **E.** 87.6 cm^2
- **R.** 333 cm^2
- **T.** 768 cm^2
- **I.** 210 cm^2
- **T.** 462.7 cm^2
- **N.** 390 cm^2
- **F.** 91.8 cm^2
- **H.** 145 cm^2

5	1	4		7	8	10	2	6	9	3	11

Start Thinking!

For use before Activity 9.3

Give a real-life example of when it would be useful to know the surface area of a cylinder.

Activity 9.3

Warm Up

For use before Activity 9.3

Find the area. Use 3.14 for π.

1. 3 in.

2. 7 ft

3. 10 cm

4. 8 cm

Lesson 9.3 Start Thinking!

For use before Lesson 9.3

Explain which cylinder has a greater surface area:

Radius: 4 cm; Height: 10 cm

Radius: 10 cm; Height: 4 cm

Lesson 9.3 Warm Up

For use before Lesson 9.3

Make a net for the cylinder. Then find the surface area of the cylinder. Round your answer to the nearest tenth.

1.

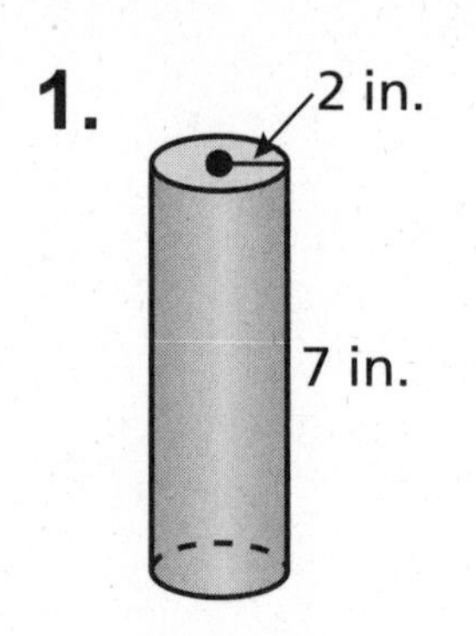

2.

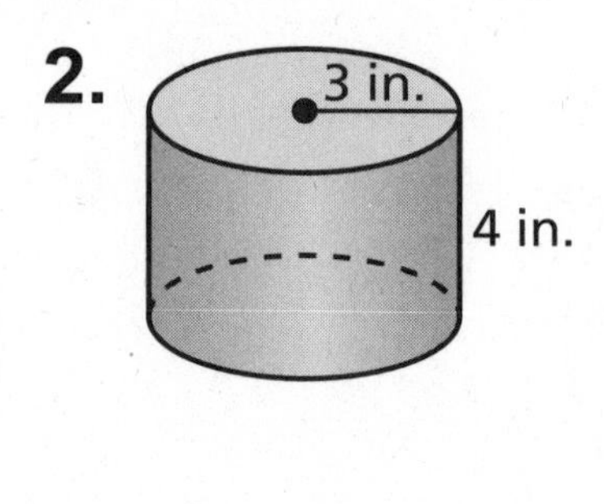

3.

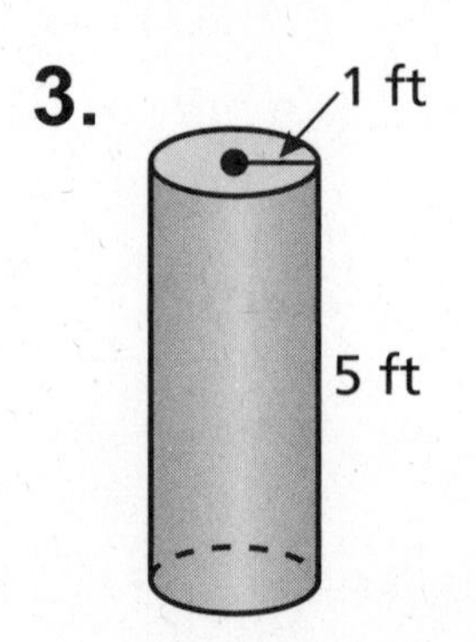

4.

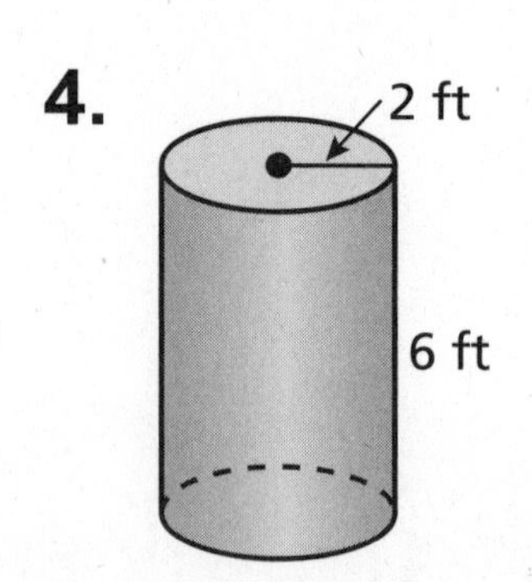

Name____________________ Date__________

9.3 Practice A

Make a net for the cylinder. Then find the surface area of the cylinder. Round your answer to the nearest tenth.

1.

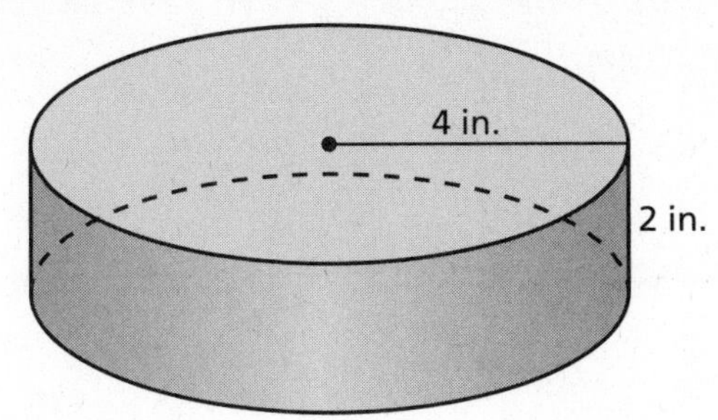

2.

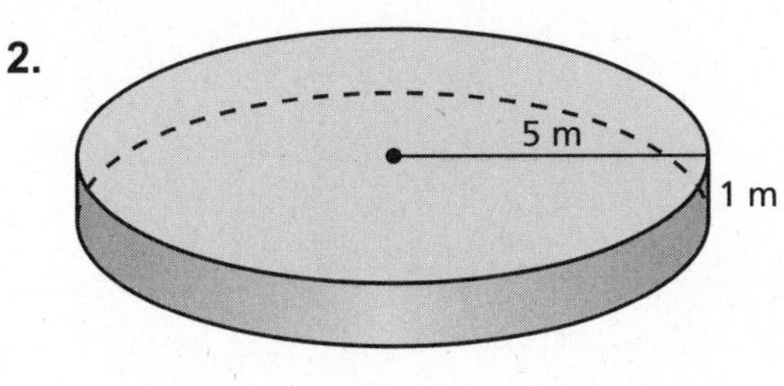

Find the surface area of the cylinder. Round your answer to the nearest tenth.

3.

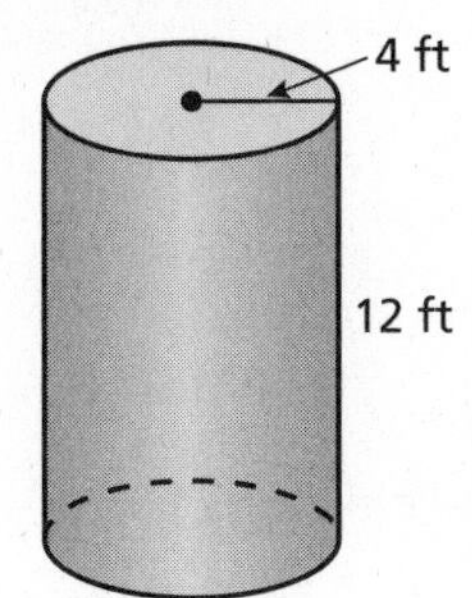

4.

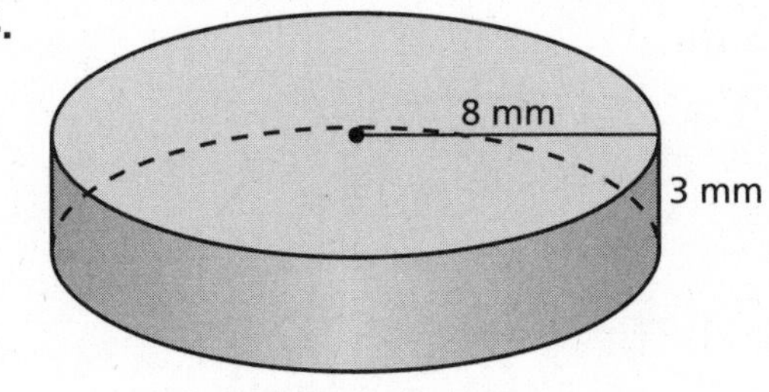

Find the lateral surface area of the cylinder. Round your answer to the nearest tenth.

5.

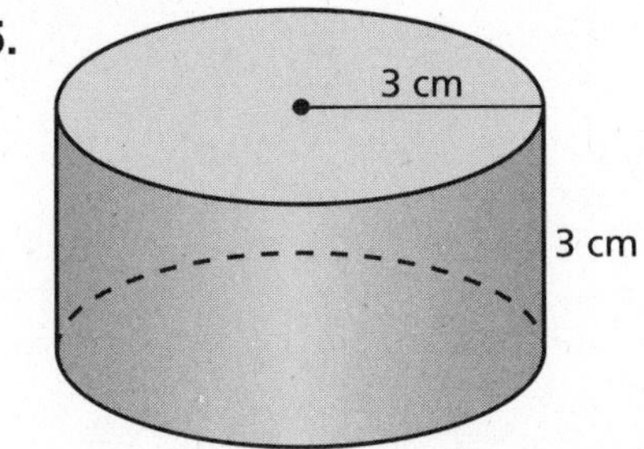

6.

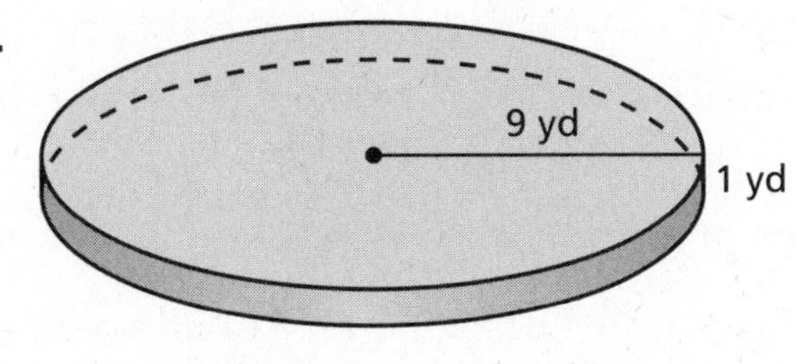

7. A deep dish pizza has a radius of 6 inches and a height of 1 inch. Find the surface area of the pizza. Round your answer to the nearest tenth.

Name ______________________________ Date ________

9.3 Practice B

Find the surface area of the cylinder. Round your answer to the nearest tenth.

1.

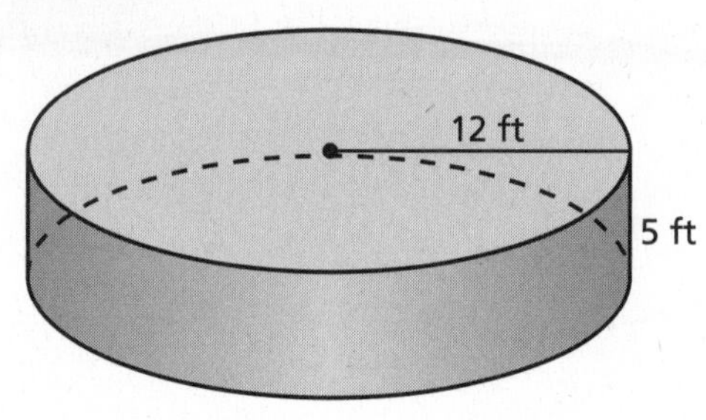

2.

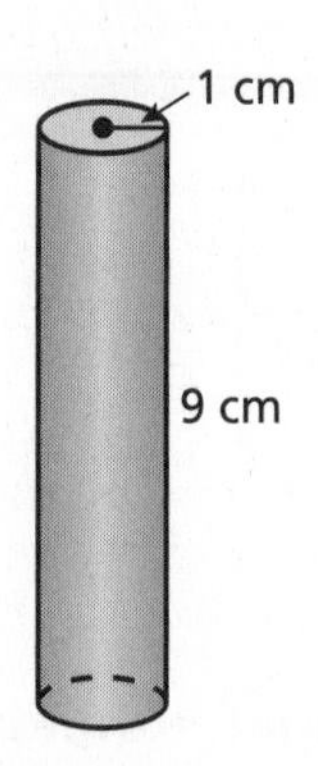

Find the lateral surface area of the cylinder. Round your answer to the nearest tenth.

3.

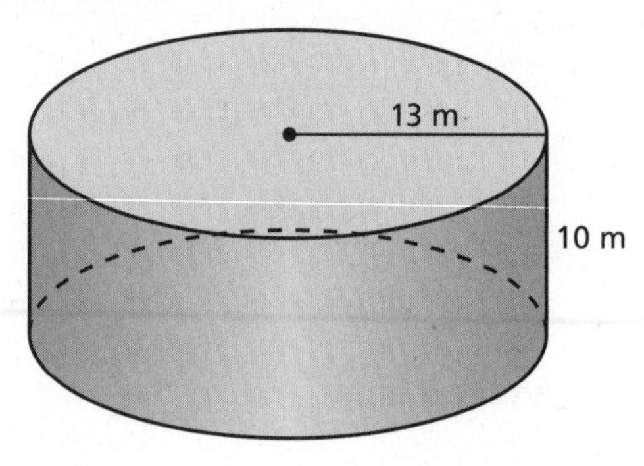

4.

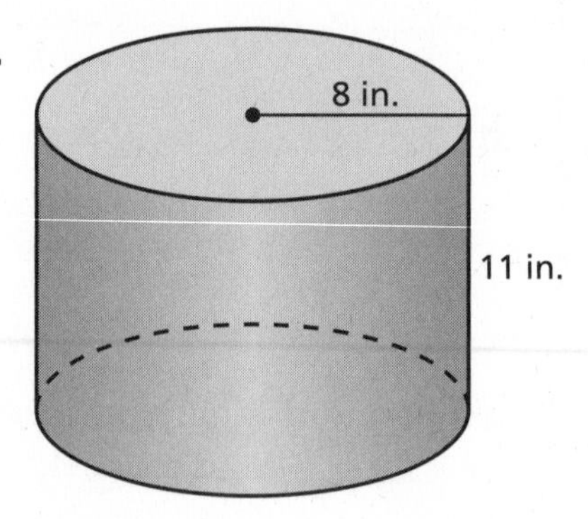

5. A quarter is worth \$0.25 and a half dollar is worth \$0.50.

a. A quarter has a diameter of $\frac{15}{16}$ inch and a height of $\frac{1}{16}$ inch. Find the surface area of a quarter. Round your answer to the nearest hundredth.

b. A half dollar has a diameter of $\frac{9}{8}$ inches and a height of $\frac{3}{32}$ inch. Find the surface area of a half dollar. Round your answer to the nearest hundredth.

c. Show that the value of the coin is not proportional to the surface area of the coin.

d. If the values of the coins were proportional to the surface areas of the coins, what would be the surface area of the half dollar? Round your answer to the nearest hundredth.

Name______________________________ Date__________

9.3 Enrichment and Extension

The Icing on the Cake

Answer the following questions. As you calculate the surface area that icing covers, keep in mind that the bottom of the cake does not get any icing.

1. A cylindrical cake is made in a pan that has a diameter of 9 inches and a height of $1\frac{1}{2}$ inches.

   a. What is the total surface area of the cake?

   b. The cake is cut into 10 equal-sized wedges. What is the total surface area of the cake now?

   c. After the cake is cut, what percent of the cake's surface is covered with icing?

   d. Cake does not stay as moist after it has been cut into pieces. Use surface area to explain this.

2. José has decided to make a heart-shaped cake using a square pan that is 9 inches by 9 inches and a circular pan with a diameter of 9 inches. Both pans are $1\frac{1}{2}$ inches tall. The diagram below shows the top view of the cake.

   a. José knows from experience that one 8-ounce container of icing will cover a cake made from his square pan exactly the way he likes it. If he covers his heart-shaped cake the same way, how many ounces of icing will he use?

   b. How many 8-ounce containers of icing will he have to buy? How much icing will be left over?

Name ______________________________ Date __________

Puzzle Time

Did You Hear About...

A	B	C	D	E	F
G	H	I	J	K	L
M	N	O	P	Q	R

Complete each exercise. Find the answer in the answer column. Write the word under the answer in the box containing the exercise letter.

565.2 m^2	WEEK
276.3 m^2	AND
753.6 ft^2	HIS
424.3 in.^2	CATCH
401.9 m^2	BOUGHT
325.2 ft^2	DAY
439.6 cm^2	AWAY
301.4 m^2	OLD
282.6 cm^2	BOOMERANG
533.8 in.^2	THROW
100.5 ft^2	MAN

Find the combined area of *both bases* of the cylinder. Use 3.14 for π. Round to the nearest tenth.

A. $r = 2$ in.

B. $r = 4$ ft

C. $r = 5$ cm

D. $r = 8$ m

Find the area of the *lateral surface* of the cylinder. Use 3.14 for π. Round to the nearest tenth.

E. $r = 3$ ft; $h = 6$ ft

F. $r = 8$ in.; $h = 7$ in.

G. $r = 9$ cm; $h = 5$ cm

H. $r = 4$ m; $h = 11$ m

Find the surface area of the cylinder. Use 3.14 for π. Round to the nearest tenth.

I. $r = 1$ in.; $h = 7$ in.

J. $r = 5$ cm; $h = 3$ cm

K. $r = 6$ m; $h = 9$ m

L. $r = 2$ ft; $h = 8$ ft

M. $r = 4$ m; $h = 4$ m

N. $r = 5$ in.; $h = 12$ in.

O. $r = 10$ ft; $h = 2$ ft

P. $r = 3$ m; $h = 13$ m

Q. A cylindrical cookie jar has a height of 9 inches. The radius of its base is 4 inches. What is its surface area?

R. A cylindrical coffee can has a height of 14 centimeters. The radius of its base is 5 centimeters. What is the area of its label?

113.0 ft^2	A
25.1 in.^2	THE
35.6 m^2	STORE
201.0 m^2	TO
50.2 in.^2	SPENT
326.6 in.^2	ONE
187.4 cm^2	BUY
125.6 ft^2	TRYING
157 cm^2	WHO
251.2 cm^2	A
351.7 in.^2	NEW

Do two-dimensional figures have volume? Explain.

Do three-dimensional figures have volume? Explain.

Activity 9.4

Warm Up

For use before Activity 9.4

Multiply.

1. $7 \times 5 \times 8$

2. $12 \times 7 \times 8$

3. $(13)(10)(7)$

4. $11 \bullet 15 \bullet 3$

5. $(14)(20)(4)$

6. $12 \bullet 16 \bullet 21$

Start Thinking!

For use before Lesson 9.4

You are buying decorative sand for art projects. Explain how volume would be helpful in figuring out which size box to use to store the decorative sand.

Warm Up

For use before Lesson 9.4

Find the volume of the prism.

1.

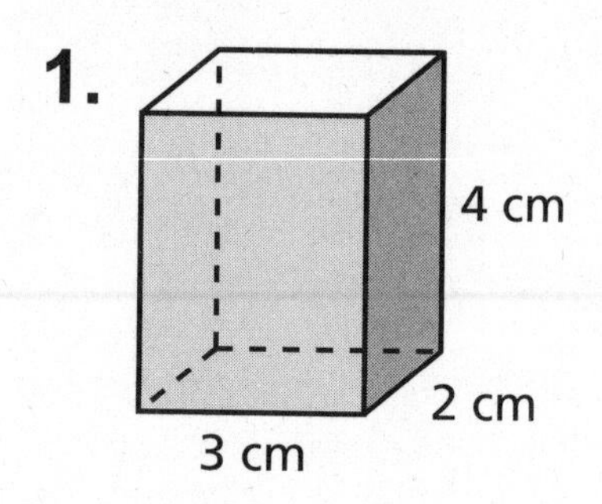

2.

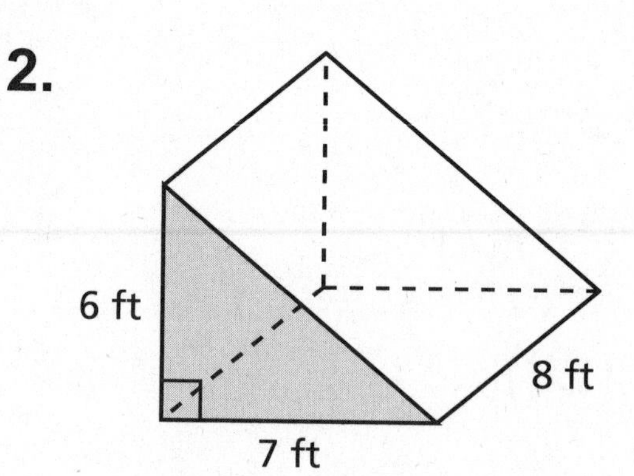

3.

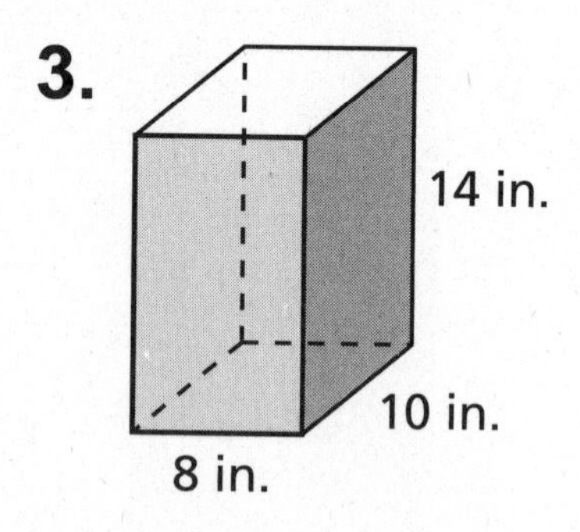

4.

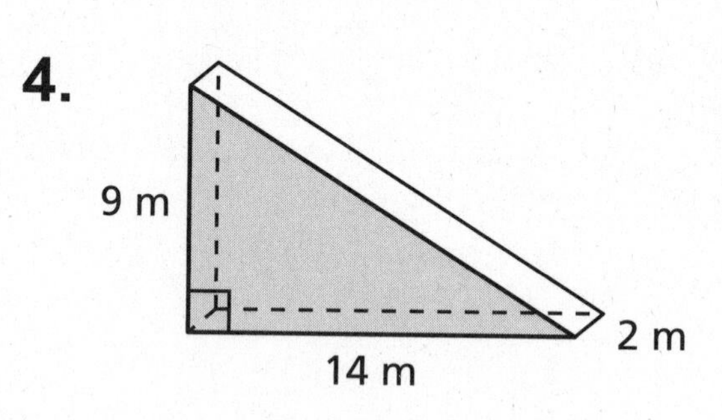

Name________________________________ Date__________

9.4 Practice A

Find the volume of the prism.

1.

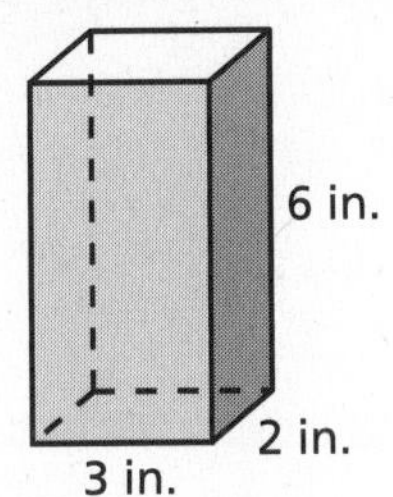

2.

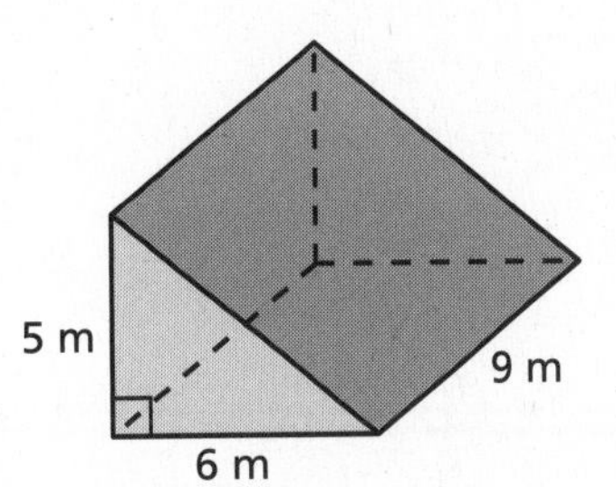

3.

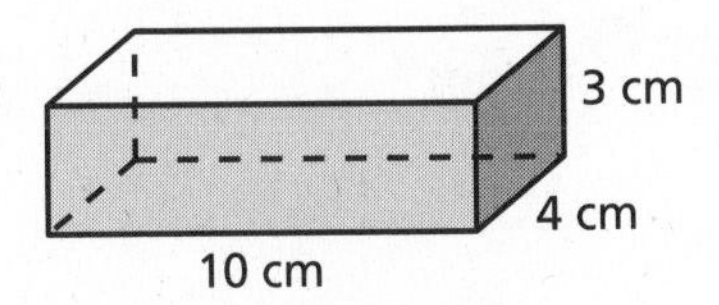

4.

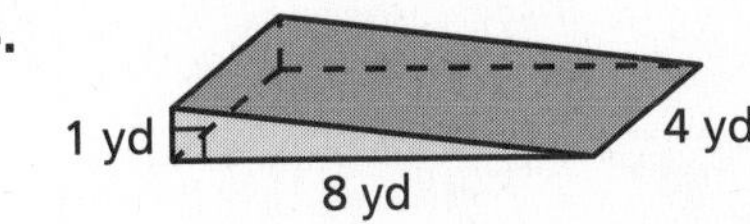

5.

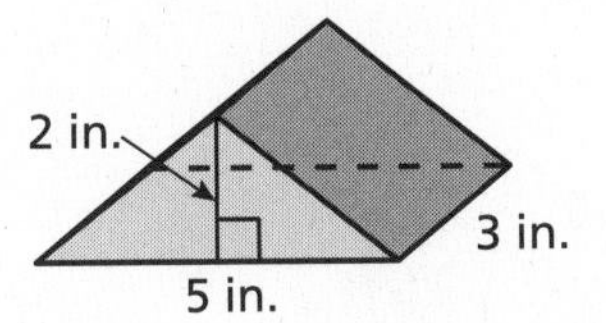

6.

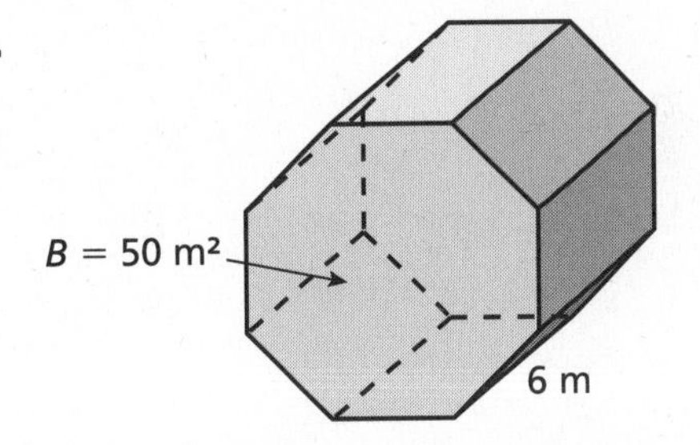

7. A cell phone is in the shape of a rectangular prism, with a length of 4 inches, a width of 2 inches, and a height of 1 inch. What is the volume of the cell phone?

8. A recycle bin is in the shape of a trapezoidal prism. The area of the base is 220 square inches and the height is 24 inches. What is the volume of the recycle bin?

9. A water jug is in the shape of a prism. The area of the base is 100 square inches and the height is 20 inches. How many gallons of water will the water jug hold? $\left(1 \text{ gal} = 231 \text{ in.}^3\right)$ Round your answer to the nearest tenth.

Name ____ Date __________

9.4 Practice B

Find the volume of the prism.

1.

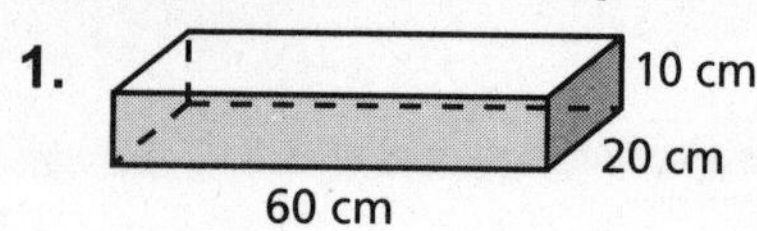

2.

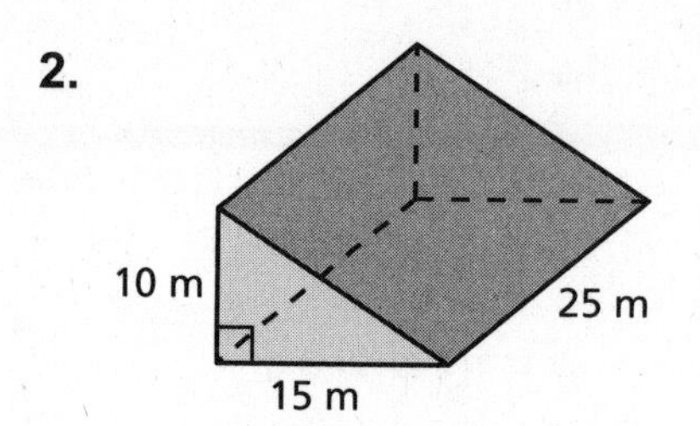

3.

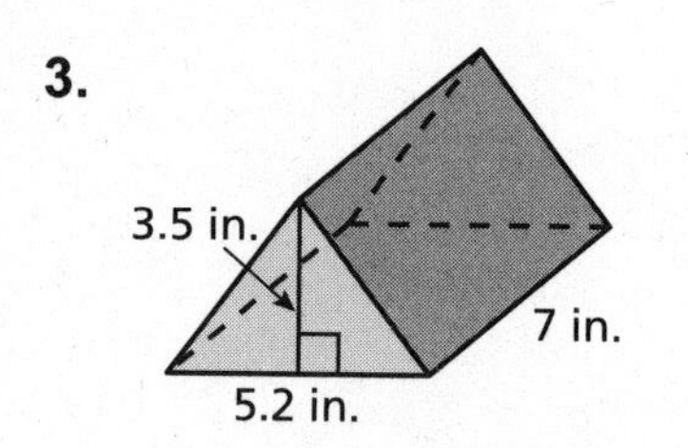

4.

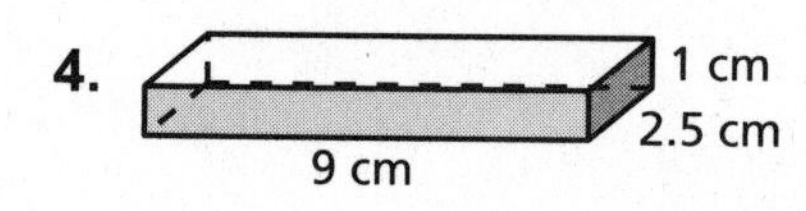

5.

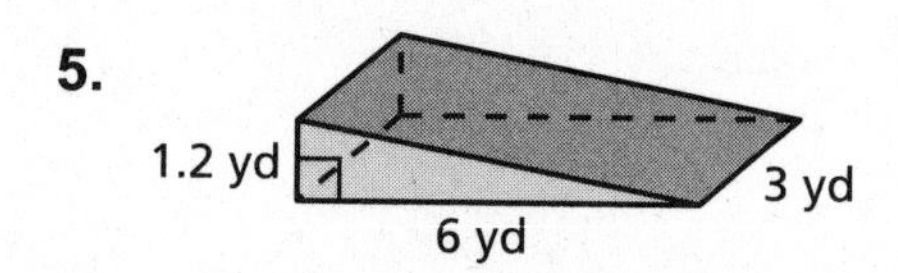

6.

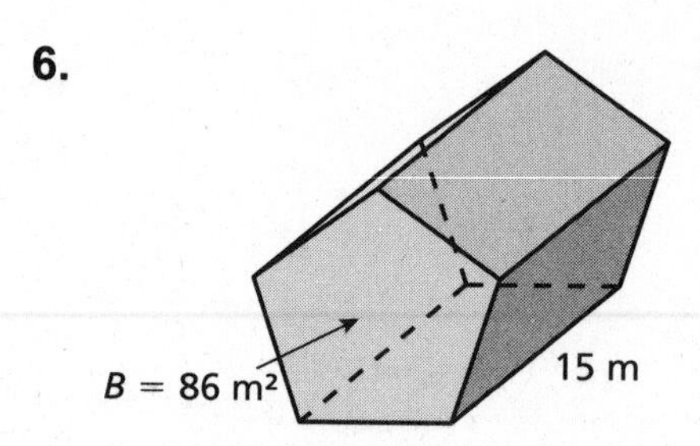

7. A mailbox is in the shape of a prism. The area of the base is 52 square inches and the height is 18 inches. What is the volume of the mailbox?

8. A chicken broth container is in the shape of a rectangular prism, with a length of 9.5 centimeters, a width of 6 centimeters, and a height of 16.5 centimeters. The container is 90% full. How many liters of chicken broth are in the container? $\left(1 \text{ L} = 1000 \text{ cm}^3\right)$ Round your answer to the nearest hundredth.

9. How many cubic feet are in a cubic yard? Use a sketch to explain your reasoning.

Name____ Date__________

9.4 Enrichment and Extension

Brain Buster Boxes

As you work, keep in mind that prisms with the same dimensions in a different order do not count as different prisms. (For example, a rectangular prism that is 5 inches long, 10 inches wide, and 10 inches tall is the same as a rectangular prism that is 10 inches long, 10 inches wide, and 5 inches tall. The prism has just been rotated.)

1. Make a list of all the different rectangular prisms you can make using only side lengths of 5 inches, 10 inches, and 15 inches.

    a. Find the volume and surface area of each of your prisms.

    b. Copy and complete the table by putting the volumes in order from least to greatest.

Length	Width	Height	Volume	Surface Area

2. Make a list of all the different rectangular prisms you can make using only side lengths of 5 inches, 8 inches, 15 inches, and 50 inches.

    a. Find the volume and surface area of each of your prisms.

    b. Use a table similar to Exercise 1 part (b) to put the volumes in order from least to greatest.

3. Look at both tables. As the volume increases, describe what happens to the surface area. Describe any patterns that you see.

4. How would you change the dimensions of a rectangular prism box so that it holds more while using less cardboard to make?

Name ______________________________ Date __________

Puzzle Time

What Game Do Bakers Like To Play?

Write the letter of each answer in the box containing the exercise number.

Find the volume of the prism.

1. a rectangular prism measuring 9 inches by 4 inches by 15 inches
2. a rectangular prism that measures 7 centimeters by 5 centimeters by 12 centimeters
3. a rectangular prism that measures 6 feet by 8 feet by 7 feet
4. a triangular prism with bases that have a base of 8 inches and a height of 12 inches; The height of the prism is 5 inches.
5. a triangular prism with bases that have a base of 7 feet and a height of 4 feet; The height of the prism is 9 feet.
6. a triangular prism with bases that have a base of 11 centimeters and a height of 8 centimeters; The height of the prism is 4 centimeters.
7. a pentagonal prism with a base area of 92 square inches; The height of the prism is 6 inches.
8. a hexagonal prism with a base area of 81 square centimeters; The height of the prism is 9 centimeters.
9. A baking dish shaped like a rectangular prism measures 9 inches by 12 inches by 2 inches. What is the volume of the dish?
10. A salt shaker is shaped like a pentagonal prism. Its base area is 18 square centimeters. The height of the prism is 9 centimeters. What is the volume of the salt shaker?
11. A flower garden box is shaped like an octagonal prism. The base area is 48 square feet. The height of the prism is 2 feet. What is the volume of the flower garden box?

Answers

G. 96 ft^3

C. 162 cm^3

A. 540 $in.^3$

H. 420 cm^3

T. 552 $in.^3$

U. 126 ft^3

D. 729 cm^3

I. 176 cm^3

O. 336 ft^3

C. 216 $in.^3$

T. 240 $in.^3$

4	6	10		7	1	9		8	3	5	11	2

Activity 9.5 Start Thinking!

For use before Activity 9.5

Explain the difference between the slant height of a pyramid and the height of a pyramid.

Which do you use for volume? Which do you use for surface area?

Activity 9.5 Warm Up

For use before Activity 9.5

Multiply.

1. $\frac{2}{3} \times 15$

2. $\frac{3}{4} \times 8$

3. $\frac{7}{10} \times 6$

4. $\frac{1}{3} \times 18$

5. $\frac{5}{9} \times 30$

6. $\frac{4}{13} \times 72$

Give a real-life example of how knowing the volume of a pyramid would be beneficial.

Find the volume of the pyramid.

1. 9 in., 5 in., 3 in.

2. 25 ft, 18 ft, 18 ft

3. 6 m, 4 m, 7 m

4. 18 cm, $B = 30 \text{ cm}^2$

Name____ Date__________

9.5 Practice A

Find the volume of the pyramid.

1.

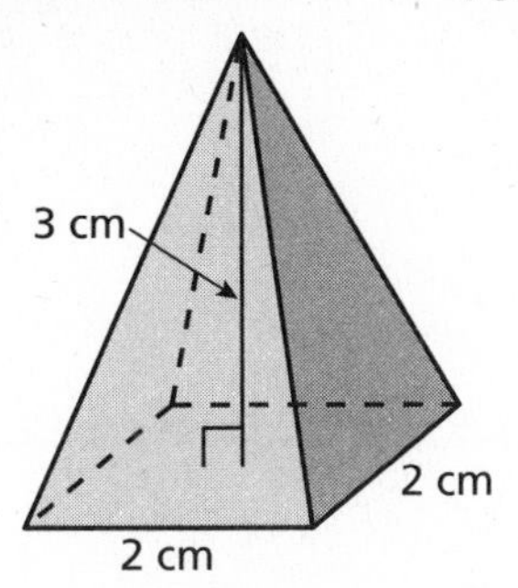

2.

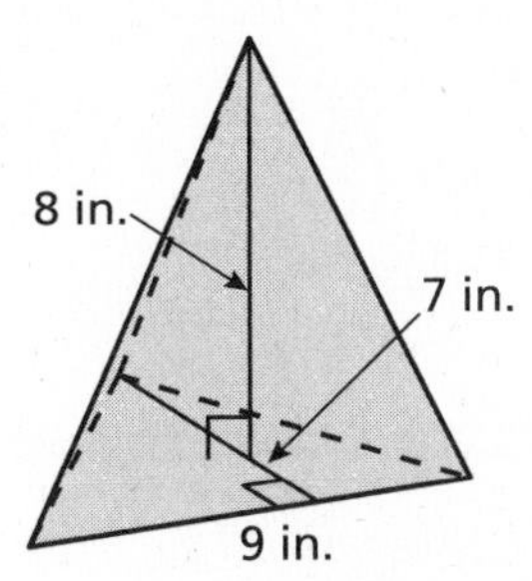

3.

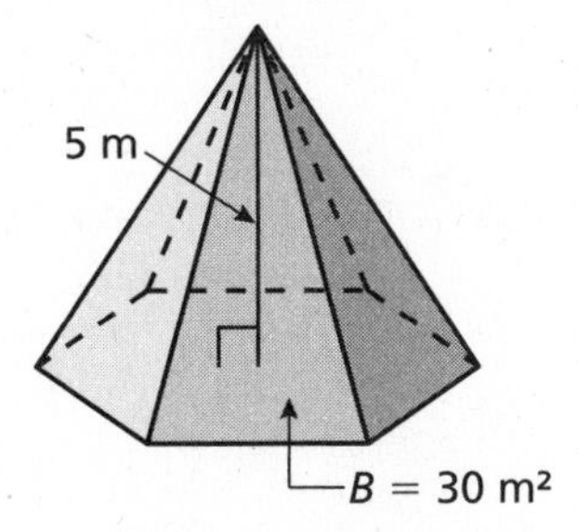

4.

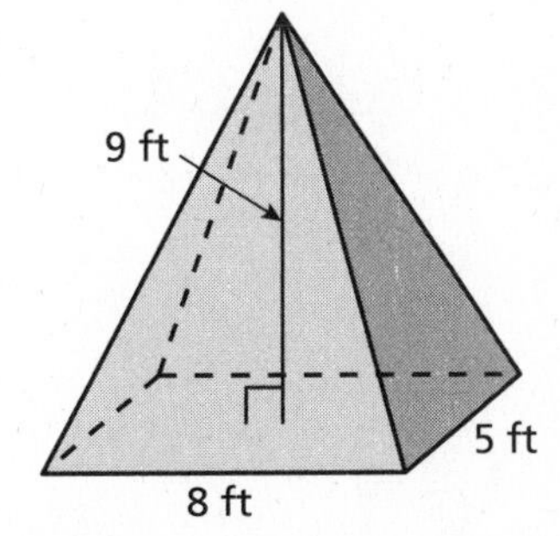

5. A tent is in the shape of a pyramid. The base is a rectangle with a length of 12 feet and a width of 10 feet. The height of the tent is 8 feet. Find the volume of the tent.

6. A sign made of solid wood is in the shape of a pyramid. The base is a triangle with a base of 6 feet and a height of 4 feet. The height of the sign is 7 feet. The wood costs $3 per cubic foot. What is the cost of the sign?

7. Two pyramids with square bases have the same volume. One pyramid has a height of 6 centimeters and the area of the base is 36 square centimeters.

a. What is the volume of the pyramids?

b. The base of the other pyramid has a side length of 3 centimeters. What is the height of this pyramid?

8. How does the volume of a pyramid change when the height is halved?

Name ______________________________ Date __________

9.5 Practice B

Find the volume of the pyramid.

1.

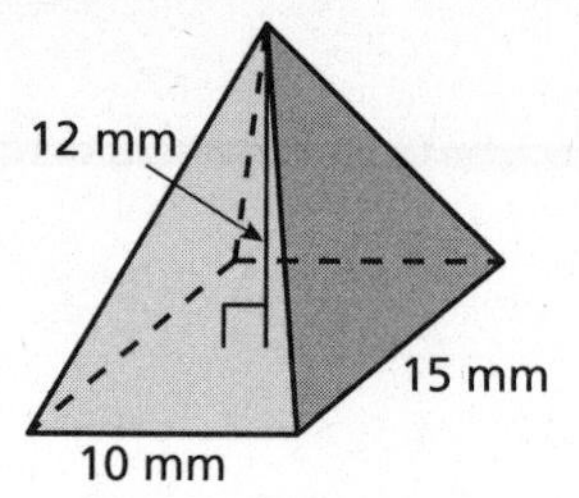

2.

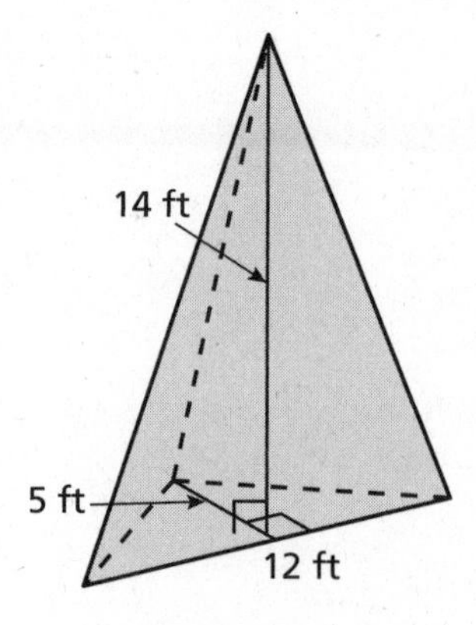

3.

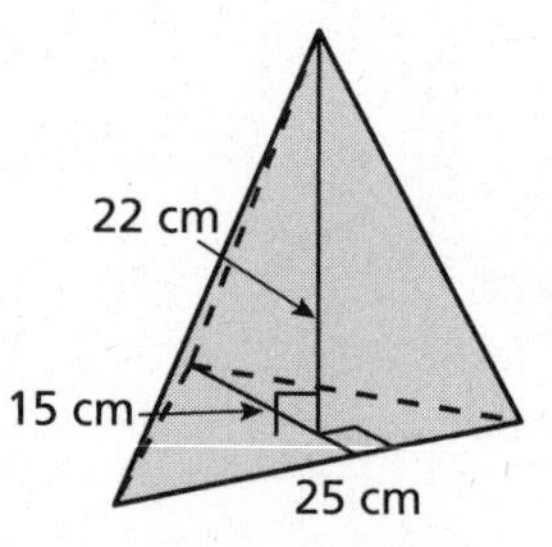

4.

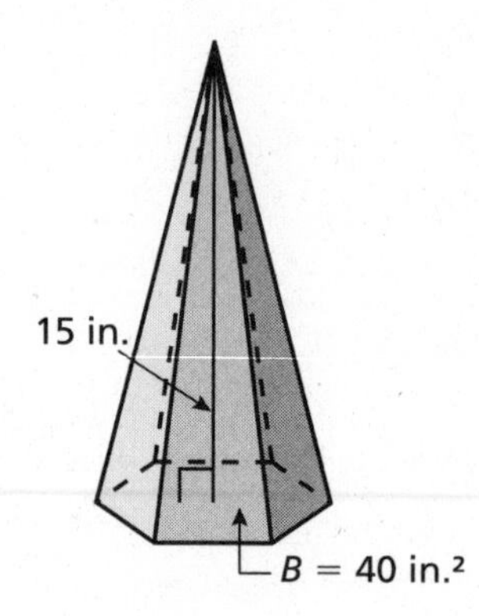

5. A pyramid has a rectangular base with length of 15 feet and a width of 8 feet. The height of the pyramid is 10 feet.

a. Find the volume of the pyramid.

b. When the pyramid was being built, the original base was dropped and it split in two pieces. Each piece was a triangle with a base of 15 feet and a height of 8 feet. Pyramids were made with these two bases, each with a height of 10 feet. Find the combined volume of the two pyramids.

c. Is the combined volume *greater than*, *less than*, or the *same as* the volume of the pyramid?

Name____ Date__________

9.5 Enrichment and Extension

Measuring Up with Sand

Children at the beach often spend a lot of time pouring sand and water from one bucket to another bucket.

1. You have a bucket that is a cube with a side length of 4 inches that you fill with sand. You pour its contents into a bucket that is a square pyramid with a base length of 4 inches and a height of 10 inches. Will the sand spill over? If so, how much sand will spill out? If not, how high will the sand be in the pyramid bucket?

2. Your friend has a bucket that is a square pyramid with a base length of 6 inches and a height of 9 inches that is full of sand. Your friend pours the sand into a cylindrical bucket with a diameter of 6 inches and a height of 9 inches. Will the sand spill over? If so, how much sand will spill out? If not, how high will the sand be in the cylindrical bucket?

3. Another friend has two buckets, one that is a triangular pyramid and one that is a triangular prism. The bases of both buckets are triangles with a base of 3 inches and a height of 4 inches. The pyramid is 6 inches tall. When you fill the pyramid with sand and pour it into the prism, it fills the prism exactly. What is the height of the prism?

4. You borrow a cylindrical bucket. You fill the cube bucket from Exercise 1 with water and pour the entire contents into this cylindrical bucket. The water is 7 inches high. Does this cylindrical bucket have a larger or smaller diameter than the cylindrical bucket from Exercise 2? Explain your reasoning.

5. You enter a sand castle building contest. You are given a bucket that is a square prism with base length of 6 inches and height of 10 inches that is full of red sand. According to the rules for the competition, you can use only this sand and nothing more, but you do not have to use all the sand. How would you design your sand castle? Draw a picture and give the names, dimensions, and volumes of the shapes you would use in your castle. Assume that you can form whatever shapes and sizes you choose.

Name ____________________ Date ________

Puzzle Time

What Do You Get If You Add Two Bananas To Three Apples?

Write the letter of each answer in the box containing the exercise number.

Find the volume of the pyramid with area of base *B* and height *h*.

1. $B = 16 \text{ in.}^2;\ h = 9 \text{ in.}$

G. 38 in.^3 **H.** 46 in.^3 **I.** 48 in.^3

2. $B = 168 \text{ cm}^2;\ h = 13 \text{ cm}$

L. 728 cm^3 **M.** 752 cm^3 **N.** 768 cm^3

3. $B = 54 \text{ ft}^2;\ h = 7 \text{ ft}$

Q. 96 ft^3 **R.** 126 ft^3 **S.** 148 ft^3

4. $B = 67 \text{ m}^2;\ h = 18 \text{ m}$

A. 402 m^3 **B.** 424 m^3 **C.** 468 m^3

5. $B = 55 \text{ cm}^2;\ h = 6 \text{ cm}$

C. 90 cm^3 **D.** 110 cm^3 **E.** 130 cm^3

6. $B = 63 \text{ mm}^2;\ h = 13 \text{ mm}$

R. 203 mm^3 **S.** 243 mm^3 **T.** 273 mm^3

7. $B = 78 \text{ ft}^2;\ h = 11 \text{ ft}$

F. 286 ft^3 **G.** 206 ft^3 **H.** 196 ft^3

8. $B = 311 \text{ yd}^2;\ h = 15 \text{ yd}$

A. 1555 yd^3 **B.** 1015 yd^3 **C.** 1225 yd^3

9. A grocery store has a display of tuna cans stacked to form a rectangular pyramid that is 5 feet tall. The base is 9 feet by 7 feet. What is the volume of the display?

R. 95 ft^3 **S.** 105 ft^3 **T.** 135 ft^3

10. Building a campfire, you start by stacking kindling wood to form a pentagonal pyramid that is 27 centimeters tall. The base area is 965 square centimeters. What is the volume of the campfire pyramid?

U. 8685 cm^3 **V.** 8852 cm^3 **W.** 9285 cm^3

7	3	10	1	6		9	4	2	8	5

Extension 9.5 Start Thinking!

For use before Extension 9.5

List examples of solids you see around the classroom.

Extension 9.5 Warm Up

For use before Extension 9.5

Identify the solid.

1.

2.

3. 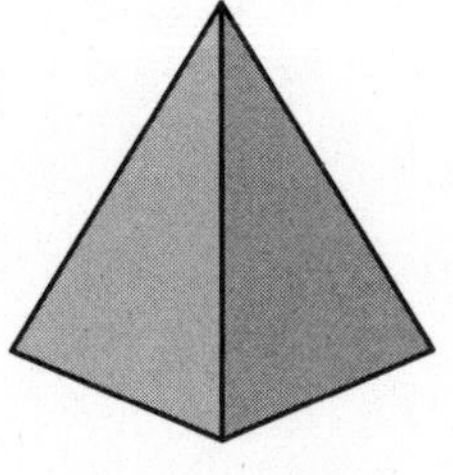

4. 

Name ______________________________ Date __________

Extension 9.5

Practice

Describe the intersection of the plane and the solid.

1.

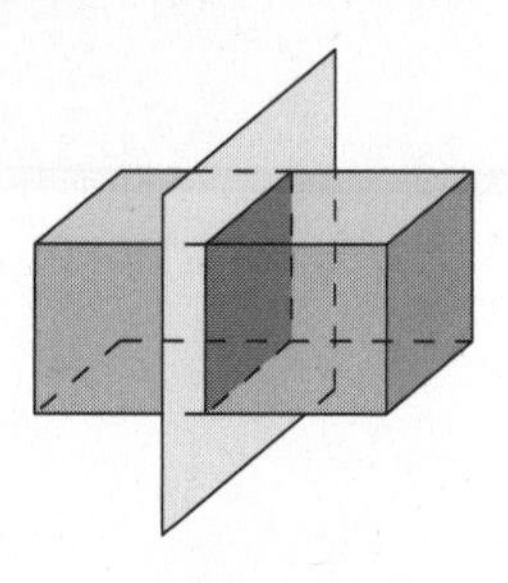

2.

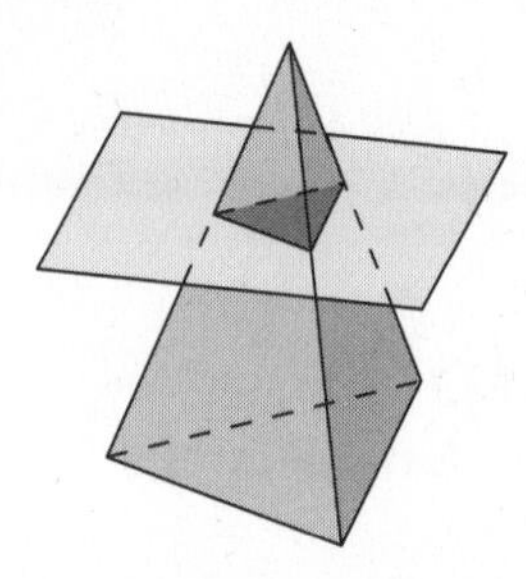

3.

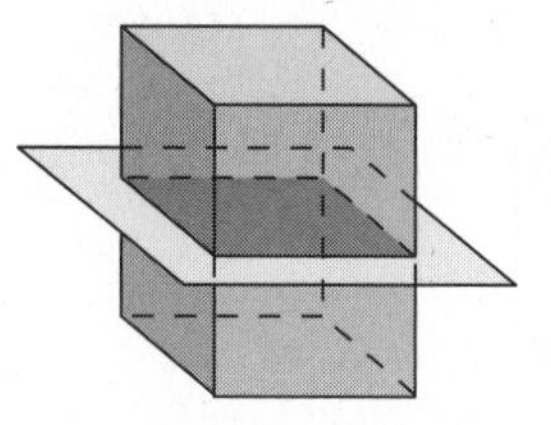

4.

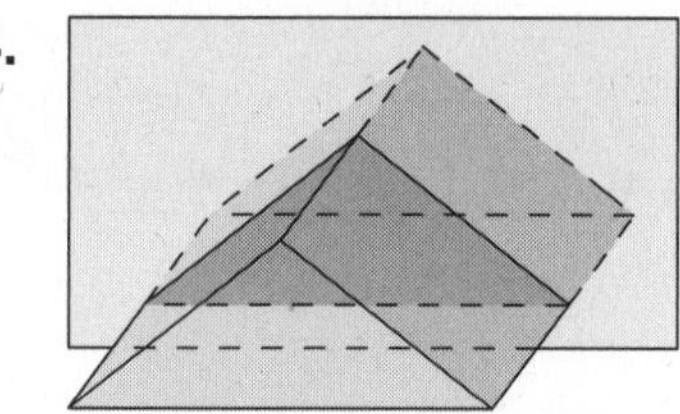

5.

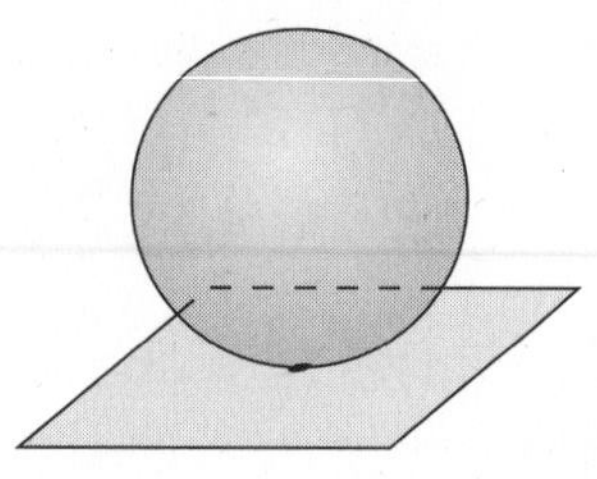

6.

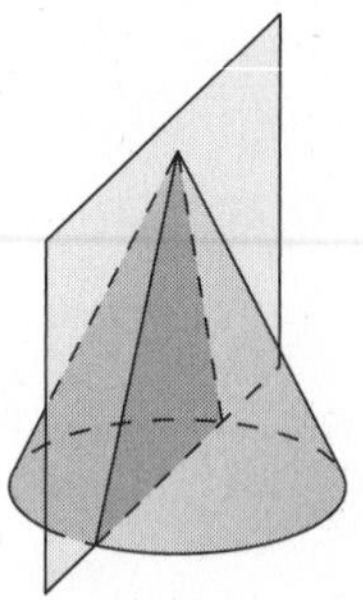

7.

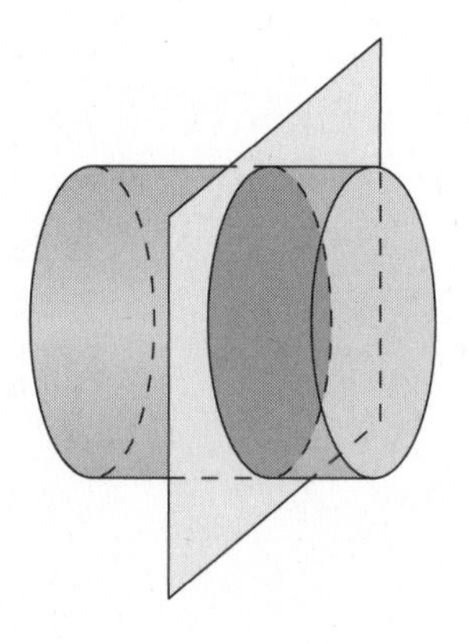

8.

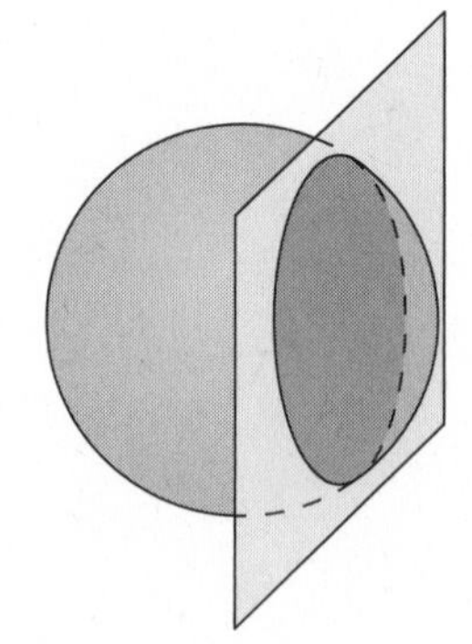

Name______________________________ Date__________

Technology Connection

For use after Section 9.1

Changing the side length of a cube

You can use a spreadsheet to find out how changing the side length of a cube affects the surface area of the cube.

EXAMPLE What happens to the surface area of a cube when its side length is doubled?

SOLUTION

Step 1 Create a spreadsheet. Put an original side length in cell A2 and the amount to multiply that length by in cell B2.

	A	B	C	D	E	F
1	Original side length	Multiplier	New side length	Original surface area	New surface area	Ratio of new to original surface area
2	5	2	=A2*B2	=6*A2^2	6*C2^2	=E2/D2
3						

Step 2 Enter a formula for the new side length in cell C2.

Step 3 Enter formulas for the surface areas in cells D2 and E2.

Step 3 You can compare the surface areas of the cubes by dividing the surface area of the new cube by the surface area of the original cube. Enter a formula for this ratio.

Step 4 Change the original value of the side length and see what happens to the ratio. Do this several times until you see a pattern.

Use a spreadsheet to help you answer each question.

1. What happens to the surface area of a cube when the side length is doubled?

2. What happens to the surface area of a cube when the side length is tripled?

3. What happens to the surface area of a cube when the side length is quadrupled?

4. What happens to the surface area of a cube when the side length is multiplied by n? Explain your answer.

Chapter 10

Name____ Date__________

Chapter 10 Probability and Statistics

Dear Family,

When you plan an outdoor event like a picnic, you cannot be certain that the weather will cooperate. Almost immediately, you begin to wonder—will it be warm or cool? sunny or cloudy? dry or rainy? There is no way to be certain, so you turn to the weather forecast to find out what is likely. When weather forecasters say there is a 60% chance of rain, do you ever wonder how they know? The weather report introduces you to the concept of probability.

The National Weather Service keeps track of daily conditions. They record the temperature, humidity, air pressure, and other data, including the weather produced by those conditions. The forecasters compare this historical data with current conditions and may see that out of 100 days with similar conditions, 60 of them were rainy days.

In probability, a *favorable outcome* is the result you are looking for, such as the number of rainy days. The ratio of the favorable outcome to the total number of outcomes is the probability.

$$\frac{\text{number of favorable outcomes}}{\text{total number of outcomes}} = \frac{\text{days with rain}}{\text{total days}} = \frac{60}{100} = 0.6 = 60\%$$

The next time you are relying on good weather, you may want to do your own research. You and your student can think about these topics:

- *The Farmer's Almanac* provides historical weather information, such as the number of times it rained on a given date. Use this information to determine the probability that it will rain on the date of your event.
- For some events, like a pool party or a picnic by the lake, you may want to get a sense of what the temperature will be. What is the probability that the temperature will be above 70 degrees the day of your event?
- What other conditions and probabilities do you want to know?

You might revise your plans if the conditions aren't favorable.

Pick a favorable day and then enjoy your picnic! Remember to watch for ants—they are almost certain to attend!

Nombre ______________________________ Fecha ________

Probabilidad y estadística

Estimada Familia:

Cuando planea un evento al aire libre, como por ejemplo un picnic, no se puede saber si el clima va a cooperar. Casi inmediatamente, uno empieza a preguntarse—¿habrá frío o calor?, ¿estará soleado o nublado?, ¿seco o mojado? No hay modo de estar seguro, por lo que uno observa el pronóstico del tiempo para ver cómo estará. Cuando los pronosticadores del tiempo dicen que hay 60% de probabilidades de lluvia, ¿alguna vez se han preguntado como lo saben? El informe del tiempo lo introduce al concepto de la probabilidad.

El Servicio de Pronósticos Nacionales hace un seguimiento de las condiciones diarias. Registran la temperatura, humedad, presión del aire y otros datos, incluyendo el tiempo producido por tales condiciones. Los pronosticadores comparan estos datos históricos con condiciones actuales y observan que por cada 100 días con condiciones similares, 60 de ellos fueron días lluviosos.

$$\frac{\text{número de resultados favorables}}{\text{número total de resultados}} = \frac{\text{días lluviosos}}{\text{días totales}} = \frac{60}{100} = 0.6 = 60\%$$

La próxima vez que tengan que contar con un buen clima, querrán hacer su propia investigación. Usted y su estudiante pueden pensar acerca de estos temas:

- *El Almanaque del Granjero* proporciona información histórica del clima, como por ejemplo el número de veces que llovió en una fecha dada. Usen esta información para determinar la probabilidad de lluvia en la fecha de su evento.
- Para algunos eventos, como por ejemplo una fiesta al lado de la piscina o un picnic por el lago, querrán saber cómo será la temperatura. ¿Cuál es la probabilidad de que la temperatura esté por encima de los 70 grados el día de su evento?
- ¿Qué otras condiciones y probabilidades desea saber?

Querrán revisar sus planes si las condiciones no son favorables.

¡Elijan un día favorable y luego disfruten su picnic! Recuerden revisar que no haya hormigas—¡por lo general suelen asistir también!

Activity 10.1 Start Thinking!
For use before Activity 10.1

If you flip a penny, how many possible results are there?

If you flip a penny and a nickel, how many possible results are there?

If you flip two pennies, how many possible results are there?

Activity 10.1 Warm Up
For use before Activity 10.1

Simplify the fraction.

1. $\frac{12}{50}$

2. $\frac{14}{28}$

3. $\frac{16}{20}$

4. $\frac{5}{25}$

5. $\frac{18}{30}$

6. $\frac{24}{42}$

Design a spinner for a board game with different colored regions in which all of the results for spinning are equally likely. Discuss the possible results.

You spin the spinner shown.

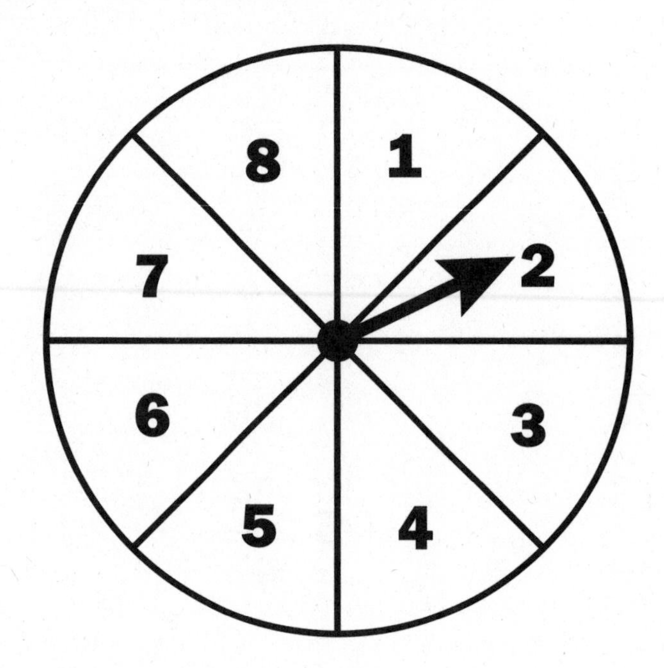

1. How many possible results are there?

2. Of the possible results, in how many ways can you spin an even number? an odd number?

Name______________________________ Date__________

10.1 Practice A

You randomly choose one of the tiles shown below. Find the favorable outcomes of the event.

1. Choosing a 4
2. Choosing an even number
3. Choosing a number less than 2
4. Choosing an odd number greater than 6
5. Choosing a number divisible by 2
6. Choosing a number greater than 10

You randomly choose one shape from the bag.
(a) Find the number of ways the event can occur.
(b) Find the favorable outcomes of the event.

7. Choosing a triangle
8. Choosing a star
9. Choosing *not* a square
10. Choosing *not* a circle
11. A beverage cooler contains bottles of orange juice and apple juice. There are 44 bottles in the cooler.
    a. You are equally likely to randomly choose a bottle of orange juice or a bottle of apple juice from the cooler. How many of the bottles are apple juice?
    b. Two of the bottles of orange juice are replaced with apple juice. How many ways can you randomly choose a bottle of apple juice from the cooler?
12. Three girls and four boys made the final round of the spelling bee.
    a. How many ways can you randomly choose a girl to be the first contestant?
    b. Given that part (a) occurred, how many ways can you randomly choose a girl to be the second contestant?

Name ______________________________ Date __________

10.1 Practice B

You randomly choose one of the tiles shown below. Find the favorable outcomes of the event.

1. Choosing an 8
2. Choosing an even number less than 7
3. Choosing a 5 or a 7
4. Choosing a number divisible by 11
5. Choosing a number that begins with the letter T
6. Choosing a number that doesn't contain line segments

You randomly choose one shape from the bag.
(a) Find the number of ways the event can occur.
(b) Find the favorable outcomes of the event.

7. Choosing a triangle
8. Choosing a star
9. Choosing *not* a square
10. Choosing *not* a circle
11. There are 12 cats and 7 dogs at the Humane Society.
    a. In how many ways can the first customer randomly choose a cat?
    b. In how many ways can the second customer randomly choose a dog?
    c. In how many ways can the third customer randomly choose a dog?
    d. In how many ways can the fourth customer randomly choose a dog?
    e. When the fifth customer arrives, what are the favorable outcomes of randomly choosing a dog?

Name____ Date__________

10.1 Enrichment and Extension

Counting Jasmine's Rectangular Designs

Jasmine has just been hired to work for a company that designs patios and walkways. Her first assignment is to make pictures of walkway and patio designs that can be made with their new decorative square stones. She needs to make one picture for each of the possible rectangles that can be made with 30 to 40 stones. Jasmine decides to put each rectangular design on a separate card. (*Note:* A rectangle that is 5 stones wide and 6 stones long is the same as one that is 6 stones wide and 5 stones long.)

1. How many cards will Jasmine have to make? List the dimensions of all the different rectangles that can be made with 30 to 40 stones.

2. Jasmine dropped her cards and one of them was ruined.

a. Is it more likely that the ruined card has an odd or even number of square stones in the design? Explain your reasoning.

b. Is it more likely that the ruined card has more or less than 35 square stones in the design? Explain your reasoning.

3. A customer wants to buy the number of blocks between 30 and 40 that will give her the most options for a rectangular patio or walkway. How many should she buy? Explain your reasoning.

4. Another customer wants to make a rectangular walkway that is between 2 and 5 blocks wide. He's not sure how long he wants to make it, but he wants to buy some blocks to set out in the space in order to visualize his options before making his decision. He wants to have at least 30 stones and at least six options to consider. What is the least number of blocks that he should buy? Explain your reasoning.

5. The side length of the square stones is 9 inches. A customer wants to make a patio that is 3 feet 3 inches by 5 feet 9 inches. The stones can be cut in halves, quarters, or thirds.

a. Draw a picture showing how many full and partial squares he will need to make the patio.

b. How many stones will he need to buy in order to have the least waste?

c. How many stones will he need to have cut and to what sizes? How many stone pieces will he have left over, and what sizes will they be?

Name ______________________________ Date __________

Puzzle Time

What Is Brown, Has A Hump, And Lives At The North Pole?

Circle the letter of each correct answer in the boxes below. The circled letters will spell out the answer to the riddle.

You randomly pick a card out of a deck of 52 cards. Find the number of ways the event can occur.

1. Choosing a spade

2. Choosing an ace

3. Choosing *not* a king

4. Choosing a red card

5. Choosing a heart that is a face card

6. Choosing *not* a diamond

Your teacher randomly selects among the following names to be your partner in a project: Girls—Amanda, Meredith, Erin, Gail, and Mackenzie; Boys—Scott and Peter. Find the number of ways the event can occur.

7. Choosing a girl

8. Choosing *not* a girl

9. Choosing Meredith

10. Choosing *not* Peter

You randomly choose one month to celebrate a family reunion. Find the number of ways the event can occur.

11. Choosing a month after March

12. Choosing a month before September

13. Choosing *not* December

14. Choosing *not* June or July

T	B	A	R	C	H	V	I	E	M	L	R	S	Y	D	F	O
7	32	4	41	16	36	10	22	13	21	42	2	17	8	15	33	12
G	**L**	**Y**	**O**	**J**	**S**	**T**	**W**	**A**	**C**	**D**	**A**	**R**	**M**	**E**	**S**	**L**
20	9	14	3	44	5	48	19	35	11	24	39	40	1	26	18	6

Activity 10.2 Start Thinking!

For use before Activity 10.2

You have one number cube to roll and a friend has two number cubes to roll. Explain who has a better chance of rolling a 4.

Activity 10.2 Warm Up

For use before Activity 10.2

Determine whether the fraction is in lowest terms. If not, simplify the fraction.

1. $\frac{6}{14}$ **2.** $\frac{12}{27}$ **3.** $\frac{7}{10}$

4. $\frac{24}{35}$ **5.** $\frac{16}{25}$ **6.** $\frac{22}{44}$

Start Thinking!
For use before Lesson 10.2

Explain how a weather forecaster might use probability.

Warm Up
For use before Lesson 10.2

You are playing a game using the spinners shown.

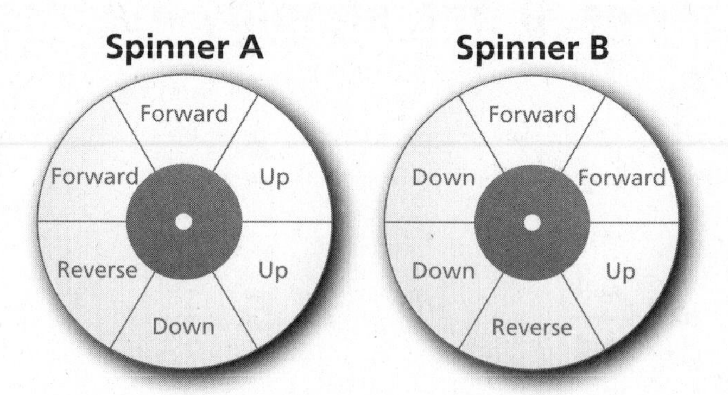

1. You want to move up. On which spinner are you more likely to spin "Up"? Explain.

2. You want to reverse. Which spinner would you spin? Explain.

Name____ Date__________

10.2 Practice A

You are playing a game using the spinners shown.

Spinner A

Spinner B

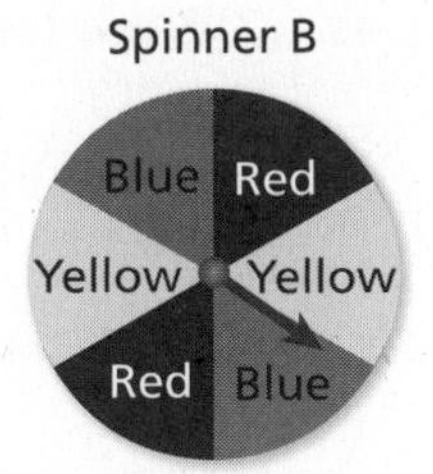

1. You want to spin red. Which spinner should you spin? Explain.

2. You want to spin yellow. Which spinner should you spin? Explain.

3. You want to spin blue. Does it matter which spinner you spin? Explain.

Describe the likelihood of the event given its probability.

4. The probability that it will snow today is zero.

5. You make a free throw 70% of the time.

6. Your band marches in $\frac{1}{6}$ of the parades.

You randomly choose one song from a collection of 4 country songs, 2 jazz songs, 3 rock songs, and 1 pop song. Find the probability of the event.

7. Choosing a jazz song

8. Choosing a pop song

9. *Not* choosing a country song

10. Choosing a blues song

11. Your football team has a 75% chance of winning a game. Your team is scheduled to play 16 games. Estimate how many games your team will win.

12. In a classroom, the probability that the teacher chooses a boy from 20 students is 0.45.

    a. How many students are *not* boys?

    b. Describe the likelihood of *not* choosing a boy.

13. A box contains ten slips of paper numbered 1 through 10. Find the probability and describe the likelihood of each event.

    a. Choosing a number greater than 2

    b. Choosing a number that is a multiple of 2

    c. Choosing a number that is less than 10

Name ________________________________ Date __________

10.2 Practice B

Describe the likelihood of the event given its probability.

1. The school bus arrives late $\frac{2}{7}$ of the time.

2. The probability that it rains during a hurricane is 1.

3. There is an 85% chance that you will go to the concert.

You randomly choose one mathematical operator from the collection. Find the probability of the event.

4. Choosing a multiplication sign

5. Choosing a plus sign

6. *Not* choosing an equal sign

7. *Not* choosing a greater than sign

8. One-half of the boxes of cereal contain a prize.

   a. Find the probability of winning a prize.

   b. Find the probability of *not* winning a prize.

   c. If you purchased two boxes of cereal, estimate the number of prizes you would receive.

9. A store has 30 blue pens, 18 black pens, and 12 red pens in stock. You buy 3 blue pens, 9 black pens, and 3 red pens. Find the probability of each event before and after your purchase. Then describe how your purchase affects the probability of each event.

   a. Randomly choosing a blue pen

   b. Randomly choosing a black pen

   c. Randomly choosing a red pen

Name ______________________________ Date __________

10.2 Enrichment and Extension

Geometric Probability

Assume that a dart thrown at the target is equally likely to hit anywhere on the target. The probability P that the dart lands in the shaded region is $P = \dfrac{\text{area of shaded region}}{\text{total area of target}}$. Find the probability that the dart lands in the shaded region.

1.

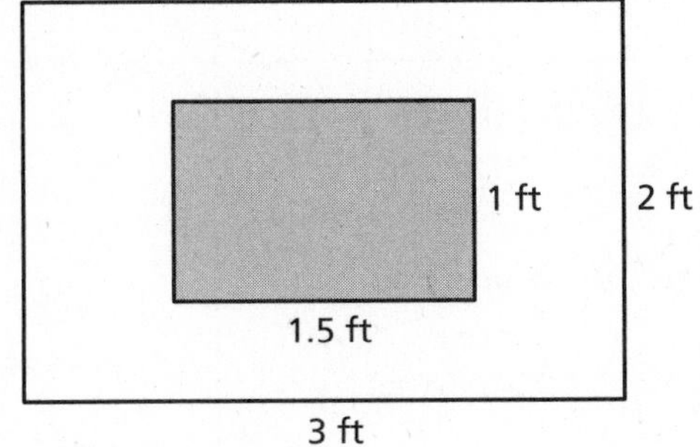

2.

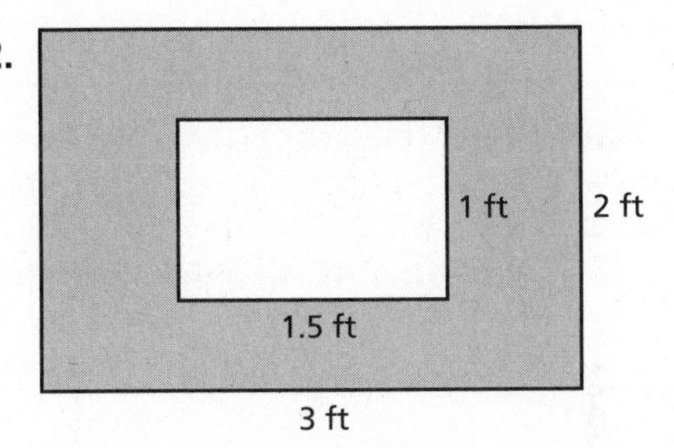

3.

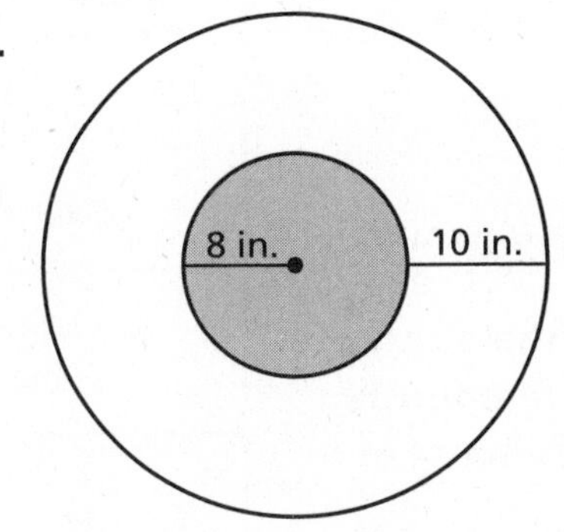

4.

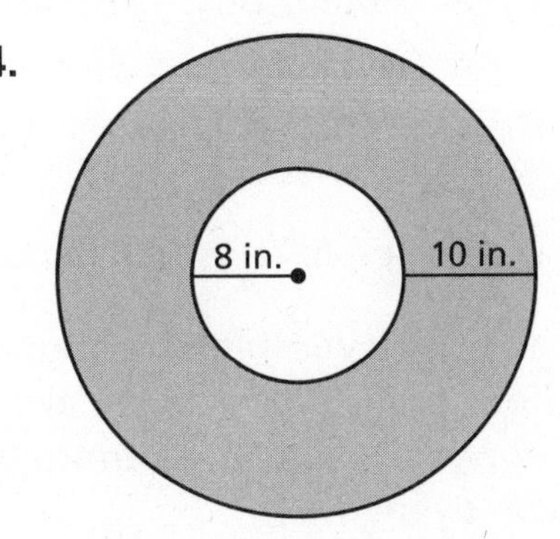

5.

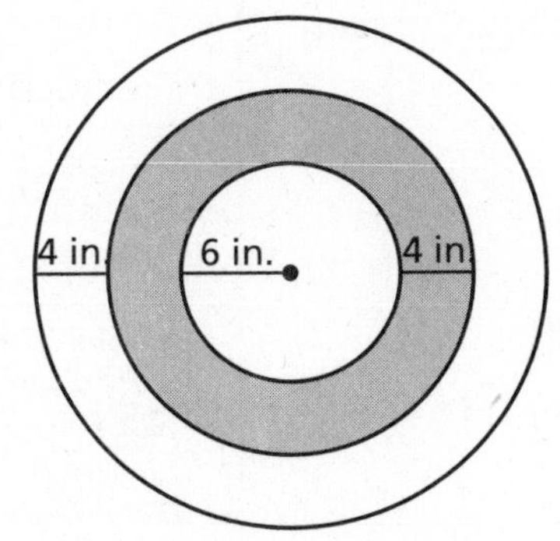

6.

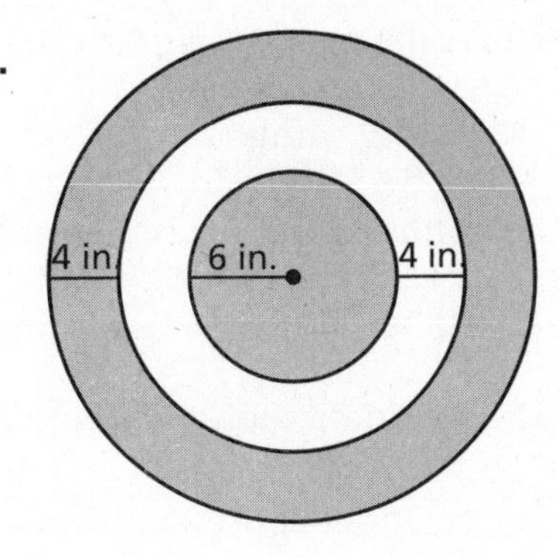

Name ______________________________ Date __________

Puzzle Time

Did You Hear About The...

A	B	C	D	E	F
G	H	I	J	K	L
M	N				

Complete each exercise. Find the answer in the answer column. Write the word under the answer in the box containing the exercise letter.

Answer	Word
$\frac{3}{4}$	SHE
Likely	WENT
$\frac{1}{4}$	BYTE
$\frac{3}{20}$	COMPUTER
$\frac{11}{20}$	GET
0	BECAUSE
Unlikely	ROCK
$\frac{13}{20}$	HE
$\frac{1}{20}$	HOW
$\frac{7}{20}$	THE

Describe the likelihood of the event given its probability.

A. You take the bus home from school $\frac{1}{4}$ of the time.

B. The probability your favorite show is on tonight is 0.

C. 50% of the time you flip a coin you flip tails.

D. Your team wins the swim meets $\frac{4}{5}$ of the time.

E. The probability that the cafeteria will have milk is 1.

An MP3 player has 60 songs stored on it. Of the songs, 21 are rock, 9 are rap, 18 are dance, and 12 are country. If songs are played randomly, find the probability of each event.

F. A rock song is played.

G. A rap song is played.

H. A dance song is played.

I. A country song is played.

J. A rock song is *not* played.

K. Either a dance song or rap song is played.

L. Either a rock song or country song is played.

M. A country song is *not* played.

N. A song is played.

Answer	Word
$\frac{1}{100}$	WHERE
$\frac{1}{5}$	SO
Impossible	STAR
1	GIG
Certain	TO
$\frac{3}{10}$	STORE
$\frac{9}{20}$	COULD
$\frac{9}{10}$	MUSIC
Equally Likely to Happen or Not Happen	WHO
$\frac{4}{5}$	A

Activity 10.3

Start Thinking!

For use before Activity 10.3

You have a bag filled with 6 red marbles, 4 blue marbles, and 8 yellow marbles. Explain to a partner how to find the probability of pulling out a red marble.

Activity 10.3

Warm Up

For use before Activity 10.3

A number cube is rolled. Determine the probability of each event.

1. Rolling a 5

2. Rolling an even number

3. Rolling a 3 or 4

4. Rolling a 6

5. Rolling a 1, 5, or 6

6. Rolling an odd number

Start Thinking!

For use before Lesson 10.3

Review with a partner how to find relative frequency. Use an example to explain.

Warm Up

For use before Lesson 10.3

You have three sticks. Each stick has one red side and one blue side. You throw the sticks 10 times and record the results. Use the table to find the relative frequency of the event.

1. Tossing 3 red

2. Tossing 1 red, 2 blue

3. Tossing 1 blue, 2 red

4. *Not* tossing all red

Outcome	Frequency
3 red	2
3 blue	4
1 red, 2 blue	0
1 blue, 2 red	4

Name____ Date__________

10.3 Practice A

You have two sticks. Each stick has one blue side and one pink side. You throw the sticks 10 times and record the results. Use the table to find the experimental probability of the event.

Outcome	Frequency
2 blue	1
2 pink	3
1 blue, 1 pink	6

1. Tossing 2 pink

2. Tossing 1 blue and 1 pink

3. *Not* tossing all pink

4. You check 15 bananas. Six of the bananas are bruised.

a. What is the experimental probability that a banana is bruised?

b. What is the experimental probability that a banana is *not* bruised?

5. Sixteen students have cell phones. Five of the cell phones have touch screens.

a. What is the experimental probability that a student's cell phone has a touch screen?

b. Out of 144 students' cell phones, how many would you expect to have touch screens?

You flip a coin twice. You repeat this process 12 times. The table gives the results.

Outcome	Frequency
2 Heads	2
1 Head, 1 Tail	7
2 Tails	3

6. Use the first table to find the experimental probability of each outcome.

7. Based upon experimental probability, which outcome is most likely?

8. The second table gives the possible outcomes of flipping a coin twice. Each of these outcomes is equally likely. What is the theoretical probability of flipping 1 tail?

1st Flip	2nd Flip
Head	Head
Head	Tail
Tail	Head
Tail	Tail

9. Compare your answers to Exercises 7 and 8.

Name ______________________________ Date __________

10.3 Practice B

You have four sticks. Two sticks have one blue side and one pink side. One stick has 2 blue sides. One stick has 2 pink sides. You throw the sticks 20 times and record the results. Use the table to find the experimental probability of the event.

Outcome	Frequency
3 blue, 1 pink	7
2 blue, 2 pink	9
1 blue, 3 pink	4

1. Tossing 1 pink and 3 blue

2. Tossing the same number of blue and pink

3. *Not* tossing 3 pink

4. Tossing at most 2 blue

5. You check 30 containers of yogurt. Seven of them have an expiration date within the next 3 days.

   a. What is the experimental probability that a container of yogurt will have an expiration date within the next 3 days?

   b. Out of 120 containers of yogurt, how many would you expect to have an expiration date within the next 3 days?

6. The plant produces 1200 packages of grapes. An inspector randomly chooses 24 packages and discovers that 8 of the packages have broken seals. How many of the 1200 packages of grapes would you expect to have broken seals?

7. You flip 3 coins 50 times, and flipping 3 tails occurs 6 times.

   a. What words above refer to the *total number of trials*?

   b. What words above refer to the *number of times the event occurs*?

   c. What words above refer to the *event*?

   d. What is the experimental probability that you flip 3 tails?

   e. How many times would you expect to flip 3 tails out of 200 trials of flipping 3 coins?

Name____________________ Date__________

10.3 Enrichment and Extension

What's the Difference?

Many card games involve making choices based on how likely it is to choose a certain card. By performing an experiment with a regular deck of cards, you will be finding the probability of certain outcomes when cards are chosen at random. A deck of cards has 4 suits. Each suit has 13 cards: a Jack, Queen, King, Ace, and the numbers 2 through 10. For this experiment, the Ace is worth 1, the Jack is worth 11, the Queen is worth 12, and the King is worth 13.

Experiment Directions: Put all of the cards face down and spread them out. Choose two cards at random. Find the absolute value of the difference between the values of the cards. Replace the two cards. Mix the cards and repeat.

Answer Exercises 1 and 2 *before* performing the experiment.

1. Make a list of all the possible outcomes and design a frequency table to record your results.
2. Make some predictions. Will all the outcomes be equally likely? If not, what outcomes will be most likely? least likely? Explain your reasoning.
3. Perform the experiment at least 60 times. Record the results in your frequency table from Exercise 1.
4. Make a bar graph of your results. Compare your results with your classmates. Were they similar? Explain.
5. Describe any patterns you notice. Did they fit your predictions? What outcomes are most likely? least likely? Explain.
6. Explain why it would be difficult to find theoretical probability for this situation.
7. What is the advantage to doing a large number of trials? Explain why doing more trials is especially important for this experiment.
8. You want to change the experiment. Instead of taking the absolute value of the difference, this time you will take the value of the first card minus the value of the second card. How would this change your results? Explain.
9. A friend asks you to play the following game. Two cards are chosen at random. If the absolute value of the difference is between 1 and 6, Player 1 gets a point. If the absolute value of the difference is between 7 and 12, Player 2 gets a point. If the difference is zero, both players get a point. Replace the cards, shuffle, and repeat. The first person to get 10 points wins. Explain why this game is not fair. Rewrite the rules to make the game more fair.

Name ______________________________ Date __________

10.3 Puzzle Time

Who Kept Tom Sawyer Cool In The Summertime?

Write the letter of each answer in the box containing the exercise number.

You randomly pick a nut from a can of mixed nuts 20 times and record the results: 5 almonds, 6 peanuts, 2 hazelnuts, 3 pecans, and 4 cashews. Find the experimental probability of the event.

1. Choosing an almond
2. Choosing a peanut
3. Choosing a peanut or cashew
4. Choosing *not* an almond
5. Choosing *not* a peanut
6. Choosing a walnut

You pour 50 nuts into a bowl. Use the results from the example above to make the following predictions.

7. How many peanuts would you expect to be in the bowl?
8. How many almonds and pecans would you expect to be in the bowl?
9. How many nuts that are *not* a peanut would you expect to be in the bowl?

You and your friends decide to play hide-and-seek. In a plastic container, there are 2 blue flashlights, 4 green flashlights, 1 red flashlight, 3 white flashlights, and 2 black flashlights. Find the theoretical probability of the event.

10. Choosing a green flashlight
11. Choosing a black flashlight
12. Choosing a red flashlight
13. Choosing a flashlight that is *not* blue
14. The theoretical probability of choosing a green marble is $\frac{1}{3}$. If there are 6 marbles in the bag, how many marbles would you expect to be green?

Answers

C. $\frac{3}{4}$	**E.** 0
B. $\frac{1}{3}$	**A.** 2
U. 15	**Y.** 20
E. $\frac{1}{4}$	**R.** $\frac{1}{6}$
K. $\frac{1}{12}$	**H.** 35
L. $\frac{3}{10}$	**F.** $\frac{5}{6}$
R. $\frac{1}{2}$	**N.** $\frac{7}{10}$

9	7	4	12	2	6	10	1	11	3	8		13	14	5

Activity 10.4 Start Thinking!

For use before Activity 10.4

Make a list of your two favorite ice cream flavors and your four favorite ice cream toppings.

You are allowed one ice cream flavor and one topping.

Make a list of your choices. How many choices are there?

Activity 10.4 Warm Up

For use before Activity 10.4

Multiply.

1. $3 \times 4 \times 5$

2. $7 \times 3 \times 6$

3. $5 \times 5 \times 4$

4. $9 \times 10 \times 12$

5. $15 \times 10 \times 9$

6. $7 \times 6 \times 12$

Lesson 10.4 **Start Thinking!**
For use before Lesson 10.4

You have 5 pairs of jeans, 4 T-shirts, and 3 pairs of shoes. Find how many outfit combinations are possible. Explain how you got your answer.

Lesson 10.4 **Warm Up**
For use before Lesson 10.4

1. A lock is numbered from 0 to 9. Each combination uses three numbers in a right, left, right pattern. Find the total number of possible combinations for the lock.

Name______________________________ Date__________

10.4 Practice A

Use a tree diagram to find the sample space and the total number of possible outcomes.

1.

Pet	
Animal	Hamster, Guinea Pig, Snake
Name	Lucky, Shadow, Smokey, Max

2.

Ice Cream	
Cone	Waffle, Sugar
Flavor	Chocolate, Vanilla, Strawberry

Use the Fundamental Counting Principle to find the total number of possible outcomes.

3.

Pizza	
Size	Small, Medium, Large
Crust	Thin, Thick, Regular

4.

Car	
Transmission	Automatic, Manual
Doors	2-door, 4-door
Color	Red, Blue, Black, White

5. You are taking a true-false test that has 10 questions. Assuming you answer every question, in how many different ways can the test be completed?

6. A game system allows players to design a personal picture. Each picture is designed by choosing from male or female, 8 face shapes, 48 eyes, 12 noses, 24 mouths, and 82 hair styles. How many different pictures are possible?

Name ______________________ Date __________

10.4 Practice B

Use a tree diagram to find the sample space and the total number of possible outcomes.

1.

Vacation	
Destination	Amusement Park, Zoo, Beach
Transportation	Car, Plane

2.

Game	
Coin	Quarter, Dime, Nickel, Penny
Card	King, Queen, Jack

Use the Fundamental Counting Principle to find the total number of possible outcomes.

3.

Computer	
Hard Drive	200 GB, 400 GB
Monitor	17-inch, 20-inch, 22-inch, 24-inch

4.

Sandwich	
Bread	Italian, Wheat
Meat	Ham, Roast Beef, Salami
Cheese	American, Provolone, Swiss

5. You need to hang seven pictures in a straight line.

a. In how many ways can this be accomplished?

b. If the picture of your great-grandfather must be in the middle, how many ways can the seven pictures be hung?

6. A license plate must contain two letters followed by four digits. How many license plates are possible? If the rule changed to five digits instead of four digits, how many more license plates would be possible?

Name______________________________ Date__________

10.4 Enrichment and Extension

Sandwich Shop

A local sandwich shop is running a sandwich special for lunch. A customer can build his or her own sandwich using the choices in the table. The customer selects one item from each category.

Use the menu board to answer the questions.

1. How many different sandwiches can you make from the choices on the menu board?

2. The sandwich shop has several customers that are vegetarians. In place of the fillings listed on the menu, the sandwich shop uses a vegetable spread on these orders. How many different vegetarian sandwiches can you make from the choices on the menu board?

3. Customers at the sandwich shop have the option of making their sandwiches into combo platters by adding their choice of chips, pretzels, or an apple.

   a. How will the addition of the side choices change the total number of combinations calculated in Exercise 1?

   b. How many combo platters are possible using the choices on the menu board?

4. A customer orders a ham sandwich on wheat bread. Draw a tree diagram that illustrates the possible sandwiches that could result from the order.

5. What is the probability of a customer ordering a ham sandwich on wheat bread with ketchup and no cheese?

Name ______________________________ Date __________

10.4 Puzzle Time

What Is An Ant Dictator?

Write the letter of each answer in the box containing the exercise number.

Use the Fundamental Counting Principle to find the total number of possible outcomes.

1. A restaurant offers five flavors of milkshakes. There are three sizes for each flavor.
2. Students were asked to schedule one from each category: chorus or band; French, Spanish, or German; art, wood shop, or physical education.
3. When ordering a birthday cake, you will need to choose one from each category: white, chocolate, or marble cake; raspberry or strawberry filling; white or buttercream frosting; $\frac{1}{4}$ or $\frac{1}{2}$ sheet cake.
4. Your seventh grade class is selling apparel with the school mascot to raise money for a class trip. The sizes are small, medium, large, or extra large. You can choose a T-shirt, a long-sleeved shirt, or a sweatshirt.
5. Each school lunch includes a choice of a main entrée, vegetable, fruit, and beverage. Today, the main entrées are spaghetti, fish sandwich, or cheeseburger. The vegetables are corn, green beans, or carrots. The fruit is an apple or a banana. The beverages are milk or juice.
6. You roll a number cube and flip a coin. What is the probability of rolling an even number and flipping tails?
7. You roll a number cube and flip a coin. What is the probability of rolling a number less than 5 and flipping tails?

Answers

A. 18

R. 12

C. $\frac{1}{2}$

Y. $\frac{1}{4}$

A. 15

T. 36

K. 20

T. $\frac{1}{3}$

N. 24

2		7	6	4	1	3	5

Start Thinking!

For use before Activity 10.5

Think about the words independent and dependent. What do they mean?

Give an example of a time in which you were independent. Give an example of a time in which you were dependent.

Warm Up

For use before Activity 10.5

You randomly choose one marble from a bag containing 5 blue marbles, 2 red marbles, 2 green marbles, and 1 purple marble. Find the favorable outcomes of the event.

1. Choosing a blue marble
2. Choosing a red marble
3. Choosing a green marble
4. Choosing a purple marble
5. *Not* choosing a blue marble
6. *Not* choosing a red marble

Lesson 10.5 Start Thinking!
For use before Lesson 10.5

You have a bag of marbles. You draw a marble, set it aside, and a draw a second marble. Your friend says the events are independent. Is your friend correct? Explain.

Lesson 10.5 Warm Up
For use before Lesson 10.5

Tell whether the events are *independent* or *dependent*. Explain.

1. You roll a number cube twice. The first roll is a 3 and the second roll is an odd number.

2. You flip a coin twice. The first flip is heads and the second flip is tails.

3. You randomly draw a marble from a bag containing 3 red marbles and 5 blue marbles. You keep the marble and then draw a second marble.

4. You randomly draw a marble from a bag containing 6 red marbles and 2 blue marbles. You put the marble back and then draw a second marble.

Name ______________________ Date __________

10.5 Practice A

Tell whether the events are *independent* or *dependent*. Explain.

1. You spin a spinner twice.

   First Spin: You spin a 2. Second Spin: You spin an odd number.

2. Your committee is voting on the leadership team.

   First Vote: You vote for a president. Second Vote: You vote for a vice president.

You spin the spinner and flip a coin. Find the probability of the compound event.

3. Spinning an odd number and flipping heads

4. *Not* spinning a 5 and flipping tails

You randomly choose one of the tiles. Without replacing the first tile, you choose a second tile. Find the probability of the compound event.

5. Choosing a 6 and then a prime number

6. Choosing two odd numbers

7. You randomly pull two bills from your wallet. What is the probability they are both $20?

You roll a number cube twice. Find the probability of the compound event.

8. Rolling two numbers whose sum is 2

9. Rolling an even number and then an odd number

Name ____________________ Date ________

10.5 Practice B

Tell whether the events are *independent* or *dependent*. Explain.

1. You throw the bowling ball at the pins. You have two throws to knock down ten pins.

First Throw: You knock down 6 pins. Second Throw: You knock down 1 pin.

2. You roll a number cube twice.

First Roll: You roll an odd number. Second Roll: You roll a number less than 2.

You spin the spinner and flip a coin. Find the probability of the compound event.

3. Spinning a 1 and flipping tails

4. *Not* spinning an even number and flipping heads

You randomly choose one of the tiles. Without replacing the first tile, you choose a second tile. Find the probability of the compound event.

5. Choosing tiles whose sum is 12

6. Choosing a 6 and then a number greater than 4

7. You randomly draw two cards from a standard deck of 52 cards. What is the probability you draw two hearts?

8. You forgot the combination for your lock. Each wheel has the numbers 0 through 9. What is the probability that you guess the combination correctly?

9. A license plate has two letters followed by three digits. What is the probability that the numbers on the license plate are all odd numbers?

Name____ Date__________

10.5 Enrichment and Extension

Winning on a Game Show

You are on a game show. You are spinning a wheel that has 20 sections, ranging from $5 to $100 in increments of $5. You win by spinning more points than your opponent in one spin or a combination of two spins. But, you lose if you exceed $100.

Your opponent spins $85 in one spin and decides not to spin again.

1. What is the probability that you tie your opponent on the first spin?

2. How many values on the wheel are greater than $85? What is the probability that you win on your first spin?

3. How many values on the wheel are less than $85? What is the probability that you spin less than your opponent on your first spin?

4. You spin $35 on your first spin.

   a. What values on the wheel would make you win on your second spin?

   b. What is the probability that the wheel lands on a winning section on your second spin?

5. You spin $60 on your first spin.

   a. What values on the wheel would make you win on your second spin?

   b. What is the probability that the wheel lands on a winning section on your second spin?

6. What do you notice about the probabilities in Exercises 2, 4(b), and 5(b)? Explain any similarity.

7. You spin less than $85 on your first spin. What is the probability that the wheel lands on a winning section on your second spin?

Your opponent's score is given. (a) Find the probability that you win after one spin. (b) Find the probability that you spin a lesser amount on your first spin, and then win on your second spin.

8. $75   9. $45   10. $90   11. $25

Name ______________________________ Date __________

Puzzle Time

What Animal Goes "Baa-Baa-Woof?"

Write the letter of each answer in the box containing the exercise number.

You roll a number cube once and flip a coin. Find the probability of the compound event.

1. Rolling a factor of 12 and flipping tails
2. Rolling a perfect square and flipping heads

You have a bag that contains 7 red marbles and 5 blue marbles. You randomly choose one of the marbles. Without replacing the first marble, you choose a second marble. Find the probability of the events.

3. Choosing a red marble and then a blue marble
4. Choosing a blue marble and then another blue marble
5. Without replacing the first and second marble, you choose a blue marble, a red marble, and then another red marble.

You are playing a treasure hunt card game that includes 8 treasure chests, 7 pirates, and 9 islands. Each player is dealt 5 cards. Before seeing any of the cards, you randomly make a guess as to which treasure chest is hidden, which pirate buried the treasure, and on which island the treasure is buried.

6. What is the probability that you got all three correct before looking at your cards?
7. You look at your cards and are able to eliminate 2 of the treasure chests, 1 of the pirates, and 2 of the islands. Now you try to guess the correct treasure chest, pirate, and island. What is the probability that you get all three correct?
8. One of your opponents looks at her cards and is able to eliminate 3 treasure chests and 2 pirates, but none of the islands. She tries to guess the correct treasure chest, pirate, and island. What is the probability that she gets all three correct?
9. Another of your opponents looks at his cards and is able to eliminate 5 treasure chests, but no pirates and no islands. He tries to guess the correct treasure chest, pirate, and island. What is the probability that he gets all three correct?

Answers

P. $\frac{5}{33}$

E. $\frac{1}{504}$

D. $\frac{1}{225}$

E. $\frac{1}{6}$

G. $\frac{7}{44}$

A. $\frac{5}{12}$

H. $\frac{35}{132}$

S. $\frac{1}{189}$

O. $\frac{1}{252}$

1		9	3	6	2	4	8	7	5

Start Thinking!

For use before Extension 10.5

A proofreader finds two mistakes in 50 papers. About how many mistakes would you expect in 1000 papers? Explain how to find the answer.

Warm Up

For use before Extension 10.5

You roll a number cube 10 times and record the results. Use the table to find the experimental probability of the event.

1. Tossing a 3

2. Tossing a 1

3. Tossing a 6

4. *Not* tossing a 4

Outcome	Frequency
1	4
2	1
3	1
4	2
5	1
6	1

Name ______________________________ Date __________

Practice

On a spinner, there is a 40% chance of spinning green and a 30% chance of spinning red. Design and use a simulation involving 100 randomly generated numbers to find the experimental probability that you will spin green on the first spin and red on the second spin.

1. Use the random number generator on a graphing calculator. Randomly generate 100 numbers from 0 to 99. The table below shows the results.

94	90	14	51	40	73	4	33	99	20
79	95	22	36	0	93	10	0	54	85
97	27	27	12	5	72	1	42	30	97
2	83	61	20	98	72	30	24	94	92
4	11	69	98	63	31	8	99	19	39
11	24	85	37	59	60	7	1	1	69
70	88	37	11	45	98	69	54	63	92
67	79	55	33	21	62	88	12	45	46
28	81	98	49	40	22	62	61	80	77
46	92	62	33	45	80	86	25	71	46

Let the digits 1 through 4 in the tens place represent green on the first spin and the digits 1 through 3 in the ones place represent red on the second spin. Any number that meets these criteria represents green on the first spin and a red on the second spin.

How many numbers meet the criteria?

2. Find the experimental probability that you spin green on the first spin and red on the second spin.

3. Try to find the theoretical probability of spinning green on the first spin and red on the second spin. What do you think happens to the experimental probability when you increase the number of trials in the simulation?

Activity 10.6 Start Thinking!

For use before Activity 10.6

Review with a partner how to determine if events are *independent* or *dependent*. Use an example.

Activity 10.6 Warm Up

For use before Activity 10.6

You flip a coin and roll a number cube. Find the probability of the event.

1. Flipping heads and rolling a 6

2. Flipping heads and rolling an odd number

3. Flipping heads and rolling a number greater than 3

4. Flipping tails and rolling a number less than 5

Lesson 10.6 Start Thinking!

For use before Lesson 10.6

You survey 20 students in your school to find their favorite summer activity. Can you make conclusions about the population of your school based on the results? Explain.

Lesson 10.6 Warm Up

For use before Lesson 10.6

Identify the population and sample.

1. residents of a city; senior residents of a city
2. members of a gym who play basketball; members of a gym
3. books in a classroom; nonfiction books in a classroom
4. travel mugs in a souvenir shop; mugs in a souvenir shop

Name____ Date__________

10.6 Practice A

Identify the population and the sample.

1. All students in a school

   30 students in the school

2. 75 strawberries in the field

   All the strawberries in the field

3. You want to know the number of students in your school who read some of the newspaper at least once a week. You survey 30 random students that you meet in the hallway between classes.

   a. What is the population of your survey?

   b. What is the sample of your survey?

   c. Is the sample biased or unbiased? Explain.

Which sample is better for making a prediction? Explain.

4.

Predict the number of residents in St. Lucie County who own a home.	
Sample A	A random sample of 100 residents in the county
Sample B	A random sample of 100 residents in the city of Fort Pierce

5.

Predict the number of people at a beach who are wearing sunscreen.	
Sample A	A random sample of 50 people at the beach
Sample B	A random sample of 5 people at the beach

Determine whether you would survey the population or a sample. Explain.

6. You want to know the average weight of the members of your family.

7. You want to know the number of grocery stores in Florida that carry your favorite cereal.

8. A survey asked 60 randomly chosen students if they eat school lunch. Forty said yes. There were 560 school lunches sold today. Predict the number of students who attend the school.

Name ______________________________ Date __________

10.6 Practice B

1. You want to know the number of fans at the Miami Dolphins and Dallas Cowboys game that think the Dolphins will win. You survey 50 fans with season tickets for the Dolphins.

   a. What is the population of your survey?

   b. What is the sample of your survey?

   c. Is the sample reasonable? Explain.

2. Which sample is better for making a prediction? Explain.

Predict the number of families in your town with two or more children.	
Sample A	A random sample of 10 families living near your home
Sample B	A random sample of 10 families living in your town

Determine whether you would survey the population or a sample. Explain.

3. You want to know the favorite clothing store of the students at your school.

4. You want to know the favorite topic of students in your history class.

5. An administrator surveys a random sample of 48 out of 900 middle school students. Using the survey results, the administrator predicts that 225 students are in favor of the new dress code. How many of the 48 students surveyed were in favor of the new dress code?

6. The table shows the results of a survey of 75 randomly chosen individuals. In the survey, each individual was asked to name his or her favorite type of music.

Music	Frequency
Rock	20
Country	23
Rap	30
Classical	2

   a. Do you think the individuals surveyed were adults or teenagers? Explain your reasoning.

   b. What other data displays could be used to show the data?

   c. If you were to repeat the survey using randomly chosen adults, would you predict that the results of the adult survey will be different if you surveyed adults in their 30s versus adults in their 70s? Explain your reasoning.

Name____ Date__________

10.6 Enrichment and Extension

The Electoral College

The President of the United States is chosen by the Electoral College. The electors usually vote for whichever presidential candidate won the popular vote in their state. The number of electors from each state is equal to the number of senators plus the number of representatives in the House. Every state has two senators, regardless of its population. The number of representatives in the House is proportional to the state's population. The District of Columbia gets three electors even though they do not have representation in Congress.

Distribution of 2004 and 2008 Electoral Votes			
State/District	**Senators**	**House Representatives**	**Electors**
California	2	53	55
District of Columbia	0	0	3
Florida	2	25	27
Louisiana	2	7	9
Montana	2	1	3
Total	**100**	**435**	**538**

1. Do the Electoral College votes represent a sample of each state's population? Explain your reasoning.

2. Your school is holding a mock election, in which all 1308 students will participate in a popular vote. Each student will be assigned to represent one of the fifty states or the District of Columbia. How many of the students in your school should represent California? District of Columbia? Florida? Louisiana? Montana?

3. The presidential election results in 2000 were delayed because the popular vote results were so close in Florida. In the end, George W. Bush won the presidency without winning the most popular votes nationwide. The presidents elected in 1824, 1876, and 1888 also did not win the most popular votes. Explain how this is possible. Research one of the four elections and summarize what you find most interesting about it.

4. Research how television stations make projections for presidential elections. Include information about pre-election polls, exit polls, and actual votes reported. Also, describe how the sample precincts are chosen. How do they ensure that the sample data is representative of the state's population?

Name ______________________________ Date __________

10.6 Puzzle Time

What Did One Tuna Say To The Other When They Were Playing Cards?

Write the letter of each answer in the box containing the exercise number.

1. You want to know what students at your school would most like to attend: a professional football, basketball, or baseball game. Which sample should you choose for your survey?

G. 5 of your friends **H.** the basketball team **I.** 25 random students

2. You survey your 22 classmates on their favorite color. Six choose green as their favorite color. There are 396 students at your school. How many students in the school do you predict would choose green as their favorite color?

S. 108 **T.** 126 **U.** 198

3. A store wants to know how good their customer service is. Who should they survey?

N. the next 3 customers **O.** 50 random customers **P.** 50 random people

4. A summer camp surveys 40 campers to see if they would take tennis next week. Twelve campers say they would. If there are 250 campers, how many campers should the counselors plan on for next week's tennis lessons?

F. 60 **G.** 65 **H.** 75

5. You want to estimate how many teens in Florida get an allowance. Who should you survey?

G. 200 random Florida teens **H.** every teen at your school **I.** every teen in Florida

6. The art teacher wants to know if her art students would like to work on pottery. Who should she survey?

E. 2 random art students **F.** all of her art class **G.** the entire school

5	3		6	1	2	4

Extension 10.6 Start Thinking!

For use before Extension 10.6

Review with a partner how to make a box-and-whisker plot.

Extension 10.6 Warm Up

For use before Extension 10.6

1. Make a box-and-whisker plot of the data.

Scores on a Science Test			
84	65	98	83
96	76	77	84
94	98	80	73

Name ______________________________ Date __________

Extension 10.6 Practice

1. Work with a partner. Mark 30 small pieces of paper with an A, a B, or a C. Put the pieces of paper in a bag. Trade bags with other students in the class.

   a. Generate a sample by choosing a piece of paper from your bag 10 times, replacing the piece of paper each time. Record the number of times you choose each letter. Repeat this process to generate five more samples. Organize the results in a table.

   b. Use each sample to make an inference about the number of As and Bs in the bag. Then describe the variation of the six inferences. Make inferences about the numbers of As, Bs, and Cs in the bag based on all the samples.

   c. Take the pieces of paper out of the bag. How do your inferences compare to the population? Do you think you can make a more accurate prediction? If so, explain how.

2. Work with a partner. You want to know the mean number of hours students in band or orchestra practice their instruments each week. Prior research indicates that the maximum number of hours of practice is 14 hours per week.

   a. Use the random number generator on a graphing calculator to simulate the hours of practice for 10 students in band or orchestra. Randomly generate 10 numbers from 0 to 14. Write down the results. Repeat this 9 more times, writing down the results each time.

   b. Find the mean of each of the 10 samples.

   c. Make a box-and-whisker plot of the sample means.

   d. Use the box-and-whisker plot to estimate the actual mean number of hours students in band or orchestra practice their instruments each week. How does your estimate compare to the mean of the entire data set?

Activity 10.7

Start Thinking!

For use before Activity 10.7

How do you find the mean of a data set? Explain by using an example.

How do you find the median of a data set? Explain by using an example.

Warm Up

For use before Activity 10.7

Find the median.

1. 5, 7, 8, 8, 12, 12, 14, 17, 21

2. 56, 57, 57, 62, 65, 65, 65

3. 23, 34, 35, 37, 41, 43, 43

4. 76, 77, 78, 78, 79, 81, 83, 85

5. 43, 45, 32, 34, 42, 38, 35

6. 65, 67, 62, 61, 69, 65, 68, 66

Lesson 10.7 Start Thinking!

For use before Lesson 10.7

Give an example of a data set that represents two populations that can be compared.

How would you compare that data of the two populations?

Lesson 10.7 Warm Up

For use before Lesson 10.7

The tables show the numbers of baskets made by two basketball teams.

Team 1				
45	52	65	56	70
56	58	49	55	64

Team 2				
52	56	65	72	49
58	49	62	63	54

1. Find the mean, median, mode, range, interquartile range, and mean absolute deviation for each data set.

2. Compare the data sets.

Name______________________________ Date__________

10.7 Practice A

1. The tables show the ages of the players on two basketball teams.

Varsity Team Ages					
18	16	17	16	18	17
19	18	18	18	18	17

Junior Varsity Team Ages					
16	17	15	16	17	15
18	14	17	16	17	17

a. Find the mean, median, mode, range, interquartile range, and mean absolute deviation for each data set.

b. Compare the data sets.

c. When comparing the two populations using measures of center and variance, would you use the mean and the MAD, or the median and the IQR? Explain.

d. Express the difference in the measures of center as a multiple of the measure of variation.

2. The double box-and-whisker plot shows the number of inches of snow per week in two cities in a 16-week period.

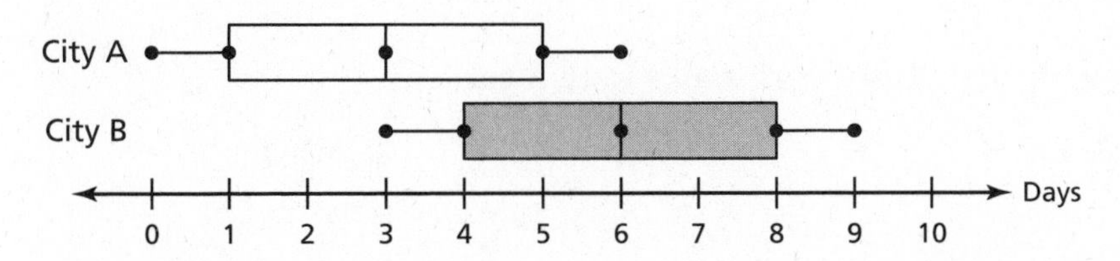

a. Compare the populations using measures of center and variation.

b. Express the difference in the measures of center as a multiple of the measure of variation.

Name ______________________________ Date __________

10.7 Practice B

1. The tables show the numbers of attendees at pep rallies for football and basketball games at a school during the year.

Football Pep Rally Attendance					
174	175	200	169	178	171
165	187	159	170	184	196
205	231	198	310	152	178

Basketball Pep Rally Attendance					
143	178	154	167	204	199
254	147	179	162	189	203
217	214	187	210	288	287

a. Find the mean, median, mode, range, interquartile range, and mean absolute deviation for each data set.

b. Compare the data sets.

c. When comparing the two populations using measures of center and variation, would you use the mean and the MAD, or the median and the IQR? Explain.

d. Express the difference in the measures of center as a multiple of the measure of variation.

2. The dot plots show the heights of corn stalks in two gardens.

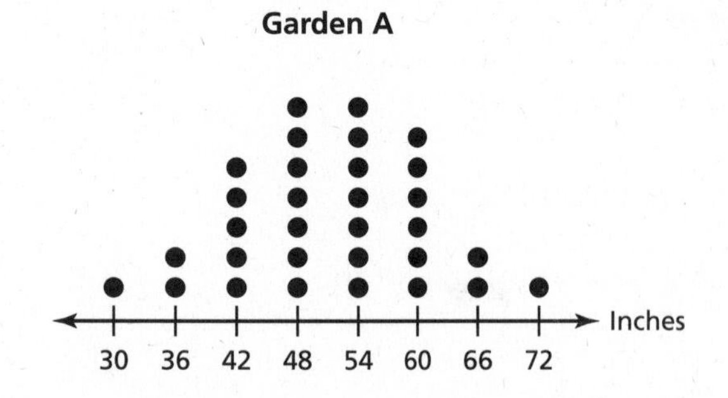

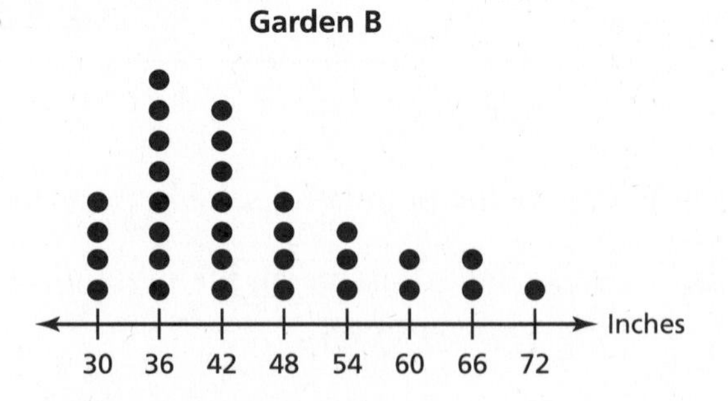

a. Compare the populations using measures of center and variation.

b. Express the difference in the measures of center as a multiple of the measure of variation.

Name______________________________ Date__________

10.7 Enrichment and Extension

Should I Keep Playing?

The **expected value** of an event is equal to the average of its outcomes—as long as all of the outcomes have an equal probability of occurring.

Example: John is playing a game where he rolls number cubes one at a time and adds the values to obtain a sum. The person scoring closest to 8 without going over wins the round. Each player can decide to continue his or her turn after two rolls. John's sum after two rolls is 5. Use the expected value of a number cube to determine if John should roll again.

$$\frac{1 + 2 + 3 + 4 + 5 + 6}{6} = 3.5$$ The average of the outcomes is the expected value.

The expected value of a number cube is 3.5, so John should not roll again because $5 + 3.5 = 8.5$. The sum 8.5 would put John over the limit of 8.

Use expected value to help each person decide what to do in the situation described.

1. Daulton is playing a game where he draws a random card from a deck with no face cards (Jack, Queen, King, Ace, or Joker). He receives points equal to the value of the card. If Daulton gets more than 4 points on his next turn he loses the game. Should Daulton draw a card or pass?

2. Ally is playing a card game with a friend. Each player draws a card, and the card with the higher value wins the round. Her friend's card is a 5. Ally has cards of 2, 5, 6, and 8 in her pile but does not know which is next. Should Ally be confident she will win the round?

3. Paxton must spin a value of 6 or greater on his next turn or he is out for the following round. The spinner has only even numbered sections from 2 through 10. Should he spin or pass?

4. Find the probability of a success in Exercises 1–3. Would using probability rather than expected value change your advice to each person? Explain.

5. Are expected value and probability the same thing? Which is better for predicting a success? Explain.

Name ______________________________ Date __________

10.7 Puzzle Time

What Kind of House Weighs The Least?

Write the letter of each answer in the box containing the exercise number.

The dot plots show the numbers of books that students read during the school year for two classes.

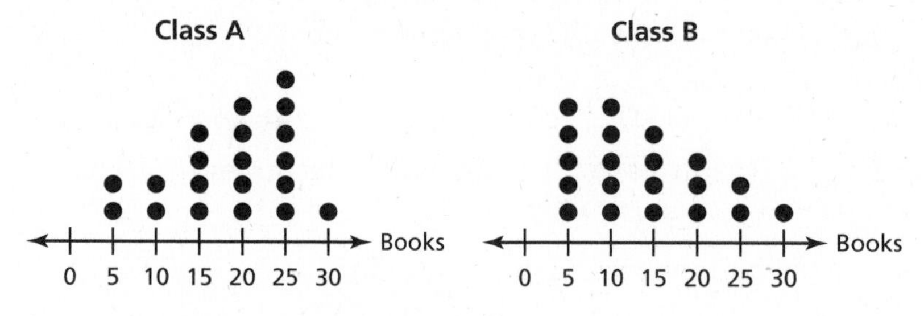

1. What is the median for Class A?

2. What is the median for Class B?

3. What is the IQR for Class A?

4. What is the IQR for Class B?

5. Compare the populations using measures of center and variation.

The tables show the numbers of books that have been signed out of the library during the school year for two classes.

Class A (Books)			
8	6	8	4
8	12	10	6
6	2	10	12
14	10	4	8

Class B (Books)			
12	12	14	9
12	12	13	9
12	7	9	13
12	6	14	10

6. What is the mean for Class A?

7. What is the mean for Class B?

8. What is the MAD for Class A?

9. What is the MAD for Class B?

10. Compare the populations using measures of center and variation.

Answers

U. 20

H. 12.5

T. 2.5

G. 2

H. 12.5

O. 8

S. 11

E. 10

I. The variation in the number of books is the same but Class A has a greater number of books.

L. The variation in the number of books is the same but Class B has a greater number of books.

10	5	9	4	8	2	6	1	7	3

Name____ Date__________

Chapter 10

Technology Connection

For use after Section 10.5

Exploring Internet-based Simulations

Although calculating the experimental probability of an event by hand may give you an intuition of the expected outcome, many times you may find that it is too time consuming to perform the simulation a high number of times. Fortunately, a computer is perfectly suited to perform these repeated simulations in a very small amount of time. The simulation for this lesson can be found at the National Library of Virtual Manipulatives in the Data Analysis & Probability section at http://nlvm.usu.edu.

EXAMPLE Find the experimental probability of rolling a 5 on a number cube using 500 trials.

SOLUTION

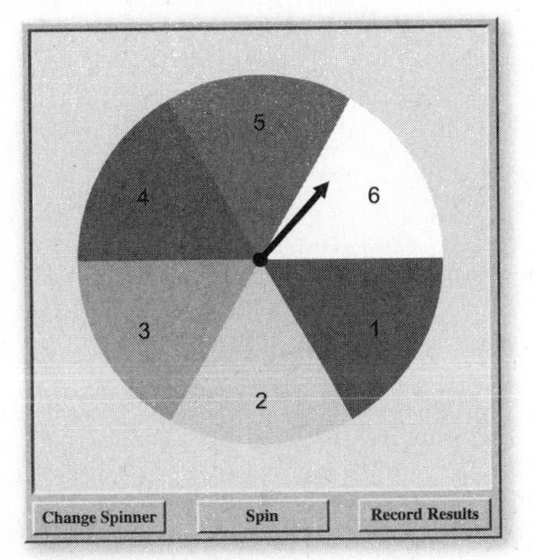

Step 1 In the Data Analysis & Probability section at the NLVM website, choose the "Spinners" activity.

Step 2 Customize the spinner by clicking the "Change Spinner" option. Then rename each category "1," "2,"…"6" to represent the numbers on a number cube. When you add the "6" category, be sure to change the spinner section from 0 to 1. Then click "Apply."

Step 3 Although you could then click the "Spin" button 500 times, it is much more efficient to click the "Record Results" button and then set the number of spins to 500. Then click "Spin."

Step 4 Your spinner will show the results of your 500 spins in a bar graph in the Results window.

Use a spinner simulation to answer the following questions.

1. Change the spinner so that it represents a coin toss consisting of two outcomes, heads or tails. Run a simulation of 1000 spins and record the results. If you divide the spinner into 4 sections (heads, tails, heads, tails), do you achieve similar results? Explain your results in terms of area.

2. Change the spinner so that it models the sums of rolling two number cubes. Weight the sections according to the theoretical probability of achieving each sum. Run a simulation of 1000 spins and compare the results to the theoretical probability of the trials.

Projects

Name____ Date__________

Project Currency Exchange

For use after Chapter 2

Objective **To convert prices from U.S. dollars to other currencies.**

Materials Newspaper or Internet access

Investigation Have you ever wanted to travel to another country? Countries around the world use different currencies. The peso, pound, yen, and euro are a few of the currencies used in other countries. You can find out the cost of an item in another country by using the current exchange rate.

Explore Think of two countries in different parts of the world that you would like to visit. Research what type of currency these countries use and what it looks like. Write a brief paragraph stating the two countries you chose, their currency, and why you would like to visit each country. Next, think of three products that you would like to purchase. Look in newspaper advertisements or on the Internet to find their prices.

Calculate

1. Look in the business section of the newspaper or on the Internet to find the exchange rates for the two countries you have chosen. You want to find the rate that converts U.S. dollars to foreign currency.

2. Use the following formula to convert the price in U.S. dollars to your chosen currencies. Money should be rounded to the nearest hundredth.

   U.S. dollars $\times$ exchange rate $=$ foreign currency equivalent

   Show your work for all calculations. Organize your data in a table.

	Cost in U.S. $	Cost in Currency 1	Cost in Currency 2
Item #1: ________			
Item #2: ________			
Item #3: ________			

3. Could you tell by the exchange rate if the price was going to be a higher number or a lower number in the foreign currency? How?

4. Why is it important to know the exchange rate if you are going to purchase something in a foreign country?

5. Explain how you could convert a price in foreign currency to U.S. dollars. Write an equation for this conversion using a product from each country.

Create a Poster Summarize your findings on a poster. Include a map to show the location of the countries you chose. You may want to include photos of the items you will purchase and the currency.

Name ____________________ Date ________

Project Student Grading Rubric

For use after Chapter 2

	Student Score	Teacher Score
Poster Information 10 points		
a. Name (4 points)	_____	_____
b. Class (2 points)	_____	_____
c. Project Name (2 points)	_____	_____
d. Due Date (2 points)	_____	_____
Explore 40 points		
a. Choose 2 countries. (5 points)	_____	_____
b. Identify the currency of two countries. (10 points)	_____	_____
c. State why you would like to visit each of the two countries. (5 points)	_____	_____
d. Find three items to purchase. (5 points)	_____	_____
e. Find the price of each item. (15 points)	_____	_____
Calculate 90 points		
a. State the exchange rate for converting U.S. dollars to foreign currency. (10 points)	_____	_____
b. Convert the prices in U.S. dollars to foreign currency. Show all work. (40 points)	_____	_____
c. Organize your data in a table. (30 points)	_____	_____
d. Present your results neatly. (10 points)	_____	_____
Poster 60 points		
a. Includes information about the country chosen, why the country was chosen, and information about the products you would purchase. (20 points)	_____	_____
b. Presents the conversion table and the answers to all the questions in a clear and readable way. (20 points)	_____	_____
c. Locates each country on a map and includes other visuals, such as images of products or currencies. (20 points)	_____	_____
FINAL GRADE	_____	_____

Project Teacher's Project Notes

For use after Chapter 2

Materials Newspaper or Internet access, poster board; optional: scissors and travel magazines, if cutting out images for the poster in class

Alternatives Students could look on the Internet to find prices in foreign currencies and convert the price to U.S. dollars. In this case, students can think of a product they would like to buy in another country to remember their visit. This can lead students to research the products for which the country is known. What might you want to buy in the countries you chose that you cannot buy in the United States?

Common Errors Students may find the exchange rate for converting a foreign currency to U.S. dollars instead of from U.S. dollars to a foreign currency.

Students may forget to round money to the nearest hundredth.

Suggestions Discuss the topic of exchange rates and do several calculations together in class before students begin the project. You may want to discuss why exchange rates can have many decimal places, but that the monetary amounts need to be rounded to the nearest hundredth.

An extensive list of exchange rates can be found on the Internet through the United States Department of the Treasury website at www.fms.treas.gov. There are also many other websites that contain current exchange rates.

Students could share their project with the class.

You may want to discuss the project with teachers in your social studies department ahead of time. This could be a nice opportunity for cross-curricular teaching. There may be some topics that students are studying in social studies that can be applied to their project.

Students could find out how the exchange rate has fluctuated in the past year, and calculate the old and new prices for an item.

Grading Rubric
For use after Chapter 2

Scoring Rubric	
A	179–200
B	159–178
C	139–158
D	119–138
F	118 or below

Poster Information 10 points

a. Name (4 points)

b. Class (2 points)

c. Project Name (2 points)

d. Due Date (2 points)

Explore 40 points

a. Choose 2 countries. (5 points)

b. Identify the currency of both countries. (10 points)

c. State why you would like to visit each of the two countries. (5 points)

d. Find three items to purchase. (5 points)

e. Find the price of each item. (15 points)

Calculate 90 points

a. State the exchange rate for converting U.S. dollars to foreign currency. (10 points)

b. Convert the prices in U.S. dollars to foreign currency. Show all work. (40 points)

c. Organize your data in a table. (30 points)

d. Present your results neatly. (10 points)

Poster 60 points

a. Includes information about the country chosen, why the country was chosen, and information about the products you would purchase. (20 points)

b. Presents the conversion table and the answers to all the questions in a clear and readable way. (20 points)

c. Locates each country on a map and includes other visuals, such as images of products or currencies. (20 points)

Name____ Date__________

Project Race Course Rates

For use after Chapter 5

Objective **Measure rates and analyze data.**

Materials Masking tape, yardstick or ruler, stopwatch or clock with a second hand, calculator, (optional: camera to take pictures of people completing the course)

Investigation Work with a partner. Use tape to mark a starting point on the floor or sidewalk. From the starting point, measure a straight line 10 feet long. Mark this distance with the tape. With the tape, label each 1-foot interval.

Think of 10 different methods you can use to "walk" the course. Some possibilities include spinning, walking backwards, crawling, hopping on one foot, etc. List the methods you choose in a table like the one below.

One partner should complete the course while the other partner measures the time it takes to complete the course to the nearest second. For each method, record the time to complete the course in the table. Then calculate the unit rate for each method. Switch roles and repeat the process.

	Partner 1		Partner 2	
Method	**Time (sec)**	**Unit rate (ft/sec)**	**Time (sec)**	**Unit rate (ft/sec)**
walk toe-to-heel				
spin				
hop on one foot				
....				

Data Analysis

- For each method, what are some things that affect the time?
- Order the rates for each person from slowest to fastest. Which method took the most time? Which method took the least amount of time? Were the slowest and fastest methods the same for both people?
- Find the mean, median, mode, and range of the data for each person and for the combined data. Are there any significant differences? Explain.
- Which measure of center best represents each person's data? the combined data? Why?
- Identify any outliers in the data sets. Does an outlier affect the mean? the median? Explain.
- Use your own data and the mean of your data. What percent of your rates were slower than your mean? faster than your mean?

Summary Make a poster to summarize your results. Include sketches, the answers to the questions above, and any other patterns or observations you made. If you took photos of someone completing the course, include them.

Name ______________________________ Date __________

Project

Student Grading Rubric

For use after Chapter 5

		Student Score	Teacher Score
Poster Information 10 points			
	a. Name (4 points)	_____	_____
	b. Class (2 points)	_____	_____
	c. Project Name (2 points)	_____	_____
	d. Due Date (2 points)	_____	_____
Introduction 20 points			
	a. Explain how you performed the experiment and collected the data, so that someone not in the class would understand what you did. (20 points)	_____	_____
Data Table 30 points			
	a. Data table is present, completely filled out, and is neat and easy to read. (10 points)	_____	_____
	b. Rates are calculated correctly. (20 points)	_____	_____
Analysis 100 points			
	a. Rates are ordered correctly and the fastest and slowest rates are correctly identified. (10 points)	_____	_____
	b. Means are correctly calculated. (10 points)	_____	_____
	c. Medians are correctly calculated. (10 points)	_____	_____
	d. Modes are correctly calculated. (10 points)	_____	_____
	e. Ranges are correctly calculated. (10 points)	_____	_____
	f. Analysis of parts (b)–(e). (20 points)	_____	_____
	g. Outliers identified and discussed. (10 points)	_____	_____
	h. Percents calculated and discussed. (20 points)	_____	_____
Poster 40 points			
	a. The poster includes a brief explanation of how you gathered the data, your table, an analysis of your data, visuals, answers to the questions above, and other observations and comparisons. (30 points)	_____	_____
	b. Present the information in a neat and organized way. (10 points)	_____	_____
FINAL GRADE		_____	_____

Project

Teacher's Project Notes

For use after Chapter 5

Materials Masking tape, yardstick or meterstick; You will need a stopwatch, watch, or wall clock that measures to the nearest second, enough for each pair of students. You may want to give students a calculator to make the calculations easier.

Alternatives Work in larger groups, or as an entire class. Create a larger data set by timing many more methods of completing the course, perhaps a different one for each student.

Or choose two or three methods and measure the time for every student to complete the course using each method. Perform data analysis for each method: for example, find the mean, median, mode, and range of the rates for rolling along the course, for toe-to-heel walking, and for hopping. Then compare the analyses for different motions—e.g. which one has the greatest range? Which has the least?

Working with a large group, you could create an obstacle course and measure each student's time to complete the course. It would also be instructive to measure each student's time on two trials, and compare the two means—does practice help?

If you have the space, you can create a longer course. If you have both slow methods (heel-to-toe walk) and fast (sprint), the range of the rates will be greater. A shorter course encourages clustering of times and rates.

Common Errors Watch for students to set up the fraction properly to find the rates: $\frac{10 \text{ ft}}{\text{time}}$.

The longest time will have the smallest rate, and the shortest time will have the greatest rate.

Many data sets will not have a mode.

Suggestions Marking each foot along the track will make it easier for students to estimate the rate, e.g. 1 foot in 2 seconds while hopping.

Before students calculate the percents, you may want to discuss what percent of data students think will be above the median (50%) versus above the mean (could be 50%, or more, or less). Then compare students' predictions with their actual results.

Project

Grading Rubric

For use after Chapter 5

Scoring Rubric	
A	179–200
B	159–178
C	139–158
D	119–138
F	118 or below

Poster 10 points

a. Name (4 points)

b. Class (2 points)

c. Project Name (2 points)

d. Due Date (2 points)

Introduction 20 points

a. Explain how you performed the experiment and collected the data, so that someone not in the class would understand what you did. (20 points)

Data Table 30 points

a. Data table is present, completely filled out, and is neat and easy to read. (10 points)

b. Rates are calculated correctly. (20 points)

Analysis 100 points

a. Rates are ordered correctly and the fastest and slowest rates are correctly identified. (10 points)

b. Means are correctly calculated. (10 points)

c. Medians are correctly calculated. (10 points)

d. Modes are correctly calculated. (10 points)

e. Ranges are correctly calculated. (10 points)

f. Analysis of parts (b)–(e). (20 points)

g. Outliers identified and discussed. (10 points)

h. Percents calculated and discussed. (20 points)

Poster 40 points

a. The poster includes a brief explanation of how the data was gathered, a completed table, an analysis of the data, visuals, and answers the questions above. (30 points)

b. The information is presented in a neat and organized way. (10 points)

Name____ Date__________

Project Composite Figures and Design

For use after Chapter 8

Objective **Use straight lines and parts of circles in a design. Then apply the formulas from the chapter to find areas and perimeters and make a model of your design.**

Materials Models may need paper, cardboard, foam, poster board, markers, fabric, wood, etc.

Investigation Many designs use circles, parts of circles, and straight lines. Design a picture, mural, garden, apartment, park, swimming pool, or other object using straight lines, circles, and parts of circles. Make a sketch of your design.

Find at least 2 perimeters and 2 areas of shapes in your design. Explain how you can apply these calculations. (For example, to know how many tiles to buy for a bathroom, or how much fencing is needed to enclose a garden, etc.)

Decide whether you know how to find all the distances in your sketch. You may be able to ask your teacher for help with finding areas of unusual shapes.

Decide how to make your model. What materials will you use?

If you have a more complex design, you may want to simplify it.

Completing the Project Turn in your sketches, your model, and a report explaining what you created. Explain why you made the design choices you did. Then give the perimeters and areas you found, and explain why you might want to know them.

Name ______________________________ Date __________

Project Student Grading Rubric

For use after Chapter 8

		Student Score	Teacher Score
Model 100 points			
	a. Name (5 points)	_____	_____
	b. Project Name (5 points)	_____	_____
	c. The sketches are neat and well-labeled. (40 points)	_____	_____
	d. The model is neat and well made. (40 points)	_____	_____
	e. The model includes both straight lines and circles or parts of circles. (10 points)	_____	_____
Report 100 points			
	a. Name (4 points)	_____	_____
	b. Class (2 points)	_____	_____
	c. Project Name (2 points)	_____	_____
	d. Due Date (2 points)	_____	_____
	e. Explanation of what the project represents. (20 points)	_____	_____
	f. Explanation of design—why various distances or shapes were used. (20 points)	_____	_____
	g. Perimeters are found correctly. (15 points)	_____	_____
	h. Explanation of why you would want to know each perimeter makes sense. (10 points)	_____	_____
	i. Areas are found correctly. (15 points)	_____	_____
	j. Explanation of why you would want to know each area makes sense. (10 points)	_____	_____
FINAL GRADE		_____	_____

Project

Teacher's Project Notes

For use after Chapter 8

Materials The materials students use for their models will depend on the type of item they design. Graph paper and rulers may be helpful for all students. For a garden or building, students may need cardboard for the base and structure, pompoms for bushes, pipe cleaners to mark borders, sugar cubes to build a fountain, etc. Encourage students to be creative.

Alternatives Students could create a full-scale project at school or at home or this could be a class project: design a maze, have the class choose one to model, determine how much material they need to lay it out (e.g. rope on the athletic field, masking tape on the floor of the gym), and then build the full-scale version.

A student could find a real-life example in a reference book or online—e.g. a garden, a circular building, a corn maze. He or she could then determine the perimeters and areas.

Common Errors Watch for students who may be over-ambitious and create a design that is too complex. Requiring each student to turn in a rough sketch before they complete the project will help you control this. Does he or she know how to find the area of an octagon? Are the shapes or design too complicated to model? Will the calculations take too long?

Suggestions You may want to introduce the project with a slide show or images showing numerous designs to help spark ideas. For example, photos of gardens with straight or curved walls and flower beds, a floor plan for a house, mosaics, geometric sculptures, etc.

Discuss how to limit the project to make it doable: A studio apartment with a rectangular common room, quarter-circle kitchen island, and square bathroom is about the right size; a 12-bedroom dream house is too complex. A mural should feature interesting geometric shapes they can work with, not a realistic rendering of dolphins. (Though dolphins in one circle, manatees in a larger circle, and turtles in a third circle could work.)

Make clear what level of effort and detail you expect in the model.

For the report, students may be unsure about what explanations to include. You may want to give students some examples. A shortened sample:

> In my design of a pool, I included a lap pool, a diving pool, and a spa. I wanted to add something more exciting, so I put in the waterfall and made the diving pool a half-circle....
>
> I wanted the lap pool to be 50 feet long so I could easily figure out how many 100-yard laps I had done. The lap pool is 6 feet wide so two people can swim at once. I tried different sizes for the spa and 8 feet across seemed to fit best....

Grading Rubric

For use after Chapter 8

Scoring Rubric	
A	179–200
B	159–178
C	139–158
D	119–138
F	118 or below

Model 100 points

a. Name (5 points)

b. Project Name (5 points)

c. The sketches are neat and well-labeled. (40 points)

d. The model is neat and well-made. (40 points)

e. The model includes both straight lines and circles or parts of circles (10 points)

Report 100 points

a. Name (4 points)

b. Class (2 points)

c. Project name (2 points)

d. Due date (2 points)

e. Explanation of what the project represents. (20 points)

f. Explanation of design—why various distances and shapes were used. (20 points)

g. Perimeters are found correctly. (15 points)

h. Explanation of why you would want to know each perimeter makes sense. (10 points)

i. Areas are found correctly. (15 points)

j. Explanation of why you would want to know each area makes sense. (10 points)

Name____ Date__________

Color Study

For use after Chapter 10

Objective **Compare experimental and theoretical probabilities using colored tiles.**

Materials One bag containing tiles of different colors, paper and pencil

Investigation In this project, you will use probability to find the chance of drawing each color tile out of a bag.

Theoretical Probability

- Count the total number of tiles in the bag.
- Separate the tiles by color and count the number of each color. Record the number of each color in a table like the one below.

Color	Number of tiles	Probability (fraction)	Probability (percent)

- Calculate the theoretical probability of drawing each color by dividing the number of each color by the total number in the bag. Write each probability as a percent. Record these amounts in the table.

Experimental Probability

- Randomly draw one tile out of the bag, record the color drawn in a table like the one above, and return the tile to the bag. Repeat this process 50 times.
- Calculate the experimental probability of drawing each color by dividing the number of each color drawn by the total number of draws. Write each probability as a percent.

Data Analysis

1. What color had the highest theoretical probability? Did the same color also have the highest experimental probability?
2. Compare the theoretical probability of drawing each color to the experimental probability of drawing each color. Are your experimental results exactly, close to, or far from your theoretical prediction? How could you change the experiment to make your results more accurate?
3. Make a histogram of the data in each table. For the table showing the theoretical frequency, use the number of tiles in the bag for the total. For the experimental frequency, use the 50 trials as the total.

Written Summary Write a summary of your results.

Name ____ Date __________

Student Grading Rubric

For use after Chapter 10

		Student Score	Teacher Score
Cover Page	10 points		
	a. Name (4 points)	______	______
	b. Class (2 points)	______	______
	c. Project Name (2 points)	______	______
	d. Due Date (2 points)	______	______
Investigation	Theoretical Probability (45 points)		
	a. Table is clear and easy to read. (10 points)	______	______
	b. Correctly calculate the theoretical probability of drawing each color. (25 points)	______	______
	c. Write each probability as a percent. (10 points)	______	______
	Experimental Probability (45 points)		
	a. Table is clear and easy to read. (10 points)	______	______
	b. Correctly calculate the experimental probability of drawing each color. (25 points)	______	______
	c. Write each probability as a percent. (10 points)	______	______
Data Analysis	70 points		
	a. Answer the questions clearly and correctly. (20 points)	______	______
	b. Construct a histogram for the theoretical frequency of each color. (25 points)	______	______
	c. Construct a histogram for the experimental frequency of each color. (25 points)	______	______
Written Summary	30 points		
	Clearly describe your results and the conclusions you drew in one or two paragraphs. (30 points)	______	______
FINAL GRADE		______	______

Project

Teacher's Project Notes

For use after Chapter 10

Materials Each student needs a bag filled with about 30–40 colored tiles, such as pattern blocks. There should be at least three different colored tiles in each bag. Four or five different colors is ideal.

Alternatives This project could also be done as an in-class activity. To expand the sample size, use the entire classes' data to calculate the theoretical probability. This will result in more accurate results and a greater chance for class discussion.

Instead of tiles, you could use squares of colored paper, a colored candy, or dry beans with the first letter of the color on the bean. (R for red, B for blue, G for green, Y for yellow, etc.)

Common Errors In the experimental probability, students may forget to return the tile to the bag after each draw. Stress that each draw is an independent event and it is important that they return the tile to the bag after each draw.

Remind students that they need to know the number of tiles in the bag for the theoretical probability and the total number of "trials" or "draws" for the experimental probability.

Suggestions You should assemble the bags with the tiles, beans, or candy ahead of time. Resealable plastic bags are good to use.

Any item with multiple colors will work. If you decide to use colored candies, remind students not to eat any of the candies until the investigation is over. (Ask how this would change their results.)

Grading Rubric

For use after Chapter 10

Scoring Rubric	
A	179–200
B	159–178
C	139–158
D	119–138
F	118 or below

Cover Page 10 points

a. Name (4 points)

b. Class (2 points)

c. Project Name (2 points)

d. Due Date (2 points)

Investigation Theoretical Probability (45 points)

a. Table is clear and easy to read. (10 points)

b. Correctly calculate the theoretical probability of drawing each color. (25 points)

c. Write each probability as a percent. (10 points)

Experimental Probability (45 points)

a. Table is clear and easy to read. (10 points)

b. Correctly calculate the experimental probability of drawing each color. (25 points)

c. Write each probability as a percent. (10 points)

Data Analysis 70 points

a. Answer the questions clearly and correctly. (20 points)

b. Construct a histogram for the theoretical frequency of each color. (25 points)

c. Construct a histogram for the experimental frequency of each color. (25 points)

Written Summary 30 points

Clearly describe your results and the conclusions you drew in one or two paragraphs. (30 points)

Credits

7 Buki Toys LTD; **8** Adapted from Singleton, Glen. *1001 Cool Jokes*. Hinkler Books Pty Ltd., 2003; **14** Singleton, Glen. *1001 Cool Jokes*. Hinkler Books Pty Ltd., 2003; **20** Adapted from Singleton, Glen. *1001 Cool Jokes*. Hinkler Books Pty Ltd., 2003; **26** *The Youth Online Club – 101 Jokes for Kids*. [Cited 12 January 2009] Available from *www.youthonline.ca/101thingstodo/jokes1-10.shtml*; **31** Buki Toys LTD; **32** Adapted from *YAHOO Kids*. [Cited 12 January 2009] Available from *www.kids.yahoo.com/jokes*; **42** *The Youth Online Club – 101 Jokes for Kids*. [Cited 12 January 2009] Available from *www.youthonline.ca/101thingstodo/jokes61-70.shtml*; **48, 54** Keller, Charles. *Best Riddle Book Ever*. Sterling Publishing Company, 1997; **60** Singleton, Glen. *1001 Cool Jokes*. Hinkler Books Pty Ltd., 2003; **70** Weigle, Oscar. *Great Big Joke and Riddle Book*. Grosset and Dunlap, 1981; **76** Keller, Charles. *Best Riddle Book Ever*. Sterling Publishing Company, 1997; **84** Singleton, Glen. *1001 Cool Jokes*. Hinkler Books Pty Ltd., 2003; **89** Buki Toys LTD; **90** *Jokes and Funny Stories*. [Cited 12 January 2009] Available from *www.thejokes.co.uk/*; **96** Singleton, Glen. *1001 Cool Jokes*. Hinkler Books Pty Ltd., 2003; **106** O'Donnell, Rosie. *Kids are Punny: Jokes Sent By Kids to The Rosie O'Donnell Show*. Warner Books, June 1997; **112** Rosenbloom, Joseph. *The World's Best Sport Riddles and Jokes*. Sterling Publishing Company, 1988; **118, 124** O'Donnell, Rosie. *Kids are Punny: Jokes Sent By Kids to The Rosie O'Donnell Show*. Warner Books, June 1997; **134** *The Youth Online Club – 101 Jokes for Kids*. [Cited 12 January 2009] Available from *www.youthonline.ca/101thingstodo/jokes61-70.shtml*; **140** Singleton, Glen. *1001 Cool Jokes*. Hinkler Books Pty Ltd., 2003; **148** *National Geographic Kids*, April 2008, 15; **154** *Adapted from Kids Jokes*. [Cited 12 January 2009] Available from *www.kidsgoals.com/kids-jokes.html*; **160** Adapted from *Buzzle.com*. [Cited 12 January 2009] Available from *www.buzzle.com/articles/funny-riddles-forkids-riddles-answers.html*; **166** Adapted from Singleton, Glen. *1001 Cool Jokes*. Hinkler Books Pty Ltd., 2003; **176** Walton, Rick. *Really, Really Bad School Jokes*. Candlewick, August 1995; **182** Weitzman, Ilana, Eva Blank and Roxanne Green. *Jokelopedia: The Biggest, Best, Silliest, Dumbest, Joke Book Ever!*. Workman Publishing Company, October 2000; **188** Rosenbloom, Joseph. *Biggest Riddle Book in the World*. Sterling Publishing Company, 1976; **194** Adapted from *Man in the Moon*. [Cited 12 January 2009] Available from *www.maninthemoon.co.uk/jokes/html*; **200** Adapted from *Ducksters*. [Cited 12 January 2009] Available from *www.ducksters.com/jokesforkids/silly.php*; **206, 212** Singleton, Glen. *1001 Cool Jokes*. Hinkler Books Pty Ltd., 2003; **222** Horsfall, Jacqueline. *Kids' Silliest Jokes*. Sterling, March 2003; **228** Hall, Katie and Lisa Eisenberg. *Buggy Riddles*. Dial Books for Young Readers, 1986; **234** Walton, Rick. Really, *Really Bad School Jokes*. Candlewick, August 1998; **242, 248** Singleton, Glen. *1001 Cool Jokes*. Hinkler Books Pty Ltd., 2003; **258** Pellowski, Michael. *Wackiest Jokes in the World*. Sterling Publishing, March 1995; **264, 270, 276** O'Donnell, Rosie. *Kids are Punny: Jokes Sent By Kids to The Rosie O'Donnell Show*. Warner Books, June 1997; **286** Rissinger, Matt and Philip Yates. *Great Book of Zany Jokes*. Sterling Publishing Company, 1994; **292** Singleton, Glen. *1001 Cool Jokes*. Hinkler Books Pty Ltd., 2003; **298** Adapted from Singleton, Glen. *1001 Cool Jokes*. Hinkler Books Pty Ltd., 2003; **304** Helmer, Marilyn and Jane Kurisu. *Funtime Riddles*. Kids Can Press, Ltd., February 2004; **310** Walton, Rick. *Really, Really Bad School Jokes*. Candlewick, August 1998; **322** Singleton, Glen. *1001 Cool Jokes*. Hinkler Books Pty Ltd., 2003; **328** Adapted from Yoe, Craig. *Mighty Big Book of Riddles*. Price Stern Sloan, July 2001; **334** Walton, Rick. *Really, Really Bad School Jokes*. Candlewick, August 1998; **340** Rosenbloom, Joseph. *Biggest Riddle Book in the World*. Sterling Publishing Company, 1976; **346** Rissinger, Matt and Philip Yates. *Totally Terrific Jokes*. Sterling, June 2001; **354** Helmer, Marilyn and Jane Kurisu. *Funtime Riddles*. Kids Can Press, Ltd., February 2004; **362** Rosenbloom, Joseph. *Biggest Riddle Book in the World*. Sterling Publishing Company, 1976

Answers

Chapter 1

1.1 Start Thinking!

For use before Activity 1.1

Sample answer: In football, there is an order to the game. For example, one team must kick-off before the other team can begin their drive, they must score a touchdown before kicking the extra point. In math problems, there is the order of operations. Students must first multiply or divide before doing addition and subtraction.

1.1 Warm Up

For use before Activity 1.1

1. $<$ **2.** $>$ **3.** $>$

4. $<$ **5.** $>$ **6.** $<$

1.1 Start Thinking!

For use before Lesson 1.1

Sample answer: Going to school is like moving the positive direction on a number line. Returning home after school is like moving the negative direction on a number line. No, you never travel a negative direction.

1.1 Warm Up

For use before Lesson 1.1

1. 15 **2.** 23 **3.** 7 **4.** 35

5. 43 **6.** 0 **7.** 39 **8.** 212

1.1 Practice A

1. 7 **2.** 12 **3.** 13 **4.** 0

5. $>$ **6.** $=$ **7.** $>$ **8.** $-5, 3$

9. $-1, |-1|, |4|, |5|, 8$ **10.** $0, |2|, |3|, |5|, 6$

11. 19 **12.** -8 **13.** -13

14. a. 2 **b.** 2 ft per sec **c.** positive **d.** 2 ft per sec

15. a. LATE **b.** TEAL **16.** *Sample answer*: -5

1.1 Practice B

1. $=$ **2.** $<$ **3.** $<$ **4.** $15, -6$

5. $-|-34|, |0|, 14, |-25|, 28$

6. $-16, 10, |-16|, |25|, |-43|$

7. 249 **8.** -183 **9.** -153

10. a. Phosphorus **b.** Oxygen

11. a. up **b.** 13 ft/sec **c.** down **d.** 17 ft/sec

12. 0

13. true; Both numbers have an absolute value of 3.

14. false; *Sample answer:* Let $x = -4$. Then $|x| = 4$ and 4 is not less than -4.

1.1 Enrichment and Extension

1. always **2.** never **3.** sometimes

4. sometimes **5.** never **6.** always

7. sometimes **8.** sometimes

9. all negative integers **10.** all positive integers

11. none **12.** all integers **13.** none

14. all integers **15.** all integers **16.** none

17. all integers **18.** none

19. all negative integers and zero

20. all negative integers and zero

21. all positive integers and zero

22. all positive integers and zero

23. all positive integers and zero

24. all negative integers and zero

25. none **26.** all negative integers

27. all positive integers **28.** all positive integers

29. none **30.** all integers

31. all positive integers **32.** all integers

33. all positive integers **34.** all integers

35. all integers **36.** all integers

Check students' picture; It should be a dog.

1.1 Puzzle Time

A TENNIS BALL

1.2 Start Thinking!

For use before Activity 1.2

Sample answer: In golf, the goal is to get below par. This would represent zero. Scores above a par on a hole would be positive integers and scores below par on a hole would be negative integers.

1.2 Warm Up

For use before Activity 1.2

1. 22 **2.** 42 **3.** 77

4. 150 **5.** 122 **6.** 221

Answers

1.2 Start Thinking!

For use before Lesson 1.2

less than; *Sample answer:* The starting temperature increases $(\text{add } +10)$ 10 degrees, then decreases $(\text{add } -12)$ 12 degrees.

1.2 Warm Up

For use before Lesson 1.2

1. 13 **2.** -6 **3.** 5

4. 4 **5.** -4 **6.** 0

1.2 Practice A

1. 10 **2.** -8 **3.** -12 **4.** 0

5. 0 **6.** 3 **7.** 3 **8.** -6

9. -17 **10.** -17 **11.** -7 **12.** -3

13. \$29

14. Use the Commutative Property to switch the positions of the terms -5 and -8. Then use the Associative Property to group the terms 8 and -8. Because they are opposites, their sum will be zero; -5

15. Use the Commutative Property to switch the positions of the terms 4 and 9. Then use the Associative Property to group the terms -4 and 4. Because they are opposites, their sum will be zero; 9

16. Use the Commutative Property to switch the positions of the terms 12 and -7. Then use the Associative Property to group the terms -5 and -7. The sum of -5 and -7 is -12, which is the opposite of 12; 0

17. 10 **18.** -9 **19.** 11

20. 4 **21.** -7 **22.** -3

23. -18 **24.** 15 **25.** -27

26. $n = 13$ **27.** $c = -4$ **28.** $k = -8$

29. $2 + (-1) + (-2)$

30.

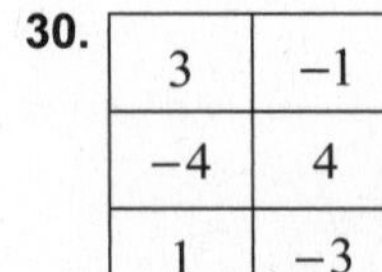

3	−1	−2
−4	4	0
1	−3	2

1.2 Practice B

1. 39 **2.** -48 **3.** 0

4. -41 **5.** 5

6. Use the Commutative Property to switch the positions of the terms -25 and -18. Then use the Associative Property to group the terms 18 and -18. Because they are opposites, their sum will be zero; -25

7. Use the Commutative Property to switch the positions of the terms 45 and -8. Then use the Associative Property to group the terms -22 and -8; 15

8. Use the Commutative Property to switch the positions of the terms -12 and 4. Then use the Associative Property to group the terms 28 and 4; 20

9. -28 **10.** 3 **11.** -27

12. 56 **13.** 18 **14.** -79

15. 37 **16.** 14 **17.** -59

18. $n = 25$ **19.** $c = 71$ **20.** $k = -80$

21. *Sample answer:* $-30, 8, 2$ and $8, 7, -5$

22. $-10°\text{F}$

23. *Sample answer:*

9	−6	−3
−1	0	1
−8	6	2

1.2 Enrichment and Extension

1.

−8	−3	−4
−1	−5	−9
−6	−7	−2

2.

−2	3	−4
−3	−1	1
2	−5	0

3.

−7	7	6	−4
4	−2	−1	1
0	2	3	−3
5	−5	−6	8

Answers

4.

−4	1	−10	3
−9	2	−3	0
5	−8	−1	−6
−2	−5	4	−7

5. Check students' work.

1.2 Puzzle Time

IN CASE HE GOT A HOLE IN ONE

1.3 Start Thinking!

For use before Activity 1.3

Sample answer: You could find the distance from where you are starting to where you are going. You could find the change in elevation.

1.3 Warm Up

For use before Activity 1.3

1. 34 **2.** 64 **3.** 77

4. 7 **5.** 45 **6.** 2

1.3 Start Thinking!

For use before Lesson 1.3

Sample answer: Rewrite the problem as an addition problem (add the opposite), then use additional rules. Do this for all 3 questions.

1.3 Warm Up

For use before Lesson 1.3

1. −3 **2.** 15 **3.** 1

4. −10 **5.** −7 **6.** −17

1.3 Practice A

1. −5 **2.** 9 **3.** −10 **4.** −3

5. 19 **6.** 8 **7.** −22 **8.** 8

9. 2 **10.** 22 **11.** 30 **12.** −45

13. −13 **14.** $7 + (-3)$ **15.** $5 - 3$

16. 2 **17.** 13 **18.** −4 **19.** 14

20. −1 **21.** 0 **22.** −18 **23.** 19

24. 26 **25.** 10 **26.** −11 **27.** −4

28. *Sample answer:* $x = -1$, $y = -3$; $x = -4$, $y = -6$

29. a. 367 ft **b.** 33 ft **c.** New Orleans

1.3 Practice B

1. −5 **2.** 29 **3.** −49 **4.** −15

5. 191 **6.** −146 **7.** −12 **8.** 18

9. $8 - (-28)$ **10.** 9

11. 17 **12.** 13 **13.** 134 **14.** −48

15. −44 **16.** −25 **17.** 103 **18.** 206

19. 71 **20.** 17 **21.** 23 **22.** 7

23. −38 **24.** −53

25. a. 94, 103, 114, 107, 84, 76, 64, 65, 75, 86, 105, 98
b. high of 99, low of −46 **c.** 145

26. For $|b| > |a|$ or a and b have different signs.

1.3 Enrichment and Extension

1. highest: 14; first dart: 7; second dart: −7

2. lowest: −14; first dart: −7; second dart: 7

3.

First Dart	7	6	5	4	3	2
Second Dart	3	2	1	0	−1	−2

First Dart	1	0	−1	−2	−3
Second Dart	−3	−4	−5	−6	−7

4.

First Dart	1	0	−1	−2	−3
Second Dart	7	6	5	4	3

First Dart	−4	−5	−6	−7
Second Dart	2	1	0	−1

5. There are 15 ways to get a score of 0. There is 1 way to get a score of 14. There is 1 way to get a score of −14.

6. You should try to land on a negative integer first and a positive integer second. Subtracting a positive integer is the same as adding a negative integer, which gives you a lower score.

Answers

7. *Sample answer:* Students may use a strategy such as aiming over 7 for the first drop and over -7 for the second drop, or if they notice their paper punches always fall in some direction, they might try to get the pieces to fall over values that will give them positive points.

1.3 Puzzle Time

NOTHING IT JUST WAVED

1.4 Start Thinking!

For use before Activity 1.4

Sample answer: A deposit is adding a positive amount of money to your account. A withdrawal is adding a negative amount to your account.

1.4 Warm Up

For use before Activity 1.4

1. 18 **2.** −14 **3.** −9

4. 15 **5.** 24 **6.** −16

1.4 Start Thinking!

For use before Lesson 1.4

Sample answer: A problem like $2(4)$ means to add 2 groups of 4.

1.4 Warm Up

For use before Lesson 1.4

1. 14 **2.** −63 **3.** −48

4. 80 **5.** −30 **6.** 60

1.4 Practice A

1. −12 **2.** −30 **3.** 16 **4.** 54

5. 0 **6.** 36 **7.** 77 **8.** −25

9. −91 **10.** −9 **11.** −24 **12.** 81

13. −35 **14.** −30 **15.** −20 **16.** −42

17. 0 **18.** 48 **19.** 100 **20.** 9

21. −9 **22.** −8 **23.** −25 **24.** −48

25. 98 **26.** −18 **27.** 22 **28.** −7

29. 96, −192 **30.** 729, −2187

31. a. −12 ft

b.

Time	3 sec	6 sec	9 sec
Height	144 ft	108 ft	72 ft

c. 15 sec **d.** −48 ft

1.4 Practice B

1. 96 **2.** −140 **3.** −84 **4.** 120

5. −384 **6.** 220 **7.** −270 **8.** 0

9. −156 **10.** 96 **11.** −729 **12.** 144

13. −144 **14.** −343 **15.** 8 **16.** −72

17. 847 **18.** 18 **19.** −19 **20.** 24

21. a. \$325

b. Because $n = 0$ represents the 1st member, the 8th member pays \$25. So, the 9th member is free.

c. \$25, the 8th member

22. a. 47 **b.** yes; −3, 16 **c.** −49

1.4 Enrichment and Extension

1. 1575 **2.** 432 **3.** −11,340 **4.** −2000

5.

$$(-3)^2 = (-3)(-3) = 9$$
$$(-3)^3 = (-3)(-3)(-3) = -27$$
$$(-3)^4 = (-3)(-3)(-3)(-3) = 81$$
$$(-3)^5 = (-3)(-3)(-3)(-3)(-3) = -243$$
$$(-3)^6 = (-3)(-3)(-3)(-3)(-3)(-3) = 729$$
$$(-3)^7 = (-3)(-3)(-3)(-3)(-3)(-3)(-3) = -2187$$
$$(-3)^8 = (-3)(-3)(-3)(-3)(-3)(-3)(-3)(-3) = 6561$$

6. The sign changes every time.
7. positive; The products are positive when the exponent is even.
8. negative; The products are negative when the exponent is odd.

1.4 Puzzle Time

DURING LEAP YEAR

1.5 Start Thinking!

For use before Activity 1.5

$+ \times +,\ + \times -,\ - \times +,\ - \times -$

1.5 Warm Up

For use before Activity 1.5

1. −50 **2.** 30 **3.** −72

4. −105 **5.** 176 **6.** −128

Answers

1.5 Start Thinking!

For use before Lesson 1.5

24, 96; Divide by 2; Multiply by 4

1.5 Warm Up

For use before Lesson 1.5

1. −4 **2.** −8 **3.** 7

4. −5 **5.** −9 **6.** 12

1.5 Practice A

1. −2 **2.** 5 **3.** −2 **4.** 0

5. −5 **6.** −3 **7.** −8 **8.** −1

9. undefined **10.** −6

11. 7 **12.** −7 **13.** 4 **14.** 3

15. −5 **16.** 10 **17.** −12 **18.** 6

19. −6, 3 **20.** 20, −4 **21.** −7

22. a. −4 **b.** 15, down **c.** down

d. Between 4 seconds and 5 seconds because the sign changed.

1.5 Practice B

1. −17 **2.** −3 **3.** 0 **4.** 16

5. −21 **6.** undefined

7. −12 **8.** 1 **9.** −11 **10.** −9

11. −52 **12.** 16 **13.** 3 **14.** −11

15. 2 **16.** 28 **17.** 50 **18.** 12

19. a. 2 **b.** 2

c. *Sample answer:* PI-Squared, because they answer both parts of more questions.

20. a. 7 hours **b.** 35 hours **c.** 8 hours 10 minutes

1.5 Enrichment and Extension

1. sometimes **2.** sometimes **3.** never

4. always **5.** always **6.** never

7. never **8.** always **9.** sometimes

10. sometimes **11.** always **12.** never

13. never **14.** always **15.** always

16. never **17.** none **18.** all integers

19. all positive integers and zero

20. all integers

21. all positive integers and zero

22. all negative integers and zero

23. all negative integers and zero

24. all positive integers and zero

25. all positive integers and zero

26. all negative integers and zero

27. all integers

28. all integers

29. all positive integers and zero

30. all positive integers and zero

31. all negative integers and zero

32. all negative integers and zero

Check students' work; It should be a person.

1.5 Puzzle Time

CATCH YOU LATER

Technology Connection

1. −8 **2.** 0 **3.** −1

4. −1 **5.** −7 **6.** −2

7. *Sample answer:* The number becomes 1 smaller, meaning it is 1 farther away from the zero. The magnitude of the number—its absolute value—becomes 1 larger, again because it moves 1 unit farther from zero.

Chapter 2

2.1 Start Thinking!

For use before Activity 2.1

you did; *Sample answer:* If you change the fractions to have common denominators, $\frac{3}{8}$ is greater than $\frac{2}{8}$ or $\frac{1}{4}$.

2.1 Warm Up

For use before Activity 2.1

1. = **2.** < **3.** >

4. > **5.** < **6.** =

Answers

2.1 Start Thinking!

For use before Lesson 2.1

The first friend is the better shooter. *Sample answer:* First write each shooter's score as a ratio (rational number). Then change each to a decimal. Finally compare the decimals to see which is the largest.

2.1 Warm Up

For use before Lesson 2.1

1. $-\frac{7}{4}, -\frac{2}{3}, -0.3, 0.6, \frac{3}{4}$ **2.** $-1.3, -\frac{6}{5}, \frac{7}{5}, 1.5, 1.65$

3. $-2.75, -0.37, \frac{5}{4}, 2.65, \frac{11}{4}$

4. $-\frac{4}{2}, -0.8, \frac{1}{8}, \frac{4}{10}, 4.5$ **5.** $-\frac{9}{3}, -0.3, \frac{6}{4}, \frac{8}{5}, 3.8$

6. $-1.5, -\frac{3}{4}, 0.6, \frac{9}{6}, \frac{7}{3}$

2.1 Practice A

1. $0.\overline{5}$ **2.** -0.375 **3.** $-0.\overline{27}$ **4.** $0.2\overline{3}$

5. $1.4\overline{16}$ **6.** $-2.\overline{3}$ **7.** $-0.\overline{590}$ **8.** $5.1\overline{6}$

9. $\frac{7}{10}$ **10.** $-\frac{3}{10}$ **11.** $-\frac{43}{100}$ **12.** $\frac{13}{25}$

13. $1\frac{1}{4}$ **14.** $-2\frac{7}{100}$ **15.** $4\frac{9}{50}$ **16.** $3\frac{1}{8}$

17. $-\frac{5}{2}, -\frac{2}{3}, -0.5, \frac{3}{2}, 1.6$

18. $-\frac{7}{4}, -1.7, 0.6, \frac{3}{4}, 1.1$

19. $-0.5, -\frac{2}{5}, 0, 0.67, \frac{7}{9}$ **20.** $-\frac{1}{3}, -0.3, 1.2, \frac{4}{3}, \frac{3}{2}$

21. a. 0.64 **b.** $\frac{16}{25}$ **22.** Ian

23. a. $77.\overline{27}\%$ **b.** yes **c.** 16

2.1 Practice B

1. 0.625 **2.** $-0.1\overline{36}$ **3.** $1.\overline{2}$ **4.** -5.075

5. $-7.\overline{45}$ **6.** $4.0\overline{6}$ **7.** $-9.\overline{1}$ **8.** $-7.8\overline{3}$

9. $\frac{17}{25}$ **10.** $-\frac{1}{100}$ **11.** $-3\frac{99}{100}$ **12.** $8\frac{149}{200}$

13. $3\frac{1}{200}$ **14.** $-13\frac{3}{250}$

15. $-9\frac{49}{50}$ **16.** $-10\frac{113}{250}$

17. your friend **18.** $>$

19. $>$ **20.** $<$

21. *Sample answer:* $-1.75, -1.\overline{6}$

22. a. $-2\frac{6}{7}, -\frac{3}{2}, 2.25, 2\frac{1}{3}$

b. 10:00 A.M. and 4:00 P.M.

c. $-2\frac{6}{7}$ feet **d.** decrease

e. increase; The tide increased at 4:00 A.M. on the given day, so it will increase again in the morning.

2.1 Enrichment and Extension

1. *Sample answer:* There are positive and negative numbers in the outer box. The larger circle has just negative integers and the smaller circle has just positive integers.

2.

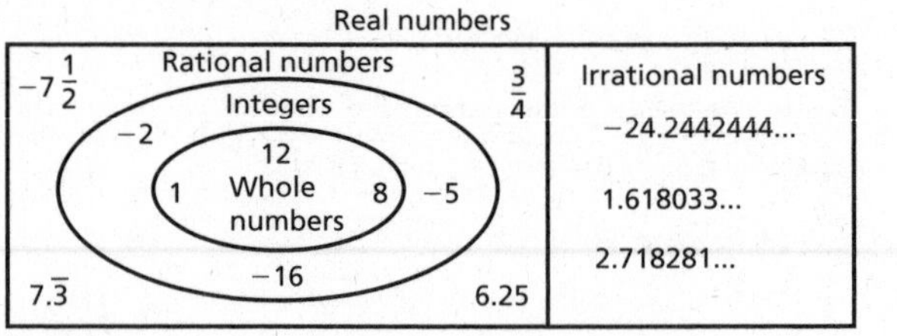

Real numbers

Rational numbers: $-7\frac{1}{2}$, $\frac{3}{4}$, $7.\overline{3}$, 6.25

Integers: -2, -5, -16

Whole numbers: 1, 12, 8

Irrational numbers: $-24.2442444...$, $1.618033...$, $2.718281...$

3. rational **4.** rational **5.** irrational

6. rational, integer, whole number

7. yes; $2 = \frac{2}{1}$; All whole numbers are rational numbers because they can be written as a fraction with a denominator of 1.

8. yes; $-14 = \frac{-14}{1}$; All integers are rational numbers because they can be written as a fraction with a denominator of 1.

9. sometimes **10.** always **11.** never

12. sometimes **13.** always

2.1 Puzzle Time

THE STUDENT WHO SAID HE COULDN'T WRITE HIS ESSAY ON GOLDFISH BECAUSE HE DIDN'T HAVE ANY WATERPROOF INK

Answers

2.2 Start Thinking!

For use before Activity 2.2

Sample answer: Add $55 + 15$, then add the answer to -20, $70 + (-20)$. Finally add the answer from the previous problem to 8, $50 + 8$. The final answer 58°. So, the temperature at the end of the day is greater than the temperature at the beginning of the day.

2.2 Warm Up

For use before Activity 2.2

1. -77 **2.** 54 **3.** 42

4. -20 **5.** 33 **6.** -99

2.2 Start Thinking!

For use before Lesson 2.2

Sample answer: Start at zero and move $\frac{4}{5}$ unit to the right. Then move $\frac{2}{5}$ unit to the left and end on $\frac{2}{5}$.

Start at zero and move 2.6 units to the left. Then move 5.8 units to the right and end at 3.2.

2.2 Warm Up

For use before Lesson 2.2

1. $2\frac{7}{9}$ **2.** $-2\frac{4}{9}$ **3.** $-\frac{11}{15}$

4. -6.9 **5.** -18.2 **6.** -2.5

2.2 Practice A

1. $-\frac{1}{8}$ **2.** $\frac{1}{3}$ **3.** $\frac{1}{6}$

4. 4.3 **5.** -2.8 **6.** -5.61

7. The second number is negative.

$$\frac{3}{10} + \left(-\frac{1}{10}\right) = \frac{3 + (-1)}{10} = \frac{2}{10} = \frac{1}{5}$$

8. $-\frac{9}{10}$ **9.** $-\frac{3}{10}$ **10.** $\frac{1}{10}$ **11.** -4.7

12. $\frac{5}{8}$ **13.** $-\frac{1}{2}$ **14.** $-1\frac{14}{15}$ **15.** 2.4

16. a. always **b.** sometimes

c. never **d.** sometimes

2.2 Practice B

1. $\frac{1}{5}$ **2.** $-1\frac{1}{12}$ **3.** $-3\frac{1}{14}$

4. -6.17 **5.** -2.34 **6.** -28.51

7. $$2\frac{5}{6} + \left(-\frac{8}{15}\right) = \frac{17}{6} + \left(-\frac{8}{15}\right) = \frac{85 + (-16)}{30}$$
$$= \frac{69}{30} = 2\frac{9}{30} = 2\frac{3}{10}$$

8. $-\frac{19}{20}$ **9.** $-\frac{1}{20}$ **10.** $\frac{11}{20}$ **11.** \$8.44

12. $\frac{13}{21}$ **13.** $-5\frac{11}{20}$ **14.** $-3\frac{7}{9}$ **15.** 20

16. When the positive number is greater than the absolute value of the negative number.

17. The snowfall for the three month period is $2\frac{1}{3}$ inches greater than the yearly average.

18. $-\$2.35$

2.2 Enrichment and Extension

1. $\boxed{3}\frac{\boxed{2}}{\boxed{5}} - 4\frac{\boxed{1}}{2} = -1\frac{1}{10}$

2. $-\boxed{2}.5\boxed{8} + \boxed{3}7.\boxed{9} = 35.32$

3. $\boxed{2}\boxed{7}.\boxed{8} - 4\boxed{3}8 = -410.2$

4. $-\frac{\boxed{3}}{\boxed{4}} - \boxed{2}\frac{\boxed{1}}{2} = -3\frac{1}{4}$

5. $-\boxed{3}.8\boxed{9} - \boxed{4}2.\boxed{3} = -46.19$

6. $\boxed{6}\frac{\boxed{7}}{8} - \boxed{8}\frac{3}{\boxed{4}} = -1\frac{7}{8}$

7. $-\boxed{6}\frac{\boxed{2}}{3} + 2\frac{\boxed{1}}{\boxed{9}} = -4\frac{5}{9}$

8. $-\boxed{3}8 + \boxed{5}\boxed{6}.\boxed{4} = 18.4$

9. *Sample answer:*

$\boxed{5}.\boxed{6}\boxed{1} - \boxed{7}.\boxed{4}\boxed{3} = -1.\boxed{8}\boxed{2}$

2.2 Puzzle Time

AT THE NORTH POLE

Answers

2.3 Start Thinking!

For use before Activity 2.3

Sample answer. One example of a sport is swimming. Each leg of a relay team is added together to get a final time. A way negative rational numbers is used in swimming is when a swimmer's time is below a record time.

2.3 Warm Up

For use before Activity 2.3

1. –70 **2.** 91 **3.** –75

4. –75 **5.** 12 **6.** 42

2.3 Start Thinking!

For use before Lesson 2.3

Sample answer: Start at zero and move $\frac{4}{5}$ unit to the left. Then move $\frac{2}{5}$ unit more to the left and end on $-1\frac{1}{5}$.

Start at zero and move 2.6 units to the left. Then move 5.8 units more to the left and end at –8.4.

2.3 Warm Up

For use before Lesson 2.3

1. $-7\frac{5}{7}$ **2.** $-13\frac{1}{2}$ **3.** $-10\frac{5}{8}$

4. –18.9 **5.** 23.8 **6.** –18

2.3 Practice A

1. –1 **2.** $1\frac{2}{3}$ **3.** $1\frac{11}{12}$ **4.** $-5\frac{1}{3}$

5. –2.46 **6.** –6.25 **7.** $10\frac{1}{4}$ **8.** 2.6

9. $3\frac{1}{15}$ **10.** $\frac{5}{12}$ **11.** $-\frac{7}{8}$ **12.** –2.57

13. a. $3\frac{2}{3}$ gal **b.** $4\frac{11}{12}$ gal **c.** $1\frac{7}{12}$ gal

14. no; If the absolute value of the second number is greater than the absolute value of the first number, then the difference is negative.

2.3 Practice B

1. $1\frac{4}{5}$ **2.** $\frac{1}{3}$ **3.** $4\frac{1}{12}$

4. $-13\frac{17}{24}$ **5.** –42.481 **6.** 3.786

7. $2\frac{8}{15}$ **8.** 13.7 **9.** 1.7

10. $7\frac{13}{16} - 9\frac{5}{8} = -1\frac{13}{16}$

11. $-5\frac{5}{6}$ **12.** –24.625 **13.** 3.975

14. $-1\frac{2}{9}$ **15.** –\$90.73 **16.** $1\frac{6}{8}$

2.3 Enrichment and Extension

1. The golfer is 4 strokes above par.

2. There is \$3.26 left in the account.

3. The team lost 2 yards.

4. negative; The amount in the bank account is \$1.87 less than the cost of the clothing.

5. positive; Distance is always positive.

6. negative; $9 - 14 = -5$.

7. zero; $6 - 3\frac{2}{3} = 2\frac{1}{3}$ and $2\frac{1}{3} - 2\frac{1}{3} = 0$.

2.3 Puzzle Time

ON A BED OF LETTUCE

2.4 Start Thinking!

For use before Activity 2.4

Gain of 2 yards (+2); Loss of 8 yards (–8);
Gain of 20 yards (+20); Loss of 5 yards (–5);
Gain of 3 yards (+3)

2.4 Warm Up

For use before Activity 2.4

1. –108 **2.** –110 **3.** 168

4. 8 **5.** –14 **6.** –27

2.4 Start Thinking!

For use before Lesson 2.4

The company shows a loss of \$2.23.

2.4 Warm Up

For use before Lesson 2.4

1. $-1\frac{4}{5}$ **2.** $-\frac{27}{125}$ **3.** 3.66

4. $\frac{6}{7}$ **5.** –1.2 **6.** –0.215

Answers

2.4 Practice A

1. negative 2. negative 3. positive 4. positive

5. $-\frac{1}{5}$ 6. $\frac{2}{3}$ 7. $-\frac{1}{21}$ 8. $-\frac{7}{10}$

9. 0.8 10. -1.5 11. $-\frac{4}{7}$ 12. $\frac{5}{6}$

13. $-3\frac{1}{3}$ 14. $1\frac{57}{64}$ 15. -7.77 16. 11.742

17. 10 pizzas 18. -2.45 in. 19. -9.46

20. $\frac{6}{11}$ 21. -14.2979 22. $2\frac{1}{2}$

23. *Sample answer:* $-\frac{5}{4}, -\frac{1}{2}$

24. -3.4

2.4 Practice B

1. $-1\frac{4}{11}$ 2. $\frac{10}{39}$ 3. $-\frac{11}{180}$ 4. $-\frac{21}{80}$

5. 8.9 6. -6.3 7. $\frac{5}{9}$ 8. $-\frac{3}{8}$

9. $-4\frac{5}{6}$ 10. $-11\frac{14}{25}$ 11. -8.525 12. 2.5654

13. 16 burgers 14. -0.52 second

15. 0.241 16. $3\frac{1}{2}$

17. -28.5939 18. $-11\frac{9}{25}$

19. **a.** \$46.25 **b.** \$72.52

2.4 Enrichment and Extension

1. 24 Zweenubs, 16 Zweedulls, 6 Zweebuds

2. \$218.32

3. 20 Zweenubs, 10 Zweedulls, 2 Zweebuds

4. \$198.88 5. $-$\$19.44 6. $-$\$97.20

7. 25 Zweenubs, 10 Zweedulls, 5 Zweebuds

8. They need to make at least \$64.21. The most profit they could earn is \$78.70.

9. No, they cannot make enough income on Zweenubs alone. Because only 25 Zweenubs are being manufactured next week, and last week's cannot be reused, there are only 13 Zweenubs left to sell. When you multiply 13 by \$4.50, you get a total possible income of \$58.50 from Zweenubs, which is less than the \$64.21 that is needed in order to break even.

2.4 Puzzle Time

WHEN HE DRIBBLES

Technology Connection

1. $\frac{9}{10}$ 2. $2\frac{3}{8}, \frac{19}{8}$ 3. $\frac{-13}{36}$

4. $3\frac{7}{75}, \frac{232}{75}$ 5. $\frac{13}{16}$ 6. $1\frac{11}{36}, \frac{47}{36}$

Chapter 3

3.1 Start Thinking!

For use before Activity 3.1

Sample answer: An expression can contain numbers, variables, and operations $\left(\text{i.e. } x + 9, 3 + 4\right)$. An equation can contain numbers, variables, operations, and has an equal sign $\left(\text{i.e. } x + 3 = 8, 1 + 2 = 3\right)$.

3.1 Warm Up

For use before Activity 3.1

1. $7 + x$ 2. $-14 - y$ 3. $-19 + n$

4. $14y$ 5. $\frac{10}{n + 6}$ 6. $6\left(\frac{x}{3}\right)$

3.1 Start Thinking!

For use before Lesson 3.1

$4x + 4$; *Sample answer:* Your brother incorrectly combined the $6x$ and the $2x$. He should have subtracted instead of added.

3.1 Warm Up

For use before Lesson 3.1

1. $14x$ 2. $7y - 3$ 3. $2x + 16$

4. $11y - 4$ 5. $10x - 15.4$ 6. $y + 5\frac{1}{2}$

3.1 Practice A

1. $-4y, 7, 9y, -3$; $-4y$ and $9y$, 7 and -3

2. $3n^2, -1.4n, 5n^2, -6.4$; $3n^2$ and $5n^2$

Answers

3. $\frac{1}{2}b^3, -b^3, 2b$; $\frac{1}{2}b^3$ and $-b^3$

4. $-6m$ 5. $14k - 8$ 6. $10.3 - 12x$

7. $13 - 8x$ 8. $28a - 7$ 9. $27x - \frac{89}{6}$

10. $-\frac{5}{8}h + 7$ 11. $-\frac{1}{4}y + 2$ 12. $18x + 16$

13. $8y + 2.1y + 3y$

The total weight carried by the runners is $8y + 2.1y + 3y = 13.1y$ ounces.

14. $65 + 35x$ kilograms 15. yes; $15a^2 - 4b$

3.1 Practice B

1. $1.3x, -2.7x^2, -5.4x, 3$; $1.3x$ and $-5.4x$

2. $10, -\frac{3}{10}m, 6m^2, \frac{2}{5}m$; $-\frac{3}{10}m$ and $\frac{2}{5}m$

3. $-\frac{35}{12}b$ 4. $180m - 40$ 5. $31.4 - 14x$

6. $-12y + 12$ 7. $-7v - 3$ 8. $\frac{17}{15}x + \frac{79}{5}$

9. $29x + 9$

10. 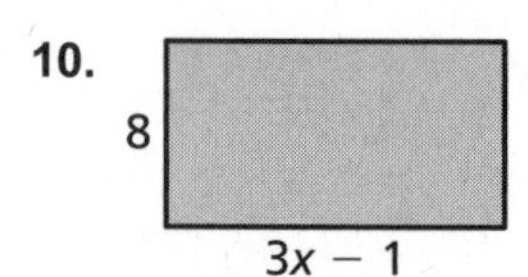

11. 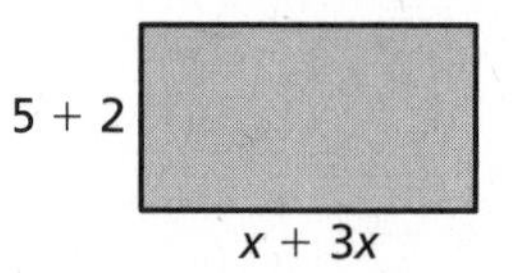

12. $2.5x + 5$ 13. $6w + 60$ 14. $15.65x$

3.1 Enrichment and Extension

1. a 2. c 3. j 4. g

5. d 6. i 7. f 8. h

9. k 10. b 11. e

12. Answers will vary. 13. Answers will vary.

3.1 Puzzle Time

THROW IT UP IN THE AIR. IT WILL COME DOWN AND SQUASH.

3.2 Start Thinking!

For use before Activity 3.2

Sample answer: like terms: $3x$, $24x$; unlike terms: $9y$, $9m$; Like terms are terms that have the same variable raised to the same exponent. Unlike terms are terms that cannot be combined further.

3.2 Warm Up

For use before Activity 3.2

1. $x + 2$ 2. $-2y - 9$ 3. $-10x + 3$

4. $3\frac{1}{2}y - \frac{2}{5}$ 5. $8.7x - 4.6$ 6. $5.1y - 3.3$

3.2 Start Thinking!

For use before Lesson 3.2

Model the expression using algebra tiles.

([+] [+] [+] [+] [+] ⊖ ⊖ ⊖ ⊖) +
([+] [+] [+] ⊖ ⊖ ⊖ ⊖ ⊖ ⊖)

Then combine the like terms. The simplified expression is $8x - 10$.

3.2 Warm Up

For use before Lesson 3.2

1. $(x - 4) + (x + 2)$; $2x - 2$

2. $(2x - 3) + (x - 3)$; $3x - 6$

3. $(x + 2) - (2x + 3)$; $-x - 1$

4. $(2x - 4) - (x + 2)$; $x - 6$

3.2 Practice A

1. $2p - 10$ 2. $2n + 3$

3. $2r + 7$ 4. $8x - 27$

5. $8.2c - 18$ 6. $28.5q - 2$

7. $6.5y + 13$ 8. $6x - 3$

9. **a.** $15t + 2$ **b.** your friend

10. $-2k + 8$ 11. $-8d - 5$ 12. $19j - 9$

13. $-12x + 26$ 14. $5t - 18$ 15. $4w + 5$

16. **a.** $4 + 0.5r$ **b.** $7.50

17. $10m + 2$

Answers

3.2 Practice B

1. $2t + 7$
2. $11k + 6$
3. $-8y + 4$
4. $13.6g - 22$
5. $-8.3s$
6. $3p + 1$
7. $-5w - \frac{5}{2}$
8. $4k - \frac{13}{5}$
9. a. $5t + 7$ b. $29t + 45$ c. $132
10. $-6u + 7$
11. $-28x + 55$
12. $11.4h - 26$
13. $4.5b + 24$
14. $2j - 8$
15. $6n + \frac{22}{5}$
16. a. $7d + 12$ b. 5 pairs

3.2 Enrichment and Extension

1. $x^3 + x^2$
2. $-2x^2 + 8x$
3. $x^5 - 4x$
4. $3x^2 - x$
5. $3x^2 - 3x$
6. $2x^2 - 2x$
7. $-16x^2 - 12x$
8. $n^2 - 4n$
9. $-3b^2 - 9b$
10. $-8w^2 - 28w$
11. $-4x^2 - 12x$
12. $-15k^2 + 68k + 18$
13. $2.2h^2 + 12.2h + 4$
14. $m^2 + 3m$
15. $-z^2 - 2z - 2$
16. $96d^9 + 18d^6$

3.2 Puzzle Time

YOU LIGHT UP MY LIFE

Extension 3.2 Start Thinking!

For use before Extension 3.2

no; The common factors of 15 and 30 are 1, 3, 5, and 15. So, the greatest common factor is 15.

Extension 3.2 Warm Up

For use before Extension 3.2

1. 2
2. 3
3. 9
4. 7
5. 6
6. 10

Extension 3.2 Practice

1. $2(4 - 11)$
2. $5(5 + 6)$
3. $3(2y + 1)$
4. $2(t - 5)$
5. $8(2p - 1)$
6. $3(7s + 5)$
7. $8(4v + 3w)$
8. $3(3b + 8c)$
9. $6(2y - 7z)$
10. $\frac{1}{2}(m + 1)$
11. $\frac{2}{3}\left(j - \frac{1}{3}\right)$
12. $1.2(k + 2)$
13. $1.5(a - 3)$
14. $3\left(f + \frac{5}{3}\right)$
15. $\frac{3}{10}(x - 2)$
16. $-\frac{1}{3}(x + 36)$
17. $-\frac{1}{6}(2x - 5y)$
18. $(3x - 2)$ in.
19. a. $6x + 9$ b. $21x + 28$ c. $15x + 19$

3.3 Start Thinking!

For use before Activity 3.3

$p - 7$ and $p + 8$; The first expression can only be written one way. The second expression could be written as $8 + p$ because addition is commutative.

3.3 Warm Up

For use before Activity 3.3

1. 42
2. −46
3. 7
4. −39
5. −6
6. 65

3.3 Start Thinking!

For use before Lesson 3.3

Sample answer: In the equation $p + 4 = 10$, subtraction will be the inverse operation used to solve this problem. By subtracting 4 from each side, the variable p will be left alone on the left side resulting in the answer 6.

3.3 Warm Up

For use before Lesson 3.3

1. $x = 5$
2. $y = 18$
3. $n = 78$
4. $p = 15$
5. $t = 57$
6. $z = 7$

3.3 Practice A

1. 7
2. −8
3. −4
4. 9.6
5. −10.7
6. $1\frac{1}{6}$
7. −18.82
8. $3\frac{5}{8}$
9. 3.94
10. $-2\frac{5}{12}$
11. $\frac{11}{14}$
12. 2.06
13. $y + 5 = -2; -7$
14. $8 + h = 12; 4$

Answers

15. $-13 = n - 4; -9$

16. $b + 2 = 9; \$7$

17. $325 = t - 75; 400°F$

18. $c - 29\frac{3}{4} = -10\frac{1}{4}; 19\frac{1}{2}$ in.

19. $76.50 = f - 31.41; \$107.91$

20. $5.2 + 8 + x = 25; 11.8$ ft

21. -4

3.3 Practice B

1. 88
2. -36
3. -6.4
4. 19.63
5. -7.143
6. 7
7. -109.04
8. $-1\frac{1}{2}$
9. 4.994
10. $-9\frac{11}{24}$
11. $6\frac{11}{15}$
12. -1.798
13. $27 = x + 12; 15$
14. $p - (-9) = 12; 3$
15. $m - 35 = -72; -37$
16. $28.12 = f - 0.14; 28.26$ seconds
17. $7\frac{3}{8} + \ell = 16\frac{1}{4}; 8\frac{7}{8}$ meters
18. $t - 183.6 = -109.3; 74.3°F$
19. \$53.30
20. 14.8, -14.8
21. 10, -14

3.3 Enrichment and Extension

1. yes
2. Karen's answer is more complete because she remembered that -7 has an absolute value of 7.
3. *Sample answer:* Plot 12 on a number line and then find the numbers that are 8 units away from 12.
4. yes; *Sample answer:* Because the absolute value of the expression $x - 5$ equals 9, then the expression $x - 5$ is equal to either 9 or -9. So, set up two equations and solve to find x.
5. Absolute value cannot be negative, but in this case x is inside the absolute value bars. So, x can be negative because you can take the absolute value of a negative number.
6. *Sample answer:* He substituted -5 for $|x|$, and the absolute value of a number is never negative.
7. Every absolute value equation does not have two solutions. In Exercise 6, by subtracting 7 from each side, you obtain $|x| = -5$. This equation has no solutions because the absolute value of a number is never negative.
8. **a.** $|x| + 6 = 6$ **b.** $|x + 3| = 5$ **c.** $|x| + 8 = 4$

3.3 Puzzle Time

LOOK MOM NO HANDS

3.4 Start Thinking!

For use before Activity 3.4

Sample answer: To determine if you have enough money to buy the items, you will need to find the total price. You can do this by adding the price together, then comparing the total with how much you have. If you are buying more than one item of the same price, you can use multiplication to find the price.

3.4 Warm Up

For use before Activity 3.4

1. -168
2. 216
3. -65
4. -32
5. -9
6. 16

3.4 Start Thinking!

For use before Lesson 3.4

Sample answer: A small bus has 13 rows. Write and solve an equation to find the number of students each seat holds if there are a total of 39 students riding the bus. A small bus carries a total of 39 students. Each seat holds 3 students. Write and solve an equation showing how many rows of seats are on the bus.

3.4 Warm Up

For use before Lesson 3.4

1. $x = 6$
2. $x = 28$
3. $x = -11$
4. $x = 60$
5. $x = 63$
6. $x = -48$

3.4 Practice A

1. 6
2. -5
3. -99
4. -160
5. 1.4
6. -5.6
7. -15
8. $4\frac{2}{3}$
9. -14.28
10. $-3\frac{1}{3}$
11. 3
12. -16
13. $\frac{1}{2}x = -\frac{5}{12}; -\frac{5}{6}$
14. $\frac{n}{0.2} = -2.6; -0.52$

Answers

15. $7.50h = 123.75$; 16.5 hours

16. $\frac{m}{4.5} = 70$; 315 miles

17. $\frac{3}{8}w = \frac{1}{2}$; $1\frac{1}{3}$ in.

18. $\frac{s}{3} = 21$; 63 students

19. $26.46 = 4s$; 6.615 in.

20. *Sample answer:* $7n = 2$

21. *Sample answer:* $\frac{y}{2} = -10$

3.4 Practice B

1. 3.75 **2.** 1.5 **3.** −36.75 **4.** −3.96

5. 7.5 **6.** −2.35 **7.** $-4\frac{4}{5}$ **8.** $0.\overline{45}$

9. −0.0084 **10.** $1\frac{1}{8}$ **11.** 9 **12.** −33

13. $12t = 0.78$; 0.065, or 6.5%

14. $72 = \frac{8}{9}p$; 9 points

15. *Sample answer:* $3n = -44.4$

16. *Sample answer:* $\frac{m}{9} = -\frac{1}{14}$

17. There are 5 students in each group and 2 students are not in a group.

18. **a.** $1.75y = 15.75$; 9 tokens

b. $12t = 17.50$; about \$1.46

c. $35 = 1.40t$; 25 free tokens

19. 6, −6

3.4 Enrichment and Extension

1. one solution; The value of x is positive because a negative must be multiplied by a positive when the result is negative.

2. one solution; The value of n is positive because a positive must be divided by a negative when the result is negative.

3. one solution; The value of g is negative because the number divided by a positive must be negative when the result is negative.

4. one solution; The value of t is negative because a negative must be multiplied by a negative when the result is positive.

5. two solutions; The value of v is either positive or negative, because the absolute value will always be positive, and a negative must be multiplied by a positive when the result is negative.

6. no solution; The absolute value of k will always be positive, and a positive divided by a negative cannot be equal to a positive.

7. no solution; The absolute value of the quotient of x and -5 will always be positive and cannot be equal to -15.

8. two solutions; The value of p is positive or negative, because the absolute value of the product of -6 and p will always be positive.

9. two solutions; The value of u is positive or negative, because the absolute value of the product of 2.7 and u will always be positive.

10. no solution; The absolute value of the product of $1\frac{1}{2}$ and b will always be positive and cannot be equal to $-13\frac{4}{5}$.

11. one solution; The value of h is positive, because the absolute value of 8 is positive, and a positive must be multiplied by a positive when the result is positive.

12. one solution; The value of y is negative, because the absolute value of -9 is positive, and a positive must be multiplied by a negative when the result is negative.

13. two solutions; The value of b is either positive or negative, because the absolute value will always be positive, and the number divided by a negative must be positive when the result is negative.

14. no solution; The absolute value of the quotient of a and -5 will always be positive and cannot be equal to -2.5.

Check students' work; It should be a fish.

3.4 Puzzle Time

THE RESTAURANT ON THE MOON THAT HAS REALLY GOOD FOOD BUT NO ATMOSPHERE

Answers

3.5 Start Thinking!

For use before Activity 3.5

Sample answer: $x + 2 = 4$; To solve this equation subtract 2 from each side of the equation. The answer is $x = 2$.

$x - 3 = 5$; To solve this equation add 3 to each side of the equation. The answer is $x = 8$.

$4x = 16$; To solve this equation divide each side of the equation by 4. The answer is $x = 4$.

$\frac{x}{5} = 3$; To solve this equation multiply each side of the equation by 5. The answer is $x = 15$.

3.5 Warm Up

For use before Activity 3.5

1. $x = 12$ **2.** $x = -9$ **3.** $x = -40$

4. $x = 12$ **5.** $x = 9.04$ **6.** $x = 4\frac{1}{6}$

3.5 Start Thinking!

For use before Lesson 3.5

Add 4 to both sides of the equation. Divide both sides by 3. The solution is $x = 5$.

3.5 Warm Up

For use before Lesson 3.5

1. $x = 8$ **2.** $x = -12$ **3.** $x = 5$

4. $x = -6$ **5.** $x = -2.65$ **6.** $x = -5.2$

3.5 Practice A

1. 4 **2.** −2.4 **3.** 3.5 **4.** 4.4

5. −3 **6.** 7.6 **7.** −1.05 **8.** −1.5

9. 17.35 **10.** −1 **11.** $-\frac{1}{2}$ **12.** −4

13. $2.5r + 4 = 21.50$; 7 rides

14. $45 + 1.99m = 68.88$; 12 movies

15. $16 + 2w = 24$; 4 feet

16. 9 **17.** $-\frac{1}{3}$ **18.** −6

19. 32 students

20. a. 10 cans **b.** $x = 10$; same answer

3.5 Practice B

1. 3 **2.** −4 **3.** 0.78125 **4.** −7

5. $2.\overline{8}$ **6.** −1.32 **7.** $-2\frac{5}{14}$ **8.** 0

9. $2.708\overline{3}$ **10.** $\frac{5}{48}$ **11.** 43 **12.** $\frac{13}{15}$

13. $132.49 + 15s = 192.49$; 4 skateboards

14. $f + 27(0.99) = 42.72$; \$15.99

15. −2.25 **16.** −9.4 **17.** $6\frac{1}{6}$

18. 16 ft, 32 ft

19. a. 56 seashells **b.** 297 seashells
c. *Sample answer:* 50, 3

3.5 Enrichment and Extension

1. 10

2. Any multiple of 10 will work because 2 and 5 will be factors of it.

3. The number 10 because it is the least common denominator of the fractions.

4. Because of the Multiplication Property of Equality

5. Multiply each term of the equation by the least common denominator of the fractions.

6. a. 46 **b.** 15 **c.** 6 **d.** $1\frac{6}{7}$ **e.** $-1\frac{2}{9}$ **f.** −6

3.5 Puzzle Time

DON'T STOP ME I'M ON A ROLL

Technology Connection

1. 47 **2.** 1.28 **3.** 2618 **4.** 5668

5. 1593 **6.** 938 **7.** 81 **8.** 3.7

9. 305.2 **10.** 5.135

Chapter 4

4.1 Start Thinking!

For use before Activity 4.1

Sample answer: For some rides at an amusement park, you must be a certain height or age to ride. The inequality $x \geq$ the minimum height or age describes the situation.

Answers

4.1 Warm Up

For use before Activity 4.1

For 1–6: $-3\frac{1}{2}$ $2\frac{1}{2}$

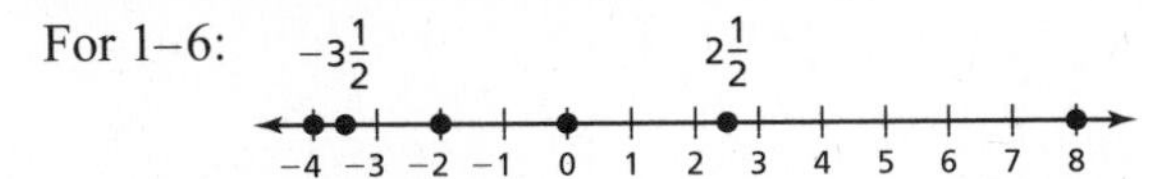

4.1 Start Thinking!

For use before Lesson 4.1

Sample answer: To move into first place, he needs to score more than 50 points; >

4.1 Warm Up

For use before Lesson 4.1

1. $x \geq -1$; all numbers greater than or equal to -1
2. $x \leq 2$; all numbers less than or equal to 2
3. $x > 4$; all numbers greater than 4
4. $x < 0$; all numbers less than 0

4.1 Practice A

1. $x < 8$; all values of x less than 8
2. $x \geq -5$; all values of x greater than or equal to -5
3. $x \leq 3$
4. $2 + y > 7$
5. $3c < -12$
6. $m - 1.5 \geq 2$
7. yes
8. no
9. no
10. yes
11. 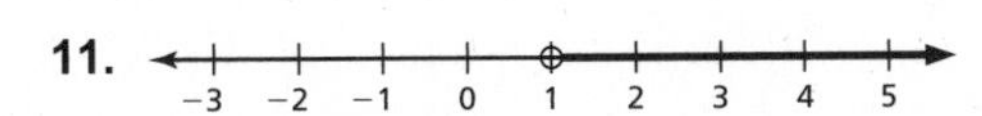
12. -2.5

13. $a \leq 12$
14. yes
15. no
16. **a.** yes **b.** no
    c. any value of x greater than -5 and less than 3

4.1 Practice B

1. $x > -4$; all values of x greater than -4
2. $x \leq 11$; all values of x less than or equal to 11
3. $x \geq 15$
4. $r + 3.7 < 1.2$
5. $\frac{h}{2} > -5$
6. $a - 8.2 \leq 12$
7. no
8. no
9. yes
10. no
11. 3.5 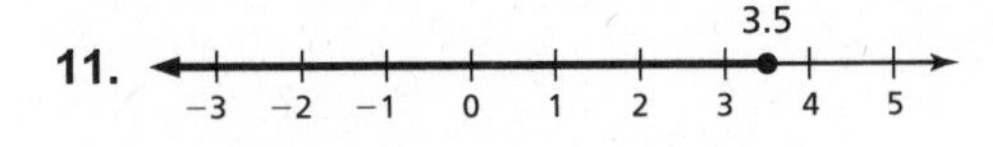
12.

13. $a \geq 16$
14. yes
15. no
16. **a.** $g \geq 3.5$;

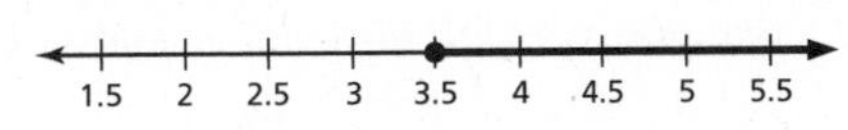

$c \geq 12$;

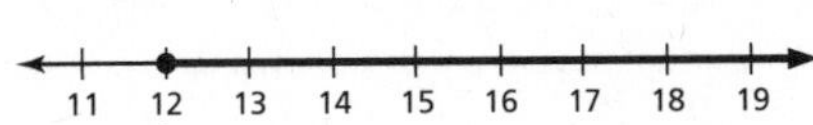

$h \geq 75$;

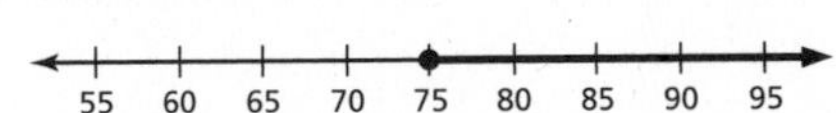

b. no; Your cousin only has 65 hours of community service and needs at least 10 more hours to meet that requirement.

4.1 Enrichment and Extension

1. **a.** $a \geq 9$ **b.** $a \leq 12$ **c.** $9 \leq a \leq 12$
2. **a.** $p \geq 0$ **b.** $p \leq 85$ **c.** $0 \leq p \leq 85$
3. $3.5 \leq n \leq 6$

4.1 Puzzle Time

A BULLDOZER

4.2 Start Thinking!

For use before Activity 4.2

$x \leq 15$; *Sample answer:* An elevator can hold up to and including 15 people.

4.2 Warm Up

For use before Activity 4.2

1. 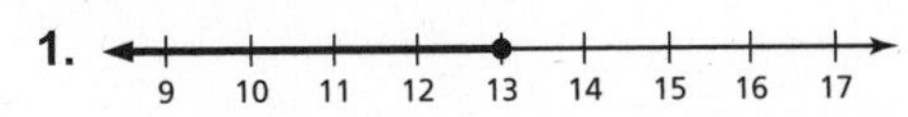
2.
3.
4.
5.
6.

Answers

4.2 Start Thinking!

For use before Lesson 4.2

Sample answer: You have \$20 to spend on souvenirs and know that you want to buy a \$5 replica of the Empire State Building. The inequality $x + 5 \leq 20$ represents the amount x that you have left to spend on souvenirs.

4.2 Warm Up

For use before Lesson 4.2

1. $x < 17$;

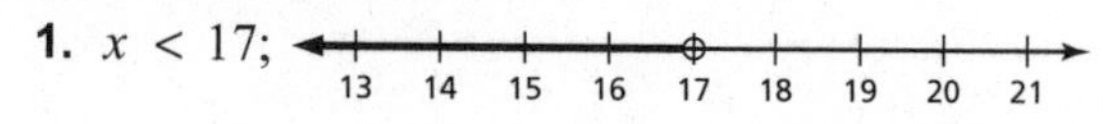

2. $h > 3$;
3. $y \leq 13$;
4. $y \geq 16$;
5. $t > 2$;
6. $x \leq 3$;

4.2 Practice A

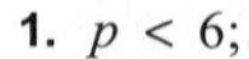

1. $p < 6$;

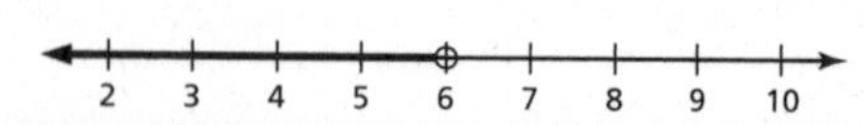

2. $s \geq -6$;
3. $k \leq 4$;
4. $n > \frac{1}{2}$;

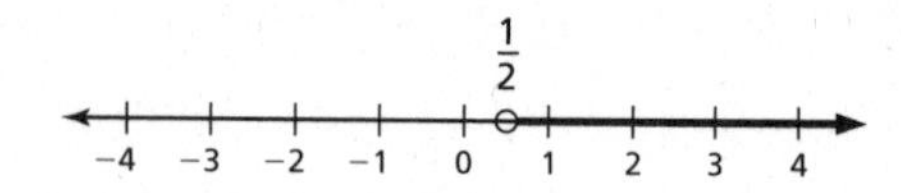

5. $z \geq 1$;
6. $t < -\frac{1}{3}$;

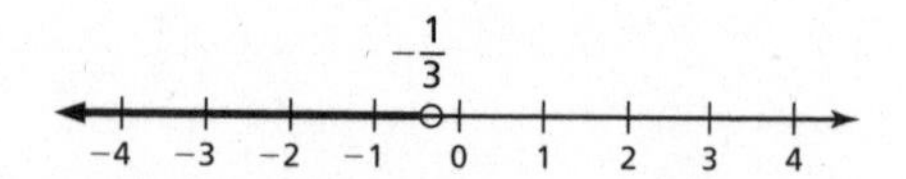

7. $d \leq -2.7$;
8. $q > 7$;

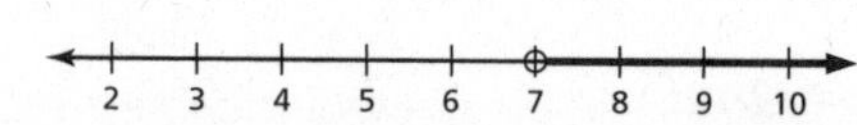

9. $a + 528 \leq 3000$; $a \leq 2472$ square feet
10. $x + 50 \leq 137.26$; $x \leq \$87.26$
11. $x + 13 < 20$; $x < 7$ meters
12. $x + 12 \geq 18$; $x \geq 6$ feet
13. a. $p + 2700 \geq 5000$; $p \geq 2300$ points

    b. $p + 3100 \geq 5000$; $p \geq 1900$ points

4.2 Practice B

1. $y \geq 5$;

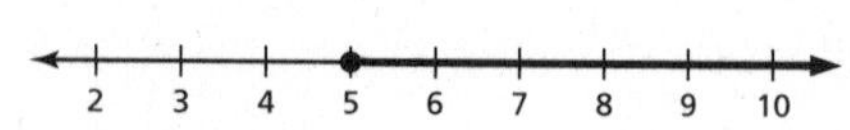

2. $w < 4.3$;

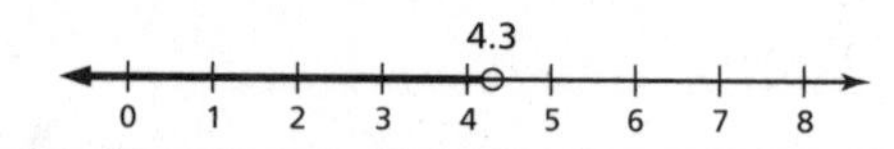

3. $v > 7\frac{2}{3}$;

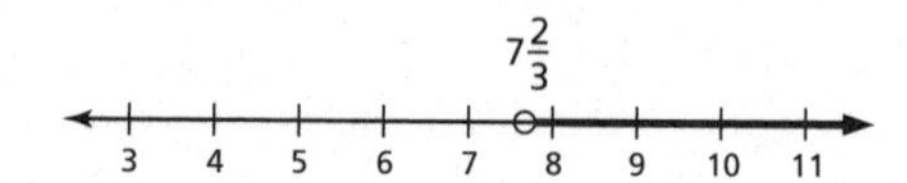

4. $k > -\frac{2}{5}$;
5. $q \geq -1$;
6. $r < 2$;

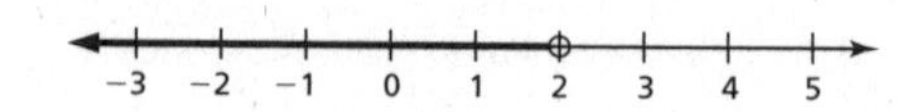

7. $c < 3.5$;

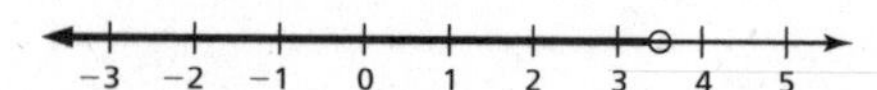

Answers

8. $p > 13.7$;

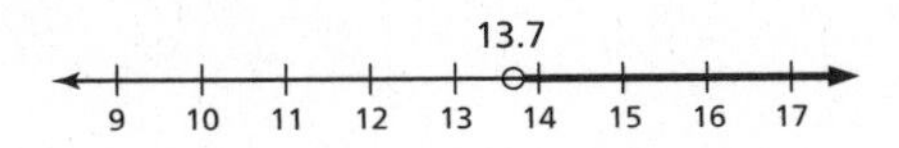

9. a. $7 + x \leq 18$; $x \leq 11$ lobsters

   b. $10 + x \leq 24$; $x \leq 14$ lobsters

   c. 6 lobsters

10. $x - 4 > 8$; $x > 12$ meters

11. $x + 35 \leq 50$; $x \leq 15$ inches

12. $c = 2$

13. all values of x greater than -5 and less than 5

14. The third side must be greater than 11 inches and less than 23 inches. Using the triangle inequality theorem, you can write and solve the inequalities $x + 6 > 17$ and $17 + 6 > x$.

4.2 Enrichment and Extension

1. 47 cm

2. a. $188 + y \geq 304.75$

   b. at least 116.75 cm

   c. at least 609.5 cm

3. a. $p - 38 \geq 162$

   b. at least 200 cm

4. Economy class will contain the greatest number of seats because they are the smallest in dimension. As a result, more of them will fit in a row of the cabin. First class will contain the least number of seats because the dimensions of the seat are so large, only a small amount of seats will fit in a row of the cabin.

4.2 Puzzle Time

THE BASEBALL GAME BETWEEN THE COLLARS AND THE SHIRTS THAT ENDED IN A TIE

4.3 Start Thinking!

For use before Activity 4.3

Sample answer: $x + 7 \leq -4$; To solve this equation subtract 7 from both sides of the equation. The answer is $x \leq -11$. $x - 6 > 8$; To solve this equation add 6 to both sides of the equation. The answer is $x > 14$.

4.3 Warm Up

For use before Activity 4.3

1. <
2. >
3. >
4. <
5. >
6. <

4.3 Start Thinking!

For use before Lesson 4.3

no; *Sample answer:* If you substitute numbers for x in each inequality, you can see that any number that makes one of the inequalities true makes the other one false.

4.3 Warm Up

For use before Lesson 4.3

1. $x < -2$
2. $x > 2$
3. $x \geq 6$
4. $x \leq -6$
5. $x > -63$
6. $x < -\frac{5}{2}$

4.3 Practice A

1. $x > 1$;

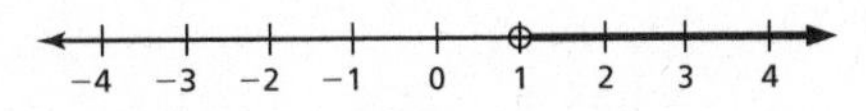

2. $r \leq 10$;

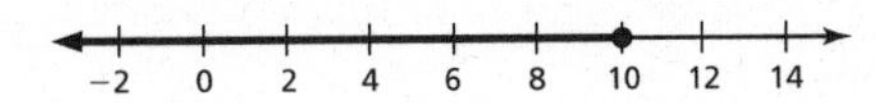

3. $h < -20$;

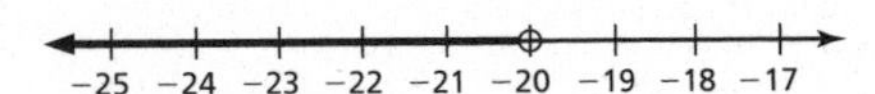

4. $u \geq 16.8$;

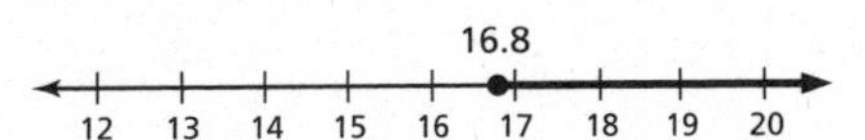

5. $j < -4.4$;

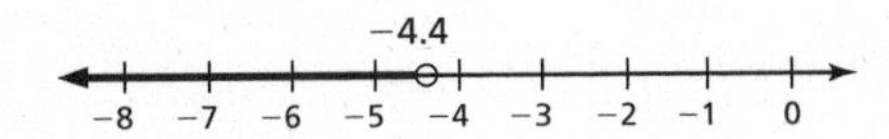

6. $x > -\frac{1}{2}$;

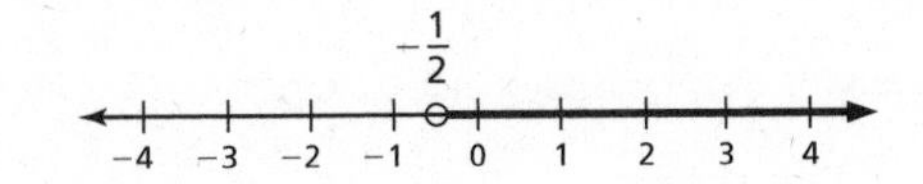

7. $5n \geq 15$; $n \geq 3$

8. $\frac{n}{4} < -1$; $n < -4$

9. $16.5g \leq 363$; $g \leq 22$ gal

Answers

10. $p \leq -5$;

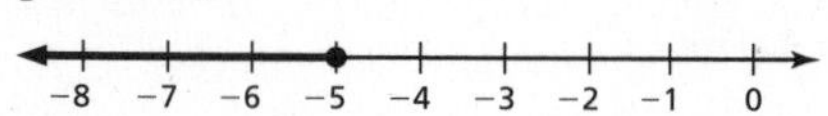

11. $v > 6$;

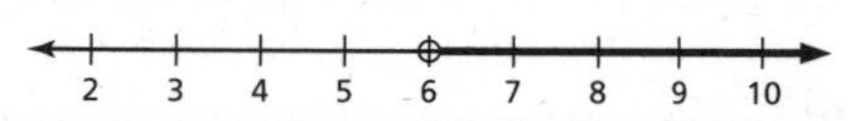

12. $g < -12.8$;

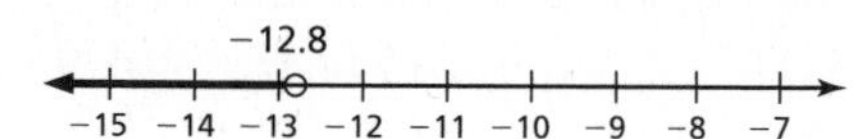

13. $y \geq -4.2$;

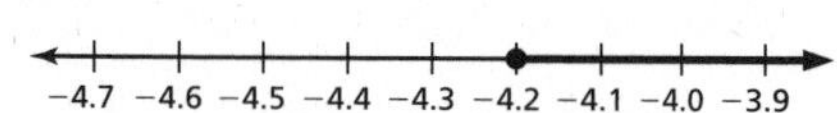

14. $h > \frac{4}{3}$;

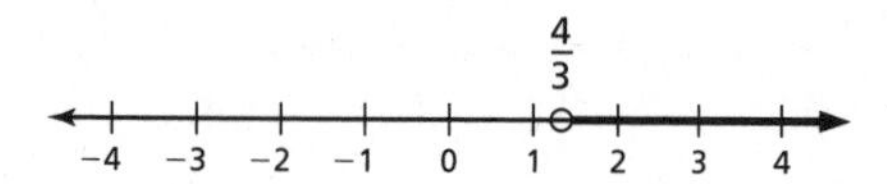

15. $a \geq 5.95$;

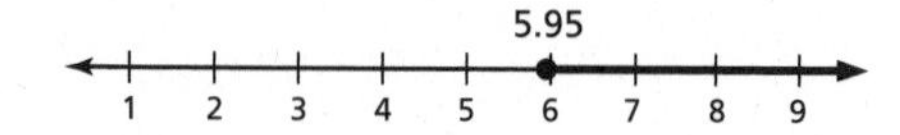

16. a. $\frac{1}{2}b \geq 20$; $b \geq 40$ beads

b. $\frac{5}{6}b \geq 20$; $b \geq 24$ beads

4.3 Practice B

1. $y \leq \frac{1}{4}$;

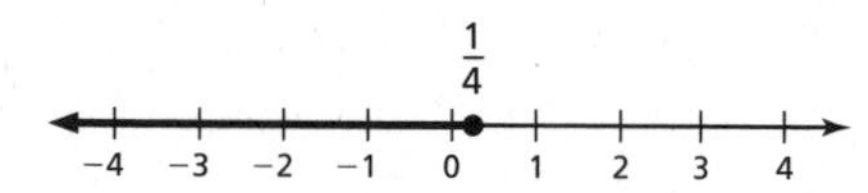

2. $p > -32$;

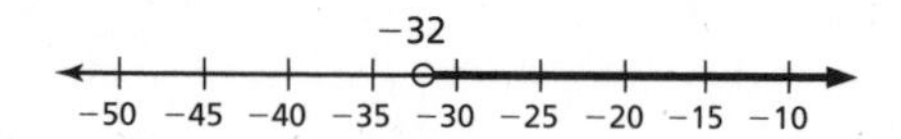

3. $g \geq 0.3$;

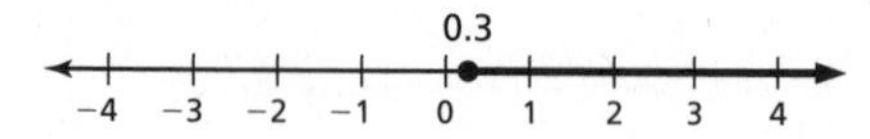

4. $k < -40$;

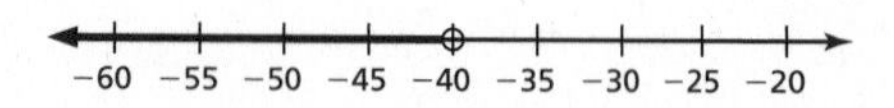

5. $s \geq 13.95$;

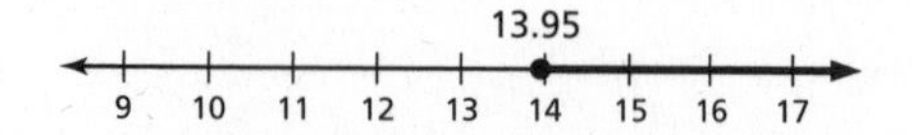

6. $x > -\frac{2}{5}$;

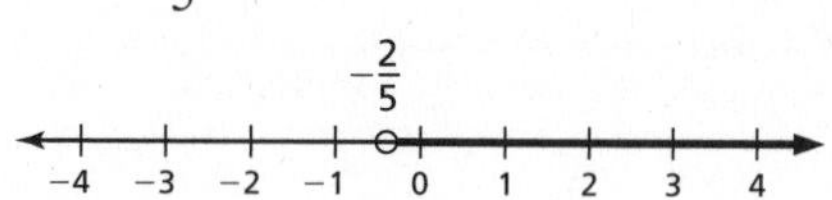

7. $\frac{n}{5} \geq 4$; $n \geq 20$

8. $2n \leq -6$; $n \leq -3$

9. $c = \frac{1}{2}$

10. $t > -4$;

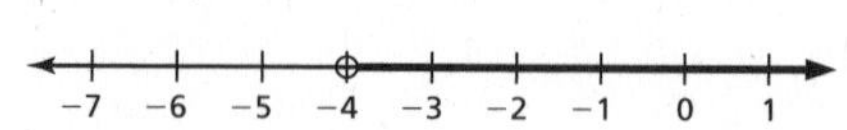

11. $u \leq \frac{2}{5}$;

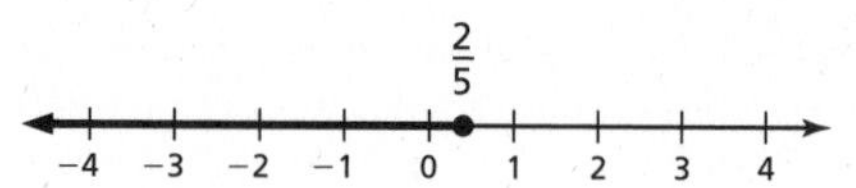

12. $q \geq -0.76$;

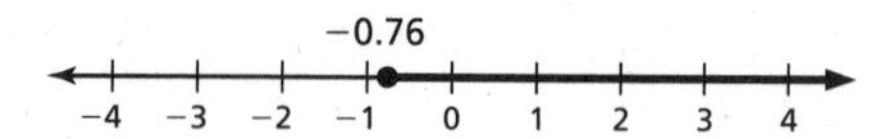

13. $d < -\frac{3}{4}$;

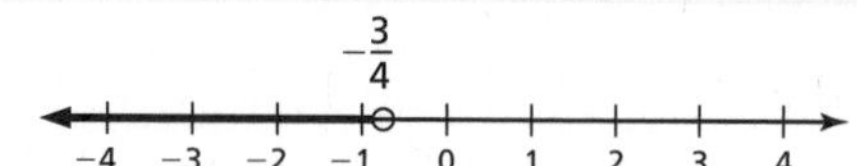

14. $r \leq 1.5$;

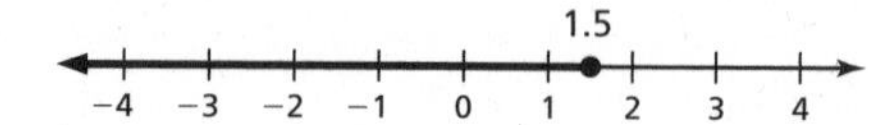

15. $j \geq 7.8$;

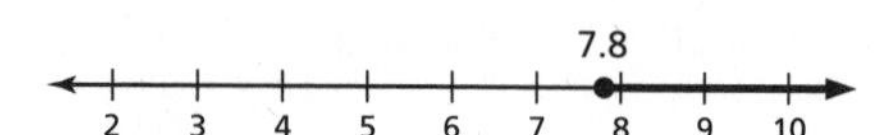

16. a. $\frac{2}{3}s \leq 10$; $s \leq 15$ shelves

b. $\frac{5}{6}x \leq 10$; $x \leq 12$ shelves

17. all values of x greater than 1 and less than 4;

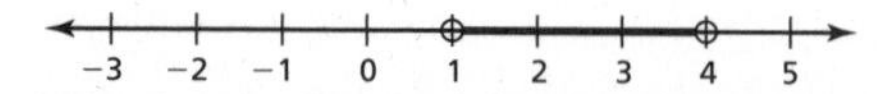

18. all values of y less than or equal to -10;

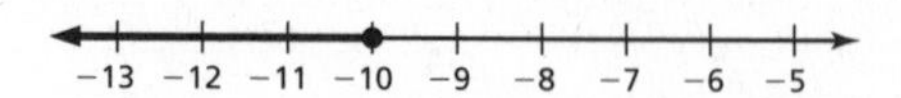

Answers

4.3 Enrichment and Extension

1. 68.2 mm
2. $e \leq 0.002$ mm
3. a. $s = 68.198$ mm, $V \approx 534{,}861.228$ mm^3
   b. $s = 68.202$ mm, $V \approx 534{,}923.972$ mm^3
   c. $68.198 \leq s \leq 68.202$
   d. $534{,}861.228 \leq V \leq 534{,}923.972$
4. a. $V \approx 534{,}901.902$ mm^3
   b. An error in the measurement of the side length will have a greater impact on the error in the volume. Because the base is a square, the error factors in the volume calculation twice.

4.3 Puzzle Time

RUN AFTER IT

4.4 Start Thinking!

For use before Activity 4.4

Sample answer: You receive a gift card for $30 to spend at an online store. The cost of shipping for your order is $5. How much can you spend without exceeding the amount on your gift card?

4.4 Warm Up

For use before Activity 4.4

1. $x < 3$
2. $x \geq 4$
3. $x > -3$
4. $x < -\frac{5}{3}$
5. $x > -4$
6. $x \geq \frac{21}{5}$

4.4 Start Thinking!

For use before Lesson 4.4

$6x + 24 > 36$; The area of the rectangle is $6(x + 4)$, or $6x + 24$. The area is more than 36 square units, so $6x + 24 > 36$. *Sample answers:* 3, 4

4.4 Warm Up

For use before Lesson 4.4

1. B
2. A

4.4 Practice A

1. $m < 3$;

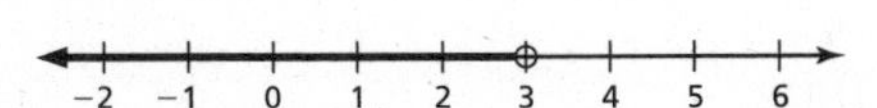

2. $r \leq 3$;

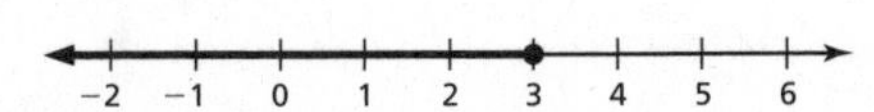

3. $k > \frac{1}{3}$;

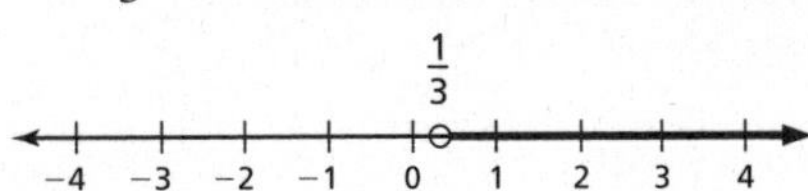

4. $c \geq -3.8$;

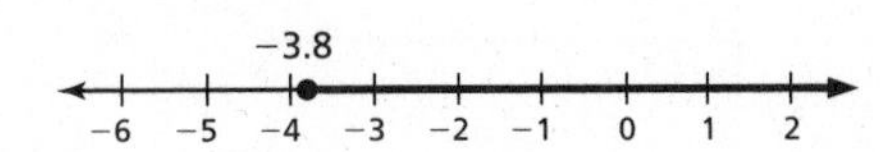

5. $20 + 0.75m \leq 65$; $m \leq 60$; at most 60 miles

6. $b > 1$;

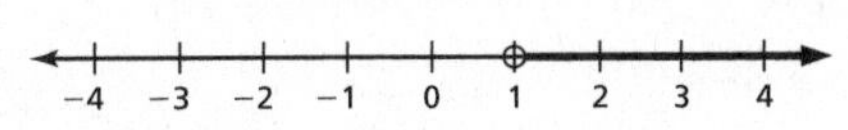

7. $p \geq -5$;

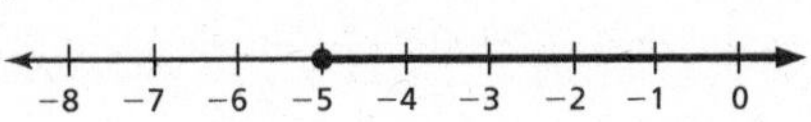

8. $d \leq 15$;

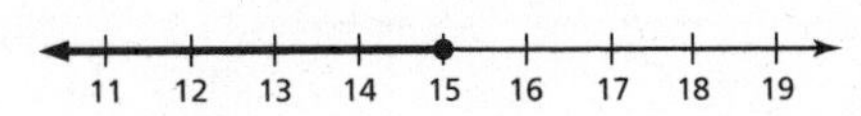

9. $a > -3.5$;

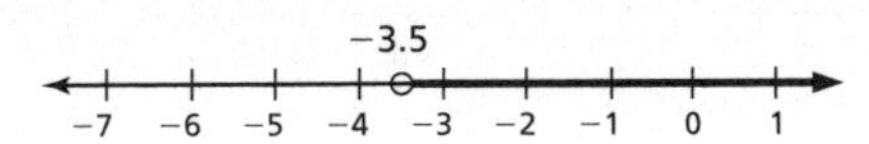

10. $5(x - 2) \geq 35$; $x \geq 9$; at least 9

11. $x > 1$;

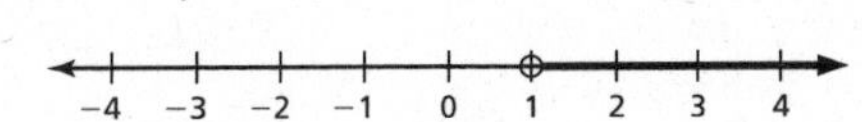

12. $w \geq 0.45$;

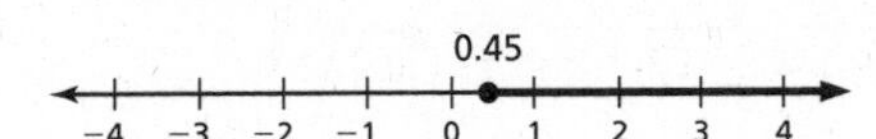

13. a. $150
    b. $150 + 20c \geq 630$; $c \geq 24$; at least 24 cell phones
    c. $150 + 20c \geq 750$; $c \geq 30$; at least 30 cell phones
    d. $150 + 20c \leq 950$; $c \leq 40$; at most 40 cell phones

4.4 Practice B

1. $q < -12$;

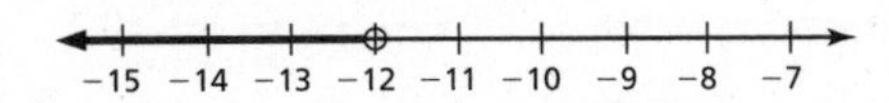

Answers

2. $v \geq -6$;

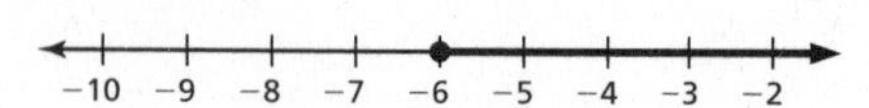

3. $k \geq -\frac{1}{2}$;

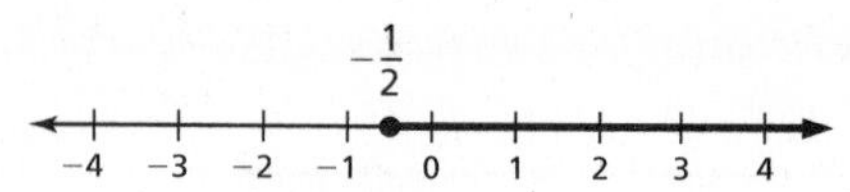

4. $n > 10$;

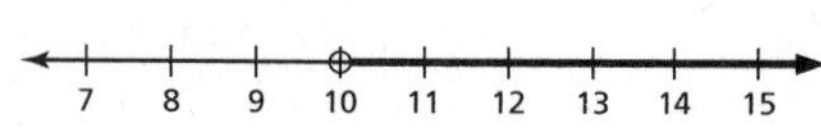

5. $300r + 2000 \geq 14{,}000$; $r \geq 40$; at least 40 occupied sites

6. $m < -4$;

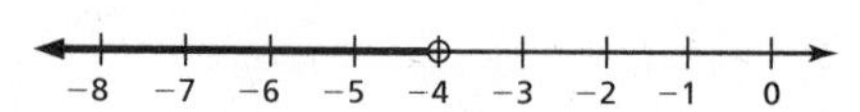

7. $f \geq -2$;

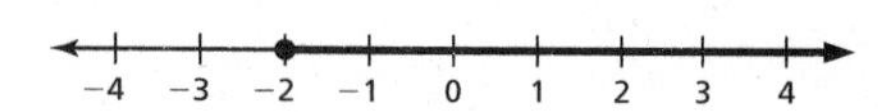

8. $p > 10$;

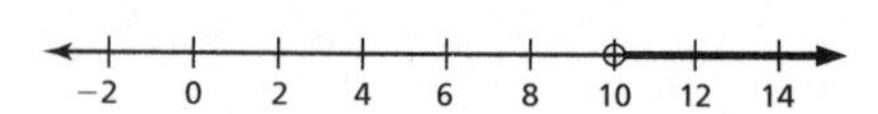

9. $w \geq 0.2$;

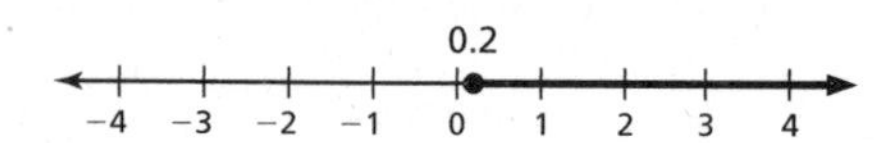

10. $4.5(x + 3) \leq 45$; $x \leq 7$

11. $x \geq 8$;

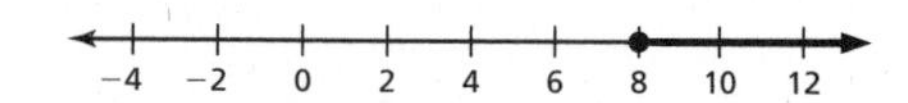

12. $v < -0.56$;

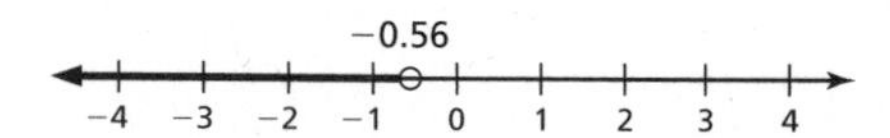

13. **a.** $750 + 6x \geq 1170$; $x \geq 70$; at least 70 animals

  b. $750 + 6x \leq 900$; $x \leq 25$; at most 25 animals

  c. \$950

4.4 Enrichment and Extension

1. $x < -2.5$;

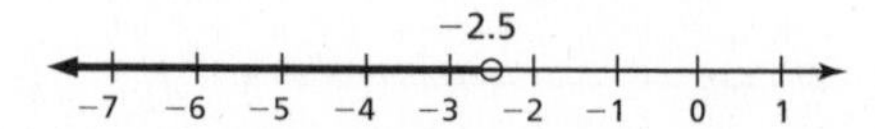

2. $x \geq 8$;

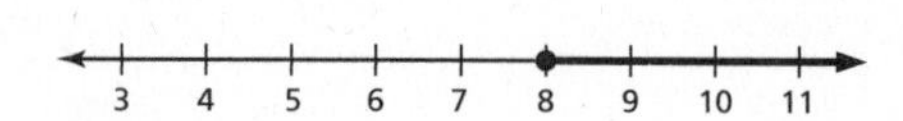

3. $x \leq -6$;

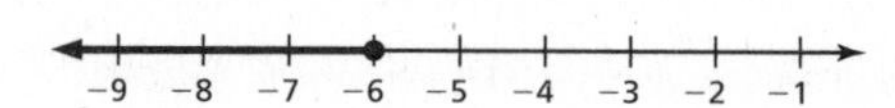

4. $x > 7$;

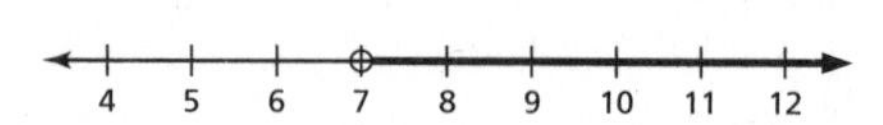

5. $x \geq 2.5$;

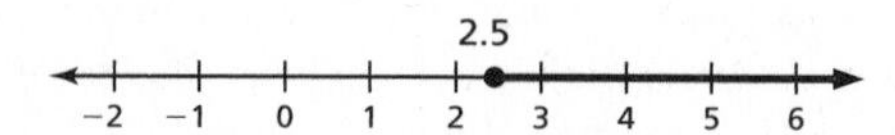

6. $x \leq 10.9$;

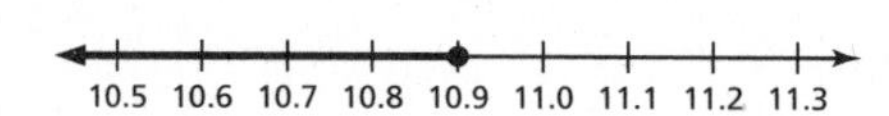

7. $x > 3$;

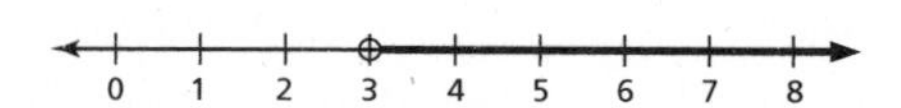

8. $x \leq -7$;

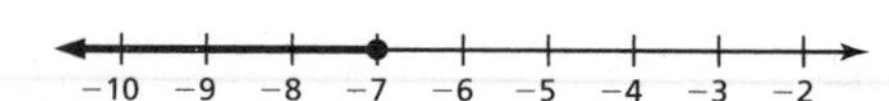

9. $x > 0$;

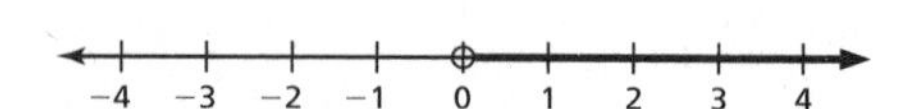

10. $\frac{85 + 91 + x}{3} \geq 90$; $x \geq 94$;

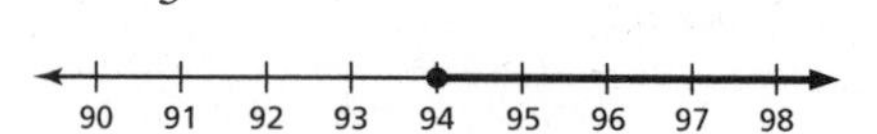

11. $\frac{x + 30 + 20}{3} \leq 20$; $x \leq 10$;

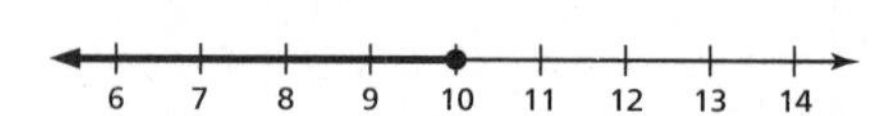

12. $5(3x + 2) > 130$; $x > 8$;

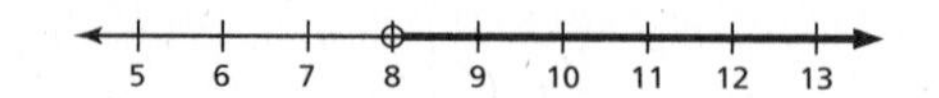

4.4 Puzzle Time

SURE BERT

Technology Connection

So, the home team earned 10 points.

Answers

Chapter 5

5.1 Start Thinking!

For use before Activity 5.1

Sample answer: For flying: You would need to consider the cost of the ticket and parking at the airport. For driving: You would need to consider the cost of gas and travel time.

5.1 Warm Up

For use before Activity 5.1

1. 0.5 h **2.** 240 min **3.** 0.25 min

4. 2.5 days **5.** 72 h **6.** 168 h

5.1 Start Thinking!

For use before Lesson 5.1

your sister; *Sample answer:* Her erasers cost $0.31 each, and your erasers cost $0.33 each.

5.1 Warm Up

For use before Lesson 5.1

1. $42 **2.** $48 **3.** 200 mi

4. $5850 **5.** $30 **6.** 210 in.

5.1 Practice A

1. $60 **2.** $3.54 **3.** 630 **4.** $\frac{4}{5}$

5. $\frac{8}{3}$ **6.** $\frac{7}{3}$ **7.** 60 mi/h

8. 3 bowlers per lane **9.** $4 per person

10. $\frac{1 \text{ lap}}{2 \text{ min}}$, or 0.5 lap per min

11. 15 grams per serving **12.** 8 min/mi

13. yes; *Sample answer:*

$$\frac{\text{blue}}{\text{orange}} = \frac{12}{9} = \frac{4}{3} \text{ and } \frac{\text{orange}}{\text{blue}} = \frac{3}{4}$$

14. 26 students per classroom

15. a. *Sample answer:* 90-pack; Often the largest package has the lowest unit rate.

b. $0.27 per pack

c. $0.18 per pack

d. $0.21 per pack

e. 60-pack; *Sample answer:* This is not what was expected.

5.1 Practice B

1. $\frac{5}{9}$ **2.** $\frac{3}{10}$ **3.** $\frac{9}{5}$

4. $\frac{500}{1}$ **5.** $\frac{29}{4}$ **6.** $\frac{3}{7}$

7. $0.225 per can **8.** $0.0645 per oz

9. about 1.82 m/sec **10.** 53 bacteria/h

11. a. B **b.** A **c.** B

d. *Sample answer:* protein **e.** calories

12. a. $6 for 4 scoops **b.** $1.50 per scoop **c.** $18

d. *Sample answer:* The graph would be a line that would be above the given graph except at point (0, 0).

e. The point would become (4, 7).

13. *Sample answer:* It costs $45 for 25 baseball tickets. What is the cost ratio?

14. *Sample answer:* You have 4 markers and 6 colored pencils. What is the ratio of markers to colored pencils?

5.1 Enrichment and Extension

1. $15.60 **2.** $30,784

3. Job A pays better. The hourly rate is $15.60, which is higher than Job B's hourly rate of $14.80. Also, Job A pays $32,448 per year, when Job B only pays $30,784 per year.

4. 180 miles **5.** 12.5 gallons **6.** 14.4 mi/gal

7. $10 per day; $50 per week; $2600 per year

8. $25

9. *Sample answer:* Sally should consider the length of time that the trips take to drive, the traffic for the different routes, road conditions, and whether or not there is public transportation available for all or part of the routes.

10. Job B's; Job A's macaroni and cheese contains 5.2 grams of fat per cup, and Job B's macaroni and cheese contains 5 grams of fat per cup.

11. Job A's; Job A's macaroni and cheese costs about $0.45 per ounce, which is less expensive than the macaroni and cheese at the cafeteria of Job B.

Answers

12. *Sample answer:* Job A pays better, has the shorter and cheaper commute, and the cheaper macaroni and cheese. It is the best choice when you consider money.

13. *Sample answer:* It is important to consider whether or not you like the type of work that you will be doing. Also, it is important to enjoy the people you work with as well as the setting in which you are working. Also, the benefits, such as health insurance, that a job offers are very important.

5.1 Puzzle Time

A FIRE-QUACKER

5.2 Start Thinking!

For use before Activity 5.2

Sample answer: Two fractions that are equivalent are $\frac{15}{18}$ and $\frac{5}{6}$. To show that they are equivalent, reduce $\frac{15}{18}$ by dividing the numerator and denominator by 3.

The reduced fraction equals $\frac{5}{6}$.

Two fractions that are not equivalent are $\frac{9}{14}$ and $\frac{13}{21}$.

They are not equivalent because $\frac{9}{14}$ and $\frac{13}{21}$ are both in simplest form.

5.2 Warm Up

For use before Activity 5.2

1. $\frac{1}{2}$ **2.** $\frac{4}{3}$ **3.** $\frac{5}{4}$ **4.** $\frac{3}{7}$

5. $\frac{5}{7}$ **6.** $\frac{10}{41}$ **7.** 12 **8.** $\frac{10}{7}$

5.2 Start Thinking!

For use before Lesson 5.2

Sample answer: You determine which is the better buy by comparing how much cereal you get for the price.

5.2 Warm Up

For use before Lesson 5.2

1. yes **2.** no **3.** no **4.** yes

5.2 Practice A

1. yes **2.** yes **3.** no **4.** yes

5. no **6.** yes **7.** yes **8.** no

9. yes **10.** no **11.** yes **12.** yes

13. no

14. yes; *Sample answer:* Each unit rate is equivalent to 2.5 minutes per lap.

15. yes **16.** yes **17.** no

18. **a.** $4.50 per chore

b. $4.50 per chore

c. yes; The rates are equivalent.

19. no; 5 tickets for $93.75 or 4 tickets for $75.20

20. 1 potato and 3 carrots

5.2 Practice B

1. no **2.** yes **3.** yes **4.** yes

5. no **6.** no **7.** yes **8.** yes

9. no

10. yes; $\frac{15}{12} = \frac{5}{4}$ and $\frac{10}{8} = \frac{5}{4}$

11. no; 3 fluid ounces of vinegar

12. **a.** $13 per pair

b. $12 per pair

c. $11.50 per pair

d. no; The cost per pair is different for each number of pairs of sandals that are purchased.

e. The buyer will purchase 30 pairs and get 3 free for $379.50, purchase 5 pairs for $65, and buy 2 pairs for $30. The total cost is $474.50.

13. 9 **14.** 15

5.2 Enrichment and Extension

1. Add 5 quarts of blue paint.

2. Add 1 quart of blue paint and 6 quarts of yellow paint.

3. Add 1 quart of yellow paint.

4. Add 1 pint of blue paint and 2 pints of red paint.

5. Add 2 pints of yellow paint and 1 pint of red paint.

6. Add 12 gallons of red paint.

7. Add 2 quarts of red paint and 4 quarts of blue paint.

8. Add 3 cups of yellow paint, 16 cups of red paint, and 4 cups of blue paint.

5.2 Puzzle Time

YOUR BREATH

Answers

Extension 5.2 Start Thinking!

For use before Extension 5.2

yes; Use the Cross Products Property to determine if the ratios are equal. Because $7 \times 6 = 42$ and $21 \times 2 = 42$, the ratios are equal.

Extension 5.2 Warm Up

For use before Extension 5.2

1. yes **2.** no

Extension 5.2 Practice

1. (0, 0): 0 pounds of chicken costs $0.
(3, 12): 3 pounds of chicken costs $12.
(5, 20): 5 pounds of chicken costs $20.

2. (0, 0): You earn $0 for working 0 hours.
(2, 50): You earn $50 for working 2 hours.
(4, 100): You earn $100 for working 4 hours.

3. $y = 2$ **4.** $y = 7$ **5.** $y = 6$ **6.** $y = 4$

7. a. Class A: 80%; Class B: 70%
b. $60
c. $7.50

5.3 Start Thinking!

For use before Activity 5.3

Sample answer: The ratios $\frac{9}{2}$ and $\frac{27}{6}$ are proportional because they are equivalent fractions.

5.3 Warm Up

For use before Activity 5.3

1. *Sample answer:* $\frac{2}{3}, \frac{4}{6}$

2. *Sample answer:* $\frac{22}{40}, \frac{33}{60}$

3. *Sample answer:* $\frac{3}{7}, \frac{9}{21}$

4. *Sample answer:* $\frac{2}{5}, \frac{6}{15}$

5. *Sample answer:* $\frac{3}{5}, \frac{6}{10}$

6. *Sample answer:* $\frac{2}{5}, \frac{4}{10}$

7. *Sample answer:* $\frac{16}{18}, \frac{24}{27}$

8. *Sample answer:* $\frac{7}{12}, \frac{21}{36}$

5.3 Start Thinking!

For use before Lesson 5.3

Yes, the scores are proportional because they both scored 5 points in 1 minute.

5.3 Warm Up

For use before Lesson 5.3

1. $\frac{p}{100} = \frac{85}{100}$ **2.** $\frac{p}{50} = \frac{74}{100}$

3. $\frac{p}{25} = \frac{80}{100}$ **4.** $\frac{p}{110} = \frac{90}{100}$

5.3 Practice A

1. $\frac{p}{70} = \frac{90}{100}$ **2.** $\frac{p}{30} = \frac{72}{100}$

3. $\frac{f}{15} = \frac{60}{100}$ **4.** $\frac{f}{24} = \frac{75}{100}$

5. $\frac{\text{2 hurricanes}}{\text{6 storms}} = \frac{\text{1 hurricane}}{n \text{ storms}}$

6. $\frac{w \text{ wins}}{\text{21 races}} = \frac{\text{8 wins}}{\text{12 races}}$

7. $\frac{\text{2 teachers}}{\text{45 students}} = \frac{t \text{ teachers}}{\text{315 students}}$

8. $a = 10$ **9.** $m = 77$ **10.** $d = 9$

11. a. $\frac{4}{9} = \frac{g}{36}$ **b.** 16 pints **c.** $6\frac{1}{2}$ gallons

12. a. 30 violin players **b.** 12 viola players **c.** $\frac{2}{5}$

13. *Sample answer:* $p = 4$, $q = 10$ and $p = 8$, $q = 20$

5.3 Practice B

1. $\frac{s}{32} = \frac{75}{100}$ **2.** $\frac{s}{80} = \frac{95}{100}$

3. The ratio of the length and height for each day should be set equal; $\frac{3.1}{h} = \frac{15.5}{45}$.

Answers

4. $\frac{3}{16} = \frac{r}{128}$ 5. $x = 15$ 6. $y = 136$

7. $r = 0.12$ 8. $k = 120$ 9. $p = 6.5$

10. $w = 32$ 11. $\frac{1}{4}$ c

12. **a.** $\frac{6}{7}, \frac{12}{14}, \frac{18}{21}$ **b.** $\frac{6}{11}, \frac{12}{22}$

 c. 12 gray keys and 22 black keys

 d. 12 gray keys, 22 black keys, 14 blue keys, 1 yellow key, 1 green key

5.3 Enrichment and Extension

1. 60 in.; 5 ft
2. *Sample answer:* If ramps are too steep, getting up and down a ramp in a wheelchair would be more difficult.
3. *Sample answer:* People in wheelchairs could get up and down more easily and be in less danger. However, the ramp will take up more room and cost more to build.
4. 30 ft
5. It meets the requirements. The ramp has a rise to run ratio of 1 : 14, which is less than 1 : 12.
6. The first sloping part is steeper; *Sample answer:* After a person pushes up the first part, the second part is not as steep and easier to push up and down when the person is more tired.
7. The first part must be 20 feet long and the second part must be 27.5 feet long.
8. *Answer should include, but is not limited to:* The design must show a maximum rise to run ratio of 1 : 12 with platforms at the top and bottom that are at least 60 inches long. Any changes in direction must have a platform that is at least 60 inches by 60 inches.

5.3 Puzzle Time

THE ORCA-DONTIST

5.4 Start Thinking!

For use before Activity 5.4

Sample answer: enlargements of pictures, doubling or halving recipes, mixing more paint

5.4 Warm Up

For use before Activity 5.4

1. 35 students per bus 2. 121 miles per hour

3. 78 pages per hour 4. 27 seats per row

5. 341 miles per hour 6. 136 pages per day

5.4 Start Thinking!

For use before Lesson 5.4

Sample answer: The bakery will need to increase the amounts proportionally to what was made in the test batch.

5.4 Warm Up

For use before Lesson 5.4

1. $a = 5$ 2. $m = 44$ 3. $v = 12$

4. $n = 16$ 5. $t = 6$ 6. $k = 2.4$

5.4 Practice A

1. 49 2. 36 3. 5 4. 6

5. 15 6. 82.5 7. 42 oranges

8. $900 9. 7.5 10. 4 11. 20

12. false; *Sample answer:* If $\frac{p}{q} = \frac{3}{5}$, then $5p = 3q$ by the Cross Products Property and $\frac{5}{p} = \frac{3}{q}$ is equivalent to $3p = 5q$.

13. **a.** 27 cm **b.** 9 cm

14. 18 men; *Sample answer:* Because 2 out of 7 people at the lecture are men, $\frac{2}{7} = \frac{m}{63}$ where m is the number of men; $m = 18$

15. $x = 39$, $y = 25$, $z = 2.1$

16. **a.** 32 min **b.** 6.4 mi/h

5.4 Practice B

1. 64 2. 27 3. 350 4. 286

5. $42\frac{2}{3}$ 6. 57.6 7. $26.64 8. 2

9. 17 10. 2.5

11. **a.** 5 servings **b.** 8 servings

 c. 45.5 g **d.** $2\frac{1}{4}$ c

12. 121 nonmembers 13. $h = \frac{4}{3}$ or $h = -\frac{4}{3}$

14. **a.** 29.19 lb **b.** 0.6 gal

Answers

5.4 Enrichment and Extension

1. 1.53

2. 15 innings; not reasonable; He has been averaging about one and a half hits and walks per inning. It would be very unlikely to go 15 innings straight without allowing a hit or a walk.

3. 13 walks; Solving the proportion $\frac{1.28}{1} = \frac{92 + w}{82}$.

4. 13 hits and walks; reasonable; He has been averaging 1.28 hits and walks per inning. Because $1.28 \bullet 9 = 11.52$, he would be expected to allow between 11 and 12 hits and walks in 9 innings. So, allowing 13 hits and walks or less is reasonable.

5. 209 shots

6. The goalies faced different numbers of shots. In order to find their combined save percentage, you must find their combined number of saves and divide by their combined number of shots on goal. Solve the proportion $\frac{907}{1000} = \frac{204}{s}$ to find how many shots on goal the second goalie has faced. The second goalie has faced 225 shots. So, the goalies have made 395 total saves and had 434 total shots on goal, for a save percentage of 0.910.

7. 31 shots on goal; not reasonable; On average, he has faced about 35 shots per game and given up 3 goals per game. So it is not likely that he could make 31 saves without allowing a goal.

5.4 Puzzle Time

THE CLOCK IN THE CAFETERIA THAT WAS SLOW BECAUSE EVERY DAY AT LUNCH IT WENT BACK FOR SECONDS

5.5 Start Thinking!

For use before Activity 5.5

Sample answer: A judge compares the scores of all of the dancers. The dancer with the highest score is awarded first place.

5.5 Warm Up

For use before Activity 5.5

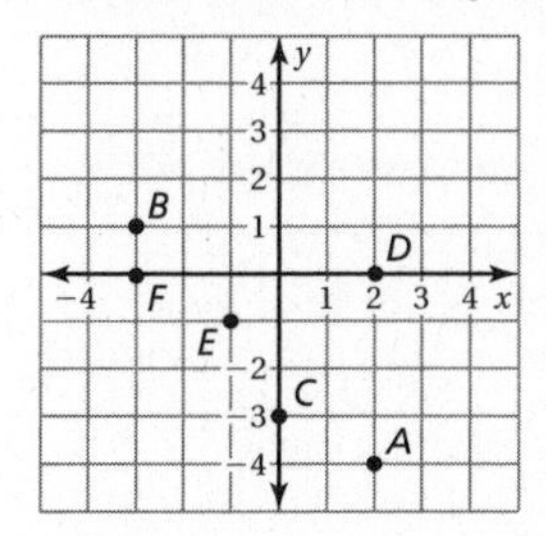

5.5 Start Thinking!

For use before Lesson 5.5

Sample answer: In auto racing, many cars travel at different speeds. They travel a distance over a period of time, which is their rate. The cars also try to maintain a consistent speed. Two or more cars can be compared on a graph to determine their speed. The car with the steepest line is traveling the fastest.

5.5 Warm Up

For use before Lesson 5.5

Time (sec)	Distance 1 (ft)	Distance 2 (ft)
0	0	0
1	12	6
2	24	12
3	36	18
4	48	24

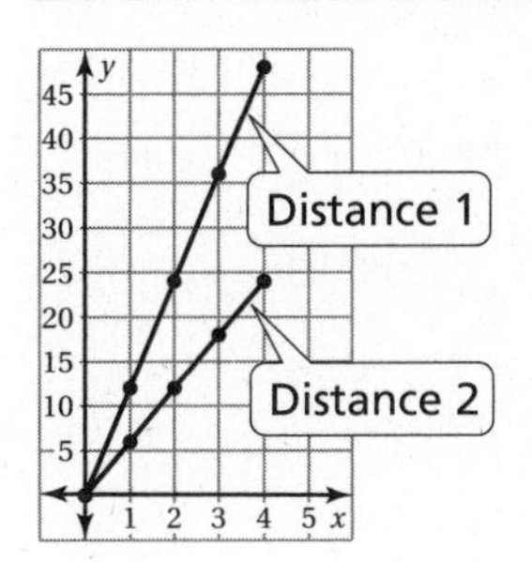

The graph for Distance 1 is steeper.

5.5 Practice A

1. 5
2. $\frac{2}{3}$
3. $\frac{3}{4}$
4. $\frac{5}{2}$

5. 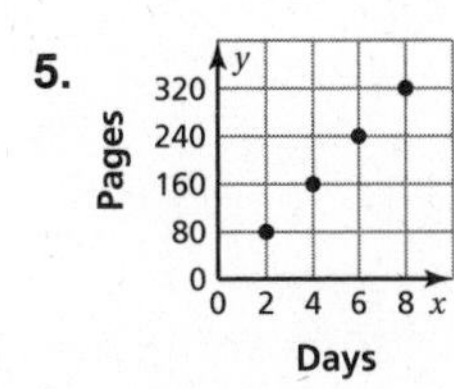

40 pages per day

6. 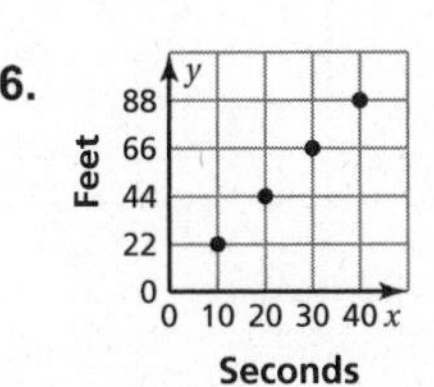

$\frac{11}{5}$ feet per second

7. 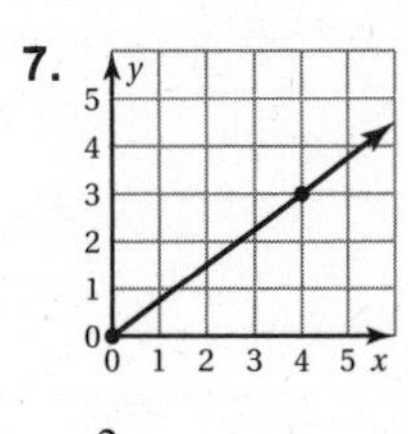

$\frac{3}{4}$

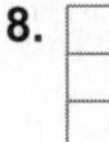

8. 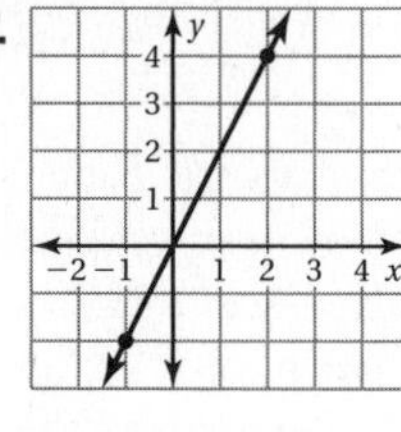

2

9. 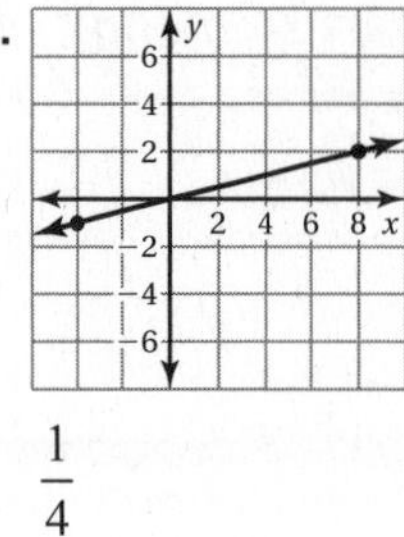

$\frac{1}{4}$

10. a. The line for the calendars is steeper than the line for the greeting cards, so a calendar is more expensive than a box of greeting cards.

b. calendar: 10; greeting cards: 6; A calendar costs \$10 and a box of greeting cards costs \$6.

c. \$6

d. 6 boxes of greeting cards or 3 calendars and 1 box of greeting cards

5.5 Practice B

1. $\frac{2}{3}$

2. $\frac{5}{2}$

3. $y = 3x$; For each value of x except 0, the value of $y = 3x$ is greater than $y = \frac{1}{4}x$.

4. $y = \frac{3}{5}x$; For each value of x except 0, the value of $y = \frac{3}{5}x$ is greater than $y = \frac{2}{5}x$.

5. 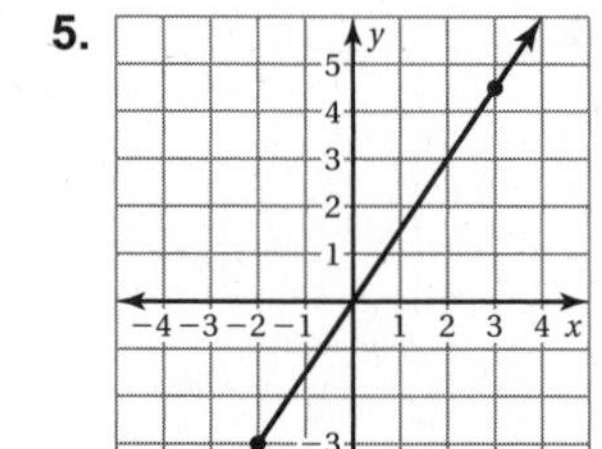

$\frac{3}{2}$

6. 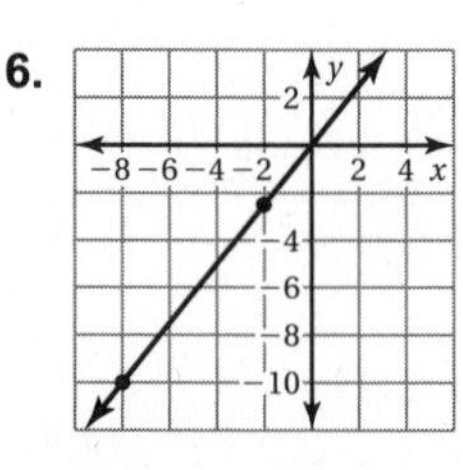

$\frac{5}{4}$

7. 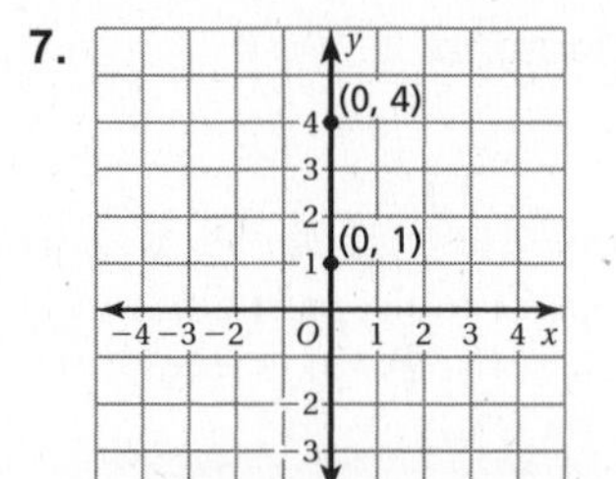

The slope is undefined because slope is the change in y divided by the change in x. The change in x is 0 in this case, and division by 0 is undefined.

8. a. 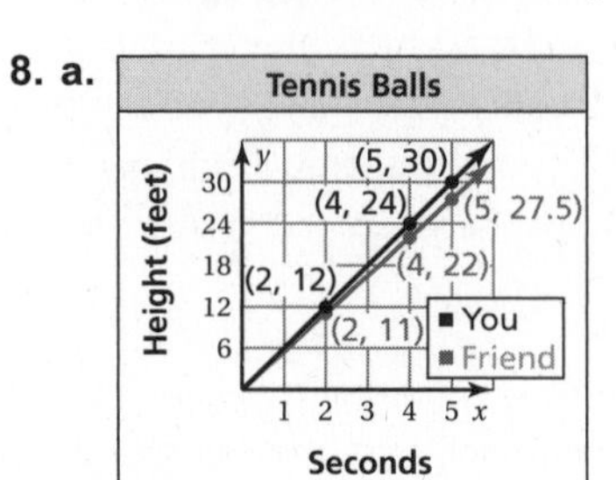

b. your ball: 6; your friend's ball: 5.5; Your ball is rising 6 ft/sec and your friend's ball is rising 5.5 ft/sec.

c. your ball; The slope is greater for the graph of the height of your tennis ball.

d. your ball: 21 ft; your friend's ball: 19.25 ft

9. 15

5.5 Enrichment and Extension

1. $\frac{1}{3}$

2. $\frac{5}{4}$

3. 1

4. $-\frac{3}{4}$

5. The slope would still be the same. You would obtain $\frac{-3}{-1}$, which is equal to 3.

6. *Sample answer:* $(6, 8), (8, 11), (10, 14)$

7. $y = -8$

8. $x = -3$

5.5 Puzzle Time

EVAPORATED MILK

5.6 Start Thinking!

For use before Activity 5.6

Sample answer: The school mascot would need to increase proportionally to your sketch. Sample dimensions for the mural could be 24 inches high and 48 inches wide.

Answers

5.6 Warm Up

For use before Activity 5.6

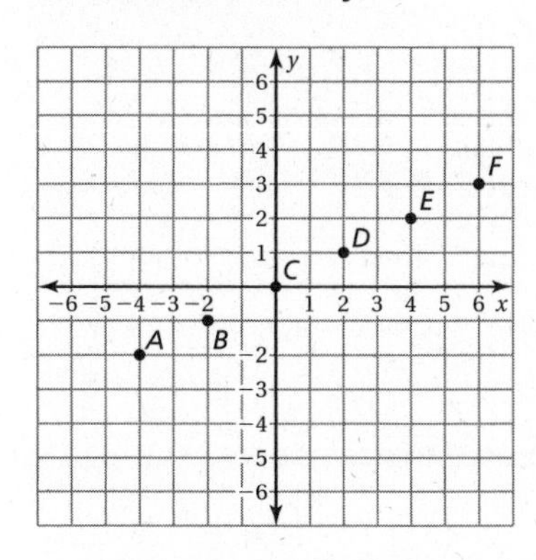

1. Start at the origin. Move 4 units left and 2 units down. Plot the point.
2. Start at the origin. Move 2 units left and 1 unit down. Plot the point.
3. Plot the point at the origin.
4. Start at the origin. Move 2 units right and 1 unit up. Plot the point.
5. Start at the origin. Move 4 units right and 2 units up. Plot the point.
6. Start at the origin. Move 6 units right and 3 units up. Plot the point.

5.6 Start Thinking!

For use before Lesson 5.6

You will need to mow your neighbor's lawn 10 times; $y = 20x$; *Sample answer:* Yes, the situation does model a direct variation because $20 stays the same (or is the constant of variation) every time you mow the lawn.

5.6 Warm Up

For use before Lesson 5.6

1. 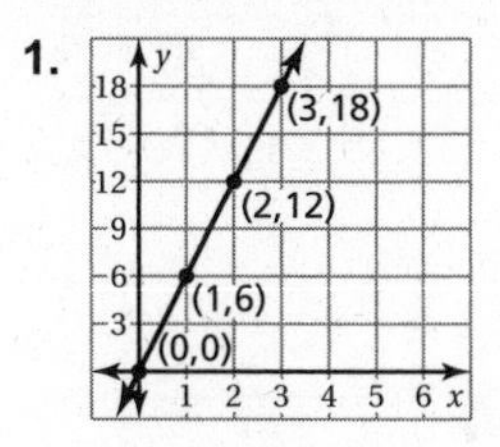

   yes; All the points lie on a line and the line passes through the origin.

2. 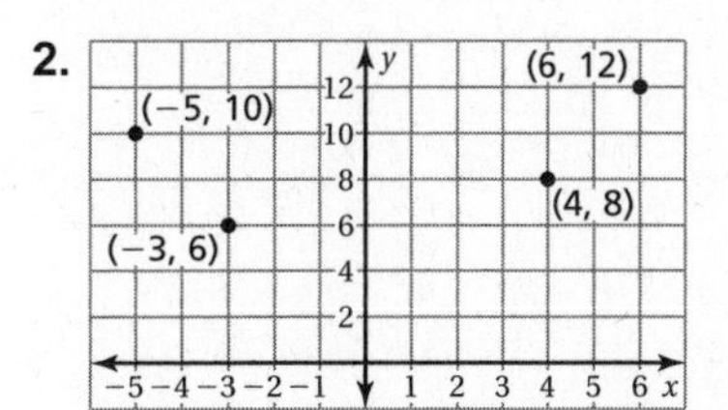

   no; The points do not lie on a line.

3. 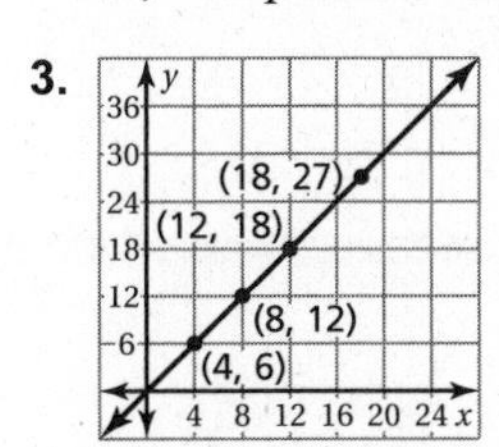

   yes; All the points lie on a line and the line passes through the origin.

5.6 Practice A

1. 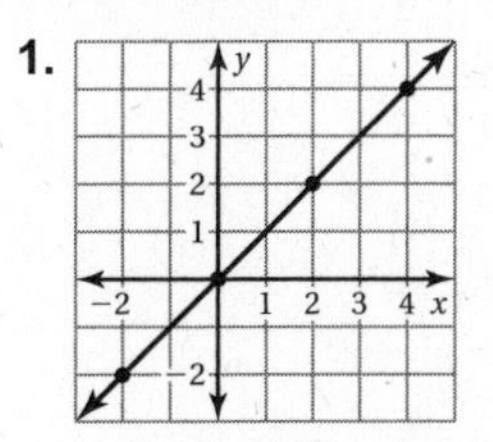

   yes; The line through the plotted points passes through the origin.

2. 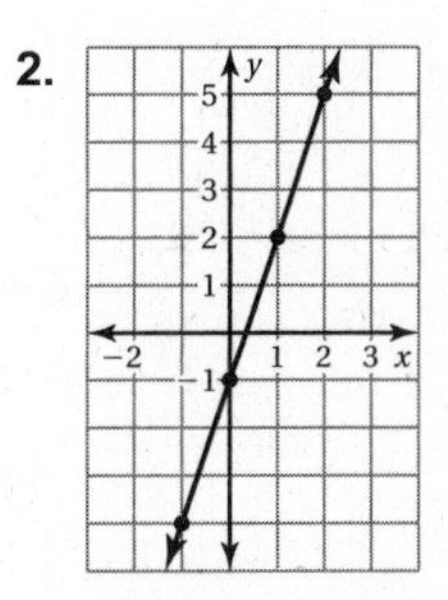

   no; The line through the plotted points does not pass through the origin.

3. no; The plotted points do not lie on a line.
4. yes; The line through the plotted points passes through the origin; $k = \frac{1}{2}$
5. yes; The equation can be written as $y = kx$; $k = 3$
6. no; The equation cannot be written as $y = kx$.
7. no; The equation cannot be written as $y = kx$.

Answers

8. 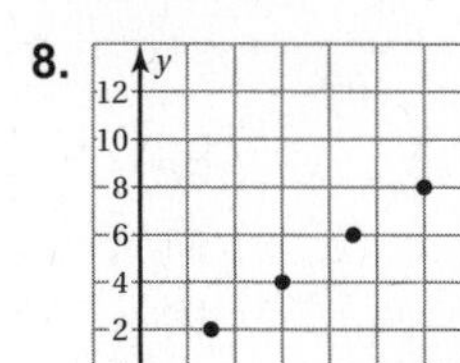

yes; $y = \frac{2}{3}x$

9. $y = 3x; k = 3$

10. $y = 5x; k = 5$

11. $y = 4x; k = 4$

12. a. 10 teaspoons **b.** $y = 2x$ **c.** 6 gallons

13. 7 tickets

14. a. $y = 0.95x$ **b.** $y = 3.8x$

c. $y = \frac{1}{0.95}x$, or $y = 1.05x$

d. They are reciprocals.

5.6 Practice B

1. 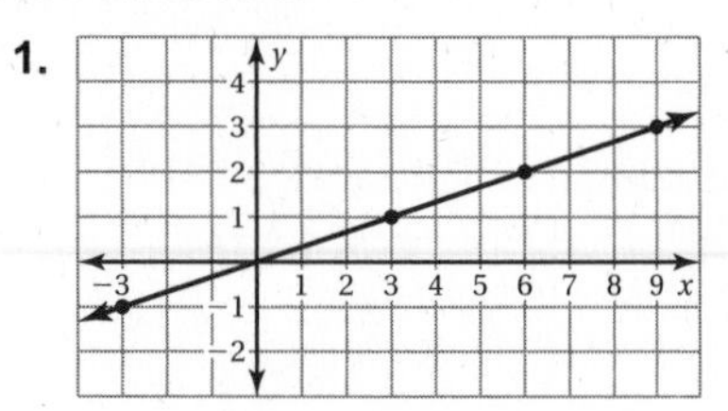

yes; The line through the plotted points passes through the origin.

2. 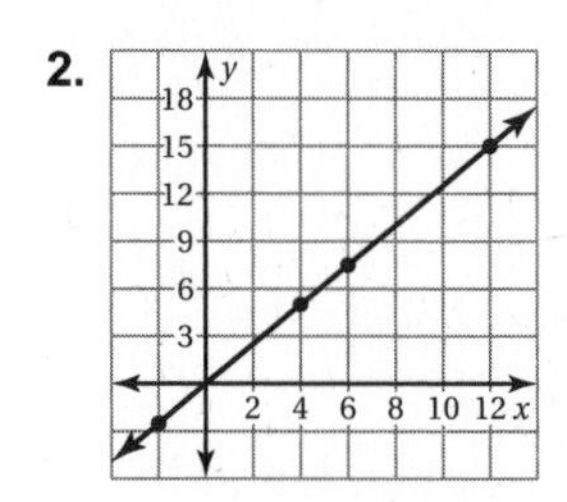

yes; The line through the plotted points passes through the origin.

3. yes; The equation can be written as $y = kx; k = 3$

4. no; The equation cannot be written as $y = kx$.

5. no; The line does not pass through the origin.

6. no; The graph is not a line.

7. a. points earned: 40, 48, 56, 64, 72, 80

b. $y = \frac{5}{4}x$

c. 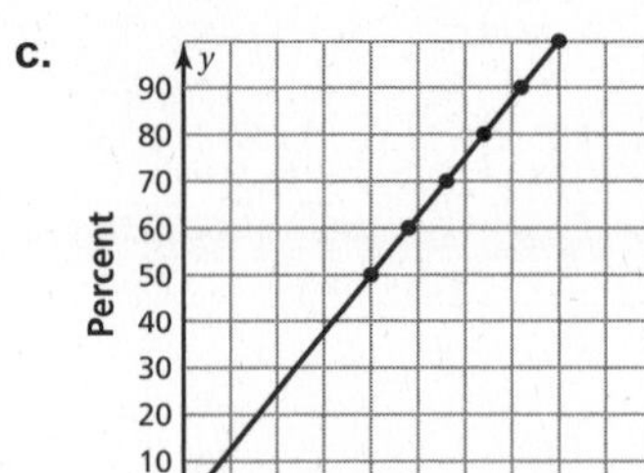

yes; The graph is a line through $(0, 0)$.

8. $y = 2x; k = 2$

9. $y = \frac{3}{2}x; k = \frac{3}{2}$

10. $y = \frac{11}{2}x; k = \frac{11}{2}$

11. friend's ramp

12. yes; Every direct variation has an equation of the form $y = kx$ where $k \neq 0$ and $(0, 0)$ is a solution of every equation of this form.

no; A graph that passes through the origin but is not a line does not represent a direct variation.

5.6 Enrichment and Extension

1. a. $P = -2.54n$; The store's profit is decreasing by \$2.54 for every Wacky Widget it purchases.

b. The profit is negative because the company is spending money to get the Wacky Widgets. Because k is negative, the graph will be a straight line that slopes downward from left to right.

c.

Items purchased, *n*	0	5	10
Profit, *P*	0	–\$12.70	–\$25.40

Items purchased, *n*	15	20
Profit, *P*	–\$38.10	–\$50.80

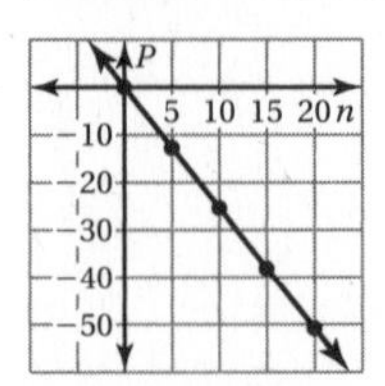

The graph makes a straight line that goes through the origin and slopes downward.

Answers

2. a. $V = 0.05p$; Tomiqua's heart pumps 0.05 liter of blood with each beat when she is at rest.

b. 4.5 L

3. a. Let m = weight on the moon, e = weight on Earth.

$$m = \frac{1}{6}e.$$

The weight on the moon is one-sixth the weight on Earth.

b. *Sample answer:* For a 150 pound person, the weight on the moon would be

$$m = \frac{1}{6}(150) = 25 \text{ lb.}$$

5.6 Puzzle Time

THEY TAKE THE BUZZ

Technology Connection

1–8. *Sample answers:* The segments could be in other positions or different lengths, but should have the same slope as those shown.

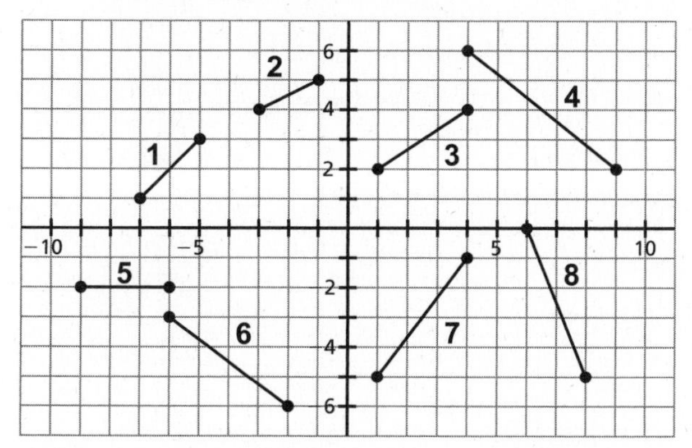

9. The slopes that are greater than 1 or less than −1 are the steepest. The slopes that are less than 1 and greater than −1 are the least steep.

10. zero

11. Vertical lines have no slope because division by zero is undefined.

12. Yes, you can move the segment around the graph without changing the slope by not moving the endpoints. The segments are parallel to one another.

Chapter 6

6.1 Start Thinking!

For use before Activity 6.1

Sample answer: The percent has the decimal point two places to the right; no; Andrew only moved the decimal point one place to the right. It should be 98%.

6.1 Warm Up

For use before Activity 6.1

1. $\frac{3}{8}$ **2.** $\frac{1}{2}$ **3.** $\frac{1}{4}$

4. $\frac{1}{8}$ **5.** $\frac{5}{6}$ **6.** $\frac{3}{8}$

6.1 Start Thinking!

For use before Lesson 6.1

Sample answer: In basketball, free throw percentage, field goal percentage, and three point percentage are often given as decimals, which makes them easier to compare.

6.1 Warm Up

For use before Lesson 6.1

1. 0.19 **2.** 0.02 **3.** 0.075

4. 89% **5.** 54% **6.** 10%

6.1 Practice A

1. 0.81 **2.** 0.78 **3.** 0.05 **4.** 0.08

5. 0.4 **6.** 0.6 **7.** 0.237 **8.** 0.1675

9. 1.5 **10.** 2.1 **11.** 1.86 **12.** 4.16

13. 1.008 **14.** 0.0517 **15.** 0.004 **16.** 0.0004

17. The decimal point is moved to the right instead of the left.

$1.475\% = 01.475\% = 0.01475$

18. 66% **19.** 32% **20.** 51% **21.** 97%

22. 1% **23.** 4% **24.** 31.2% **25.** 46.8%

26. 50% **27.** 120% **28.** 108% **29.** 116%

30. 0.3% **31.** 2.5% **32.** 2.45% **33.** 202.5%

34. The decimal point is moved from the wrong place.

$1.8 = 1.80 = 180\%$

35. 0.54 **36.** 15% **37.** 88%

Answers

38. 0.86, 86%

39. 5 times

40. $\frac{21}{100}$; 0.21

41. $\frac{3}{4}$, 0.75

42. $\frac{16}{25}$, 0.64

43. $\frac{17}{20}$, 0.85

6.1 Practice B

1. 54%
2. 37%
3. 22.2%
4. 92.9%
5. 140%
6. 250%
7. 2000%
8. 0.5%
9. $\frac{17}{25}$, 0.68
10. $\frac{9}{100}$, 0.09
11. $\frac{11}{20}$, 0.55
12. $\frac{13}{50}$, 0.26
13. $\frac{53}{125}$, 0.424
14. $\frac{92}{125}$, 0.736
15. $\frac{5}{16}$, 0.3125
16. $\frac{893}{2000}$, 0.4465
17. 64%
18. **a.** 0.48, $\frac{12}{25}$; 0.28, $\frac{7}{25}$; 0.14, $\frac{7}{50}$ **b.** 10%

c.

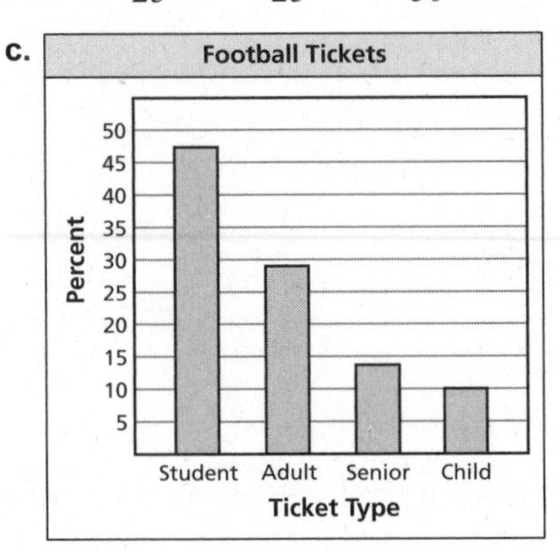

19. **a.** 33% **b.** 35% **c.** $1\frac{3}{4}$ times **d.** 27%, 0.27

20. **a–d.**

Distance	within 1 mi	1–2 mi	2–3 mi	over 3 mi
Percent	50%	28%	10%	12%

6.1 Enrichment and Extension

1. 50%; 0.5
2. 50%; 0.5
3. **a.** 25%; 0.25 **b.** 75%; 0.75 **c.** 50%; 0.5 **d.** 50%; 0.5
4. **a.** 56.25%; 0.5625 **b.** 43.75%; 0.4375
5. **a.** 4 **b.** 9; 28.125%, 0.28125

6.1 Puzzle Time

BOTH ARE SUPPOSED TO BE DUE

6.2 Start Thinking!

For use before Activity 6.2

Sample answer: You can rewrite each fraction as a decimal and compare decimals or you can write each fraction using the same denominator and compare the numerators. $\frac{7}{12}$; methods may vary.

6.2 Warm Up

For use before Activity 6.2

1. 1.4, 1.44, 3.3, 4.1, 4.3
2. 0.5, 0.7, 1.5, 1.57, 1.75
3. 0.02, 0.04, 0.2, 0.24, 0.44
4. 0.06, 0.6, 6.06, 6.6, 6.63
5. 0.02, 0.17, 0.2, 0.72, 2.27
6. 0.18, 0.81, 1.8, 1.88, 8.1

6.2 Start Thinking!

For use before Lesson 6.2

Answers will vary. Check students' work.

6.2 Warm Up

For use before Lesson 6.2

1. 45%
2. 0.4
3. $\frac{9}{17}$
4. 0.6
5. $\frac{2}{5}$
6. 240%

6.2 Practice A

1. $\frac{3}{4}$
2. 0.54
3. neither; 0.21 = 21%
4. $\frac{2}{3}$
5. 49%
6. 16%
7. neither; $\frac{12}{25} = 48\%$
8. 12%
9. 1.2
10. $\frac{31}{50}$
11. $50\frac{1}{4}$
12. $\frac{1}{8}$
13. $\frac{68}{100} = 68\%$, not 0.68%.

    So, 68% > 0.7% and $\frac{17}{25}$ is the greater number.

Answers

14. $63\%, 0.64, \frac{13}{20}$

15. $\frac{11}{25}, 45\%, 0.46$

16. $0.12, \frac{1}{8}, 13\%, 0.135$

17. $\frac{7}{8}, 90\%, 0.925, 0.93, \frac{15}{16}$

18. $3\frac{3}{5}, 362\%, 3.66, 3\frac{2}{3}, 36$

19. $0.27, 27.3\%, \frac{11}{40}, 28\%, 0.3$

20. the punch that you made

21. You made a higher percent of baskets.

22. yes; 62.5% is greater than 60%.

23. grade 7, grade 6, grade 8

6.2 Practice B

1. $\frac{1}{4}$ **2.** $\frac{5}{9}$ **3.** 3.2 **4.** 1

5. $0.3, \frac{8}{25}, 33\%, \frac{1}{3}, 33.6\%$

6. $210\%, \frac{43}{20}, 2.2, 2.\overline{2}$

7. B **8.** C **9.** A **10.** D

11. *Sample answer:* Write 31% and 32% as the decimals 0.31 and 0.32. Graph 0.31 and 0.32 on a number line. Then identify a decimal between them, such as 0.315.

12. yes; 5% of a meter $= 0.05(100 \text{ cm}) = 5$ cm;

Because 6 centimeters is greater than 5 centimeters, 6 centimeters is greater than 5% of a meter.

13. no; 6% of 1 lb $= 0.06(16 \text{ oz}) = 0.96$ oz;

Because 0.96 ounce is less than 1 ounce, 6% of a pound is less than an ounce.

14. 0.0004 of a day, 1% of an hour, $\frac{2}{3}$ of a minute

15. a. $0.45\%, 0.0082, \frac{1}{36}, \frac{1}{16}$

b. The population of Florida has grown faster than the population of the U.S.; *Sample answer:* People tend to move to Florida because it is a nice place to live.

c. *Sample answer:* no; Some day Florida will become too crowded to keep growing.

16. no;

$$\begin{aligned} 0.000002\% = 0.00000002 &= \frac{2}{100{,}000{,}000} \\ &= \frac{20}{1{,}000{,}000{,}000} \\ &= 20 \text{ parts per billion.} \end{aligned}$$

Because this is greater than 10 parts per billion, it is not an allowable amount.

6.2 Enrichment and Extension

$1\%, \frac{11}{40}, \frac{17}{60}, 31\%, \frac{9}{25}, 0.38\overline{1}, \frac{31}{75}, 41.5\%, \frac{56}{99}, 0.\overline{75}, \frac{101}{120}, 85.25\%, 100\%$

6.2 Puzzle Time

IT WAS TRULY IN A JAM

6.3 Start Thinking!

For use before Activity 6.3

Sample answer: Percents are used in banks to calculate interest rates.

6.3 Warm Up

For use before Activity 6.3

1. *Sample answer:* 260 **2.** *Sample answer:* 220

3. *Sample answer:* −140 **4.** *Sample answer:* −1200

5. *Sample answer:* −19 **6.** *Sample answer:* 500

6.3 Start Thinking!

For use before Lesson 6.3

First determine how many goals your team scored, which is 10 goals. Then draw a model. Your team scored 60% of its goals during regular play.

6.3 Warm Up

For use before Lesson 6.3

1. 12 **2.** 50% **3.** 50

4. 104 **5.** 72% **6.** 80

Answers

6.3 Practice A

1. 8 **2.** 40% **3.** 70 **4.** 138

5. $\frac{15}{w} = \frac{40}{100}$; 37.5 **6.** $\frac{24}{w} = \frac{0.6}{100}$; 4000

7. $\frac{a}{100} = \frac{27}{75}$; 36% **8.** $\frac{17}{68} = \frac{p}{100}$; 25%

9. 48 **10.** $450

11. $\frac{0.2}{16} = \frac{p}{100}$; 1.25% **12.** $\frac{19.6}{w} = \frac{24.5}{100}$; 80

13. $\frac{\frac{3}{5}}{w} = \frac{30}{100}$; 2 **14.** $\frac{a}{\frac{5}{9}} = \frac{45}{100}$; $\frac{1}{4}$

15. 7 left to complete

16. a. 60% **b.** 36% **c.** 48%

6.3 Practice B

1. $\frac{33}{w} = \frac{55}{100}$; 60 **2.** $\frac{42}{120} = \frac{p}{100}$; 35%

3. $\frac{36}{w} = \frac{0.8}{100}$; 4500 **4.** $\frac{48}{64} = \frac{p}{100}$; 75%

5. 72 **6.** $\frac{3.69}{w} = \frac{90}{100}$; $4.10

7. $\frac{6}{w} = \frac{6.25}{100}$; 96 **8.** $\frac{\frac{7}{8}}{w} = \frac{70}{100}$; 1.25

9. $\frac{7.2}{w} = \frac{250}{100}$; 2.88 **10.** $\frac{a}{\frac{3}{8}} = \frac{72}{100}$; 0.27

11. $\frac{1.4}{1.12} = \frac{p}{100}$; 1.25% **12.** $\frac{86.8}{w} = \frac{140}{100}$; 62

13. $\frac{1}{2}x$

14. a. $31.5x$ **b.** 56.25%

6.3 Enrichment and Extension

1. $x = 4$ **2.** $x = 8$ **3.** $x = 4$ **4.** $x = 17$

6.3 Puzzle Time

HOLDUP IN THE YARD WHEN TWO CLOTHESPINS HELD UP A PAIR OF PANTS

6.4 Start Thinking!

For use before Activity 6.4

36 points; *Sample answer:* The percent proportion is $\frac{a}{w} = \frac{p}{100}$. Substitute 45 for w and 80 for p. Solve for a. You need to get 36 points.

6.4 Warm Up

For use before Activity 6.4

1. 40% **2.** 55% **3.** 33

4. 18.9 **5.** 85 **6.** 1

6.4 Start Thinking!

For use before Lesson 6.4

The percent equation is correct. The number is 1.5. *Sample answer:* The student put the 30 as the part instead of the whole in the percent proportion.

6.4 Warm Up

For use before Lesson 6.4

1. 18 **2.** $33\frac{1}{3}\%$ **3.** 40

4. 96 **5.** 45% **6.** 140

6.4 Practice A

1. 40%; Methods will vary.

2. 20; Methods will vary.

3. $a = \frac{70}{100} \bullet 120$; 84 **4.** $30 = p \bullet 120$; 25%

5. $112 = \frac{56}{100} \bullet w$; 200 **6.** $128 = p \bullet 80$; 160%

7. $a = \frac{140}{100} \bullet 45$; 63 **8.** $15 = \frac{6}{100} \bullet w$; 250

9. 21 competitors **10.** 30 students

11. 75%

12. a. 38 pennies **b.** 95 rare coins **c.** 57 nickels

13. a. 6% **b.** $7.50 **c.** $12.50

14. true; To find 120% of a number, multiply it by $\frac{120}{100}$, which is a number greater than 1.

15. false; To find 0.5% of a number, multiply it by $\frac{0.5}{100}$, or $\frac{5}{1000}$.

Answers

6.4 Practice B

1. 30%; Methods will vary.

2. 140; Methods will vary.

3. $27 = \frac{0.5}{100} \bullet w$; 5400

4. $a = \frac{125}{100} \bullet 240$; 300

5. $28 = \frac{1.4}{100} \bullet w$; 2000

6. $27 = p \bullet 72$; 37.5%

7. 9.44 in.

8. 20,000 gal

9. yes

10. a. 320 students b. 80 students
 c. 32 students d. 64 students

11. 10.8

12. 1000 and 1500

13. true; $A = \frac{45}{100}B$, so $\frac{A}{B} = \frac{45}{100} = \frac{9}{20}$.

14. 42.11%

15. 131.58%

6.4 Enrichment and Extension

1. yes; You should draw again, because you are currently over 50%.

2. The lowest card you could draw is 13. The highest card you could draw is 19. You could still win with 35% of the cards.

3. The most you can get on your first three draws and still have a chance to win is 39. If you have this score after your first three draws, your only chance of winning is to get a 1. Otherwise you will be over 50%.

4. *Sample answer:* It is probably best not to draw again. You have 35%. So, only 6 out of 20 cards will increase your percent without causing you to go over. If you draw an 8 or 9 first, it would be a good idea to stop. Your current score is even harder to beat, and you are even less likely to draw a card that will increase your percent without making you go over.

5. *Sample answer:* If the target percent is higher, players will be less likely to go over. So, you will have to get closer to the target percent in order to win. If the target percent is lower, players will be more likely to go over. So, you may not have to get as close to the target percent in order to win.

6.4 Puzzle Time

THE SPORTS FAN WHO LISTENED TO A MATCH AND ENDED UP BURNING HIS EAR

6.5 Start Thinking!

For use before Activity 6.5

Sample answer: A meteorologist uses percents to tell you the chance of precipitation for a given day or the amount of humidity in the air.

6.5 Warm Up

For use before Activity 6.5

1. 45%
2. 134%
3. 54.9%
4. 108%
5. 98.5%
6. 32.25%

6.5 Start Thinking!

For use before Lesson 6.5

Sample answer: You have $250 worth of baseball cards. You are told that the value of the baseball cards will decrease 35% over the next 6 months. How much will the baseball cards be worth at the end of the 6 months?

6.5 Warm Up

For use before Lesson 6.5

1. 18 inches
2. 58 gallons
3. 126 meters
4. 39 grams
5. 114 pounds
6. 50 liters

6.5 Practice A

1. 9 dogs
2. 203 fluid ounces
3. 199 textbooks
4. 15 students
5. increase; 60%
6. decrease; 10%
7. decrease; 28.6%
8. decrease; 17.9%
9. increase; 100%
10. decrease; 50%
11. 20% increase
12. a. decrease of 20% b. 173 tickets
 c. 35.9% decrease
13. 4 musicians
14. a. 10 feet b. 25% increase c. 56.3% increase

6.5 Practice B

1. 66 employees
2. 10°
3. 45 customers
4. 4221 fans
5. decrease; 25%
6. increase; 107.3%
7. increase; 100%
8. decrease; 60%
9. 3.7% decrease
10. a. 2% decrease b. 1152 hamburgers
11. a. $36.00 b. increase of 4.2%

Answers

12. a. 14.3% increase b. 14.3% increase
    c. Troop B
    d. neither troop; The percent of change is the same.

6.5 Enrichment and Extension

1. a. 60 thousand pairs b. 75 thousand pairs
2. a. 172 thousand pairs b. 86 thousand pairs
3. You should use percent of increase because your company's percent of increase is bigger than your competitor's percent of increase, even though your competitor increased their sales by more pairs of sneakers.
4. a. 297.5 thousand pairs b. 122.5 thousand pairs
5. a. 270 thousand pairs b. 120 thousand pairs
6. You should use amount of increase because your company increased their sales by more pairs than your competitor, even though their percent of increase was greater.

6.5 Puzzle Time

SOMEBODY FRAMED IT

6.6 Start Thinking!

For use before Activity 6.6

Sample answer: A store marks up an item so they can make a profit. A store may then discount the item because either it did not sell or they need the space for new items.

6.6 Warm Up

For use before Activity 6.6

1. $a = 0.4 \bullet 238$; 95.2 2. $28 = p \bullet 70$; 40%
3. $a = 0.34 \bullet 240$; 81.6 4. $6 = 0.05 \bullet w$; 120
5. $a = 1.1 \bullet 150$; 165 6. $42 = 2.5 \bullet w$; 16.8

6.6 Start Thinking!

For use before Lesson 6.6

The jeans that cost $35 with a 30% discount are a better bargain. They cost $24.50 on sale. The other pair of jeans cost $26 on sale.

6.6 Warm Up

For use before Lesson 6.6

1. $23 2. $84.50 3. $112 4. $157.50

6.6 Practice A

1. $52.50 2. $6.30 3. $60
4. $200 5. 20% 6. 15%
7. $81.25 8. $68 9. 40%
10. a. $192 b. $134.40 c. make money; $14.40
11. $84
12. a. $63.60 b. $51
    c. $54.06 d. decrease of 15%

6.6 Practice B

1. $71.50 2. 65% 3. $72
4. $2128 5. 40% 6. $350
7. a. $31.94 b. $36.73 c. yes; $33.05
8. a. $37.99 b. 39% c. $27.99 d. 54%
9. 67%

6.6 Enrichment and Extension

1. a. $15.40 b. $5.50
   c. The coupon takes 30% off of the sale price, not the original price. So, you are saving less than 90% off the original.
2. a. In order to find the amount of markup, you find 40% of the cost. In order to find the amount of discount, you find 40% of the selling price. Because the selling price is larger than the cost, the amount of discount would be larger than the amount of markup. So, you would be discounting them by more than you marked them up, and you would lose money.
   b. $125.36 c. 29%
3. a. Yes, you would be allowed to match the competitor's price. You would be discounting the player by $9.20, which is just under a 3% discount.
   b. No, you would not be allowed to match the Internet price. The one on the Internet would cost $62.97 with shipping and handling. So, you would have to discount the router by $11.78, which is a percent of discount of 15.8%. The least amount that you can offer on the router is $63.54.

6.6 Puzzle Time

PLEASED TO EAT YOU

6.7 Start Thinking!

For use before Activity 6.7

Sample answer: Pro: Build credit; Con: Could lead to debt you cannot repay

6.7 Warm Up

For use before Activity 6.7

1. $225 2. $45 3. $1040
4. $80.75 5. $652.50 6. $549.25

Answers

6.7 Start Thinking!

For use before Lesson 6.7

Sample answer: Some factors you must consider are interest rate or if the bank charges a fee for a minimum balance.

6.7 Warm Up

For use before Lesson 6.7

1. **a.** \$45 **b.** \$795
2. **a.** \$36 **b.** \$336
3. **a.** \$280 **b.** \$1680
4. **a.** \$189 **b.** \$789
5. **a.** \$22 **b.** \$572
6. **a.** \$21 **b.** \$1221

6.7 Practice A

1. **a.** \$30 **b.** \$230
2. **a.** \$120 **b.** \$870
3. **a.** \$80 **b.** \$1680
4. **a.** \$30 **b.** \$530
5. 2%
6. 7%
7. 3 years
8. 1.5 years
9. \$367.50
10. \$1400
11. \$4200
12. 18 years
13. **a.** \$300 **b.** \$3300
14. **a.** \$3360 **b.** \$140 **c.** credit card
15. **a.** \$1296 **b.** \$1399.68

6.7 Practice B

1. **a.** \$332.80 **b.** \$2932.80
2. **a.** \$1593.75 **b.** \$76,593.75
3. 4.75%
4. 6.5%
5. 6 months
6. 3 years
7. \$35,000
8. \$7800
9. 9%
10. 10%
11. **a.** \$1208.10 **b.** \$1220.18 **c.** \$1228.11 **d.** \$28.11 **e.** 2.3%
12. 8%

6.7 Enrichment and Extension

1. If she gets \$500 off, the total cost would be \$21,320.57. If she decreases the interest rate, the total cost would be \$21,519.48. It would be better to get \$500 off.
2. The total cost would be \$22,278.63. When interest is compounded more often, the total cost increases, which means that the lender is making more money.
3. *Answer should include, but is not limited to:* The student must include the car's original price, the interest rate, how often the interest is compounded, the total cost of the loan, and the monthly payment.

6.7 Puzzle Time

ARE YOU ASLEEP

Technology Connection

1. 273
2. 10
3. 27
4. 67.5
5. 300
6. 700
7. 15.5
8. 400
9. 5
10. \$5,200

Chapter 7

7.1 Start Thinking!

For use before Activity 7.1

An acute angle is an angle whose measure is less than 90°. An obtuse angle is an angle whose measure is greater than 90° and less than 180°. *Sample answer:* Acute can mean sharp and obtuse means dull. An acute angle has a sharp point and an obtuse angle does not.

7.1 Warm Up

For use before Activity 7.1

1. right
2. obtuse
3. acute
4. straight

7.1 Start Thinking!

For use before Lesson 7.1

Sample answer:

Adjacent angles are angles that are next to each other. They share a common side and have the same vertex.

7.1 Warm Up

For use before Lesson 7.1

1. $\angle ABC$ and $\angle EBD$ measure 50°. $\angle CBD$ and $\angle ABE$ measure 130°.
2. $\angle CBD$ and $\angle ABE$

7.1 Practice A

1. *Sample answer:* adjacent: $\angle ABE$ and $\angle EBD$; $\angle ABC$ and $\angle CBD$; vertical: $\angle ABE$ and $\angle CBD$; $\angle ABC$ and $\angle EBD$
2. *Sample answer:* adjacent: $\angle NHG$ and $\angle GHI$; $\angle JHK$ and $\angle KHL$; vertical: $\angle GHI$ and $\angle KHL$; $\angle IHJ$ and $\angle NHL$
3. adjacent; 148
4. vertical; 142
5. adjacent; 18
6. vertical; 26
7.

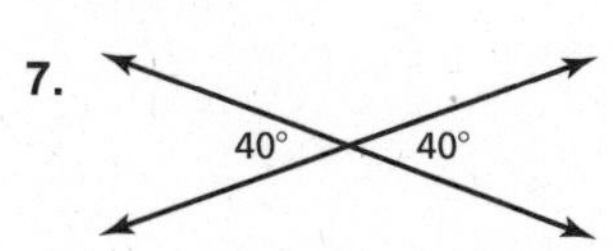

Answers

8. 9.

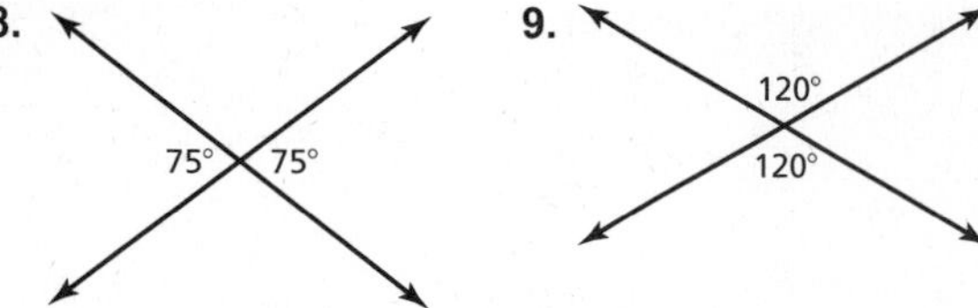

10. a.

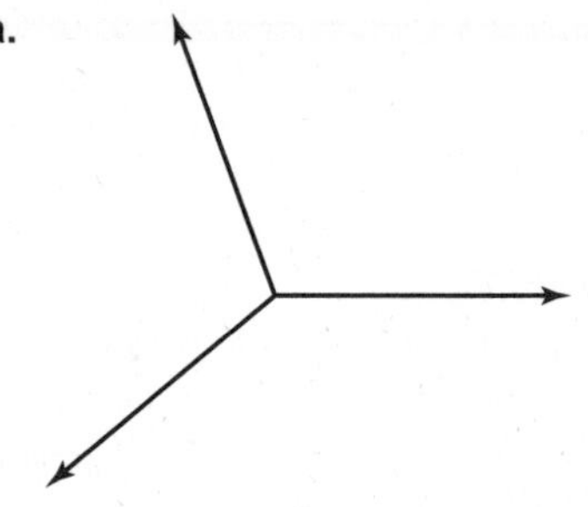

b.

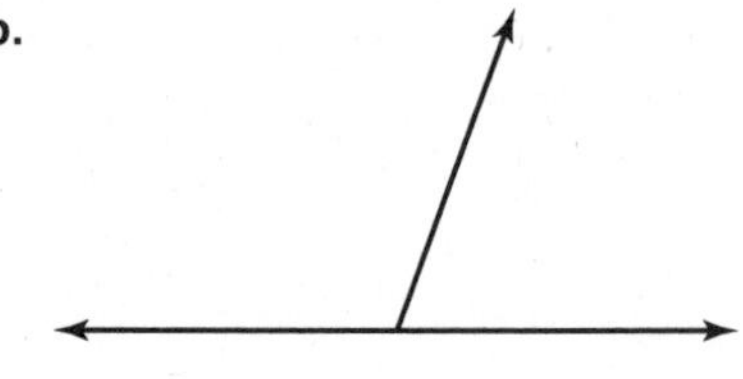

c.

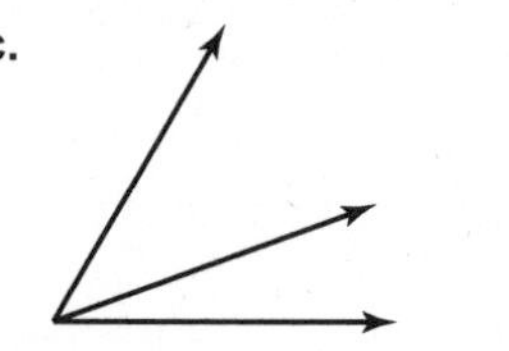

11. $\angle 1 = 48°$,

$\angle 2 = 132°$,

$\angle 3 = 48°$

7.1 Practice B

1. *Sample answer:* adjacent: $\angle IED$ and $\angle DEF$; $\angle FEG$ and $\angle GEH$; vertical: $\angle DEF$ and $\angle GEH$; $\angle FEG$ and $\angle DEH$

2. *Sample answer:* adjacent: $\angle SUT$ and $\angle SUZ$; $\angle XUY$ and $\angle WUX$; vertical: $\angle SUZ$ and $\angle WUX$; $\angle YUZ$ and $\angle VUW$

3. adjacent; 123

4. adjacent; 10

5. adjacent; 45

6. vertical; 30

7.

100°

100°

8.

9.

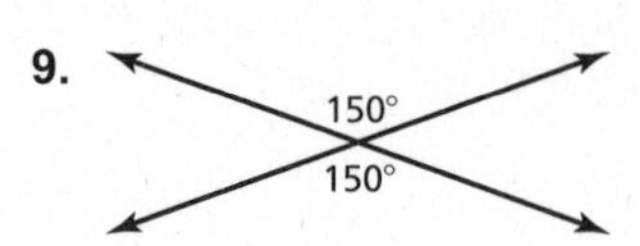

10. *Sample answer:*

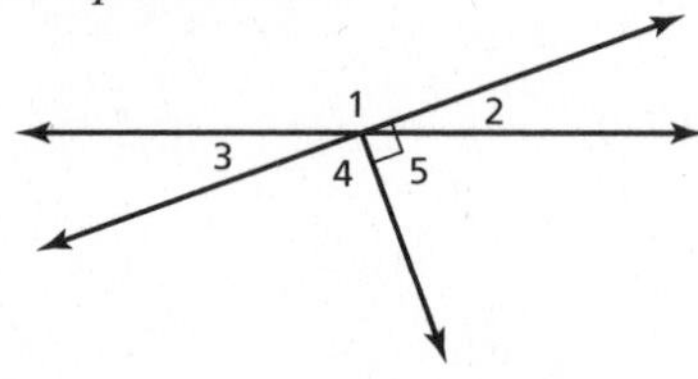

11. 75°

7.1 Enrichment and Extension

1. Angle A is an acute angle.

2–5.

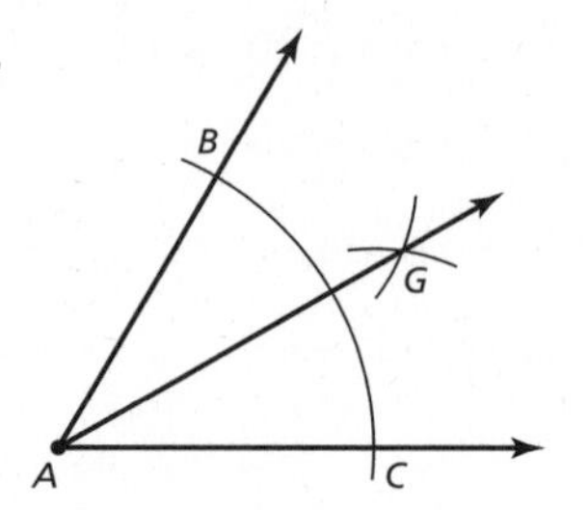

6. Angle BAC measures 60°. Each angle formed from the bisection measures 30°.

7.1 Puzzle Time

A SNEAKER

7.2 Start Thinking!

For use before Activity 7.2

Sample answer: Two angles are adjacent angles when they share common sides and have the same vertex. Two angles are vertical angles when they are opposite angles formed by the intersection of two lines. Vertical angles are congruent angles, meaning they have the same measure.

7.2 Warm Up

For use before Activity 7.2

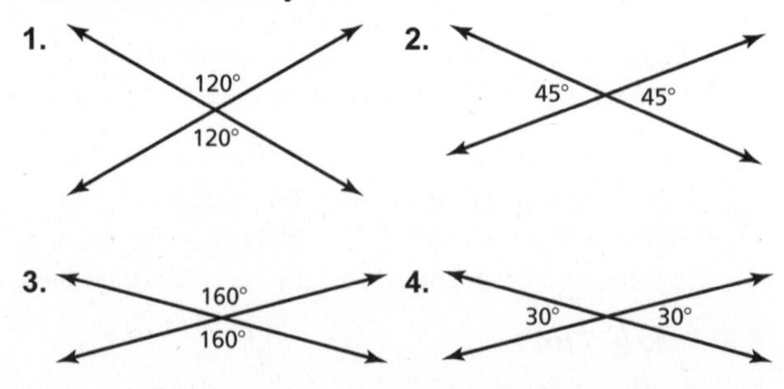

Answers

7.2 Start Thinking!

For use before Lesson 7.2

complementary; supplementary; Answers will vary. *Sample answers:* A letter S is like two Cs put together, so the sum of supplementary angles is twice as much as complementary angles. "It's *right* to give a *compliment*." C comes before S in the alphabet and 90 comes before 180 on a number line.

7.2 Warm Up

For use before Lesson 7.2

1. sometimes; x may be acute or obtuse.
2. always; y must be between $0°$ and $90°$.
3. never; If $x = 90°$ and $y < 90°$, then $x + y < 180°$.
4. sometimes; The sum $x + y$ could be any value between $90°$ and $270°$.

7.2 Practice A

1. sometimes; y could be right, acute, or obtuse.
2. never; x must be between $0°$ and $90°$.
3. complementary
4. neither
5. neither
6. supplementary
7. angle x: $52°$; angle y: $38°$
8. complementary; $x = 5$
9. supplementary; $x = 20$
10.

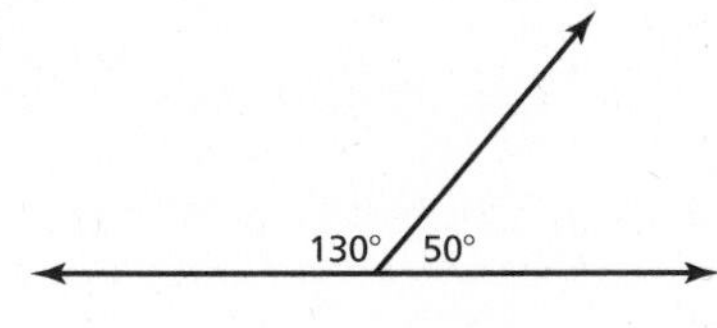

11.

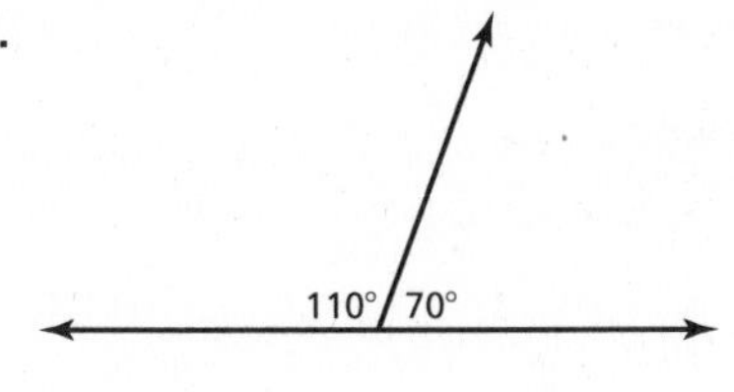

12.

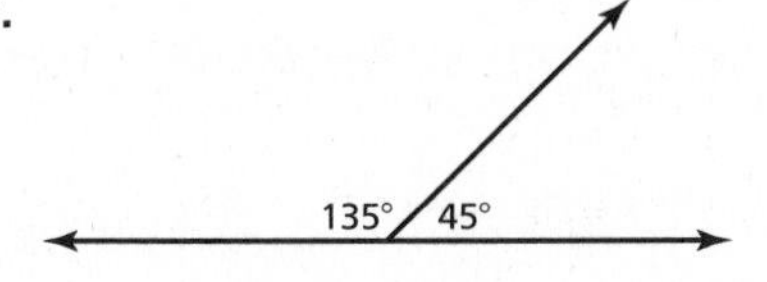

13. complementary: $45°$; supplementary: $90°$

7.2 Practice B

1. neither
2. complementary
3. supplementary
4. supplementary
5. supplementary; $x = 11$
6. complementary; $x = 6$
7. $60°$
8. $x = 5$; $y = 5$
9. **a.** $x + c = 90$ **b.** $x + s = 180$

7.2 Enrichment and Extension

1. $130°$
2. $50°$
3. $50°$
4. $30°$
5. $70°$
6. $50°$
7. $60°$
8. $40°$
9. $40°$
10. $50°$
11. $50°$
12. $130°$
13. $140°$
14. $40°$
15. $140°$

7.2 Puzzle Time

LEAP YEAR

7.3 Start Thinking!

For use before Activity 7.3

Sample answer: A yield sign is an equilateral triangle.

7.3 Warm Up

For use before Activity 7.3

1–6. Check students' work.

7.3 Start Thinking!

For use before Lesson 7.3

$90°$; *Sample answer:* Begin by drawing the angle with measure $30°$. Then draw the second angle of $60°$ off one of the rays. Measure the third angle. The third angle measure is $90°$.

7.3 Warm Up

For use before Lesson 7.3

1–4. Check students' work.

7.3 Practice A

1. equiangular, equilateral, acute
2. obtuse, scalene
3. right, scalene
4. isosceles, acute

Answers

5. 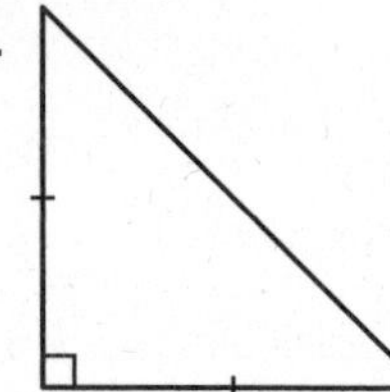

6. 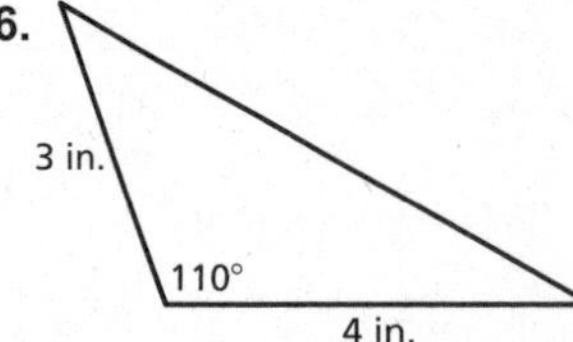

7. a. 45, 90, 45
   b. Every triangle has one $90°$ angle and two $45°$ angles.
   c. An isosceles right triangle has two $45°$ angles.

7.3 Practice B

1. acute, isosceles
2. obtuse, scalene

3. 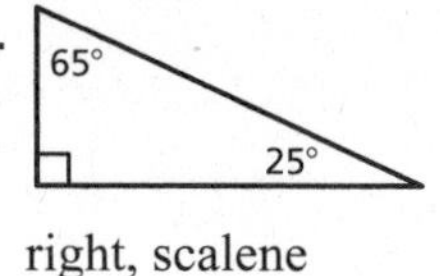

right, scalene

4. 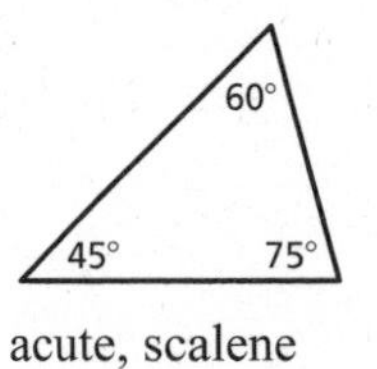

acute, scalene

5. 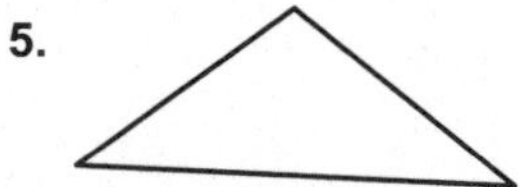

6. 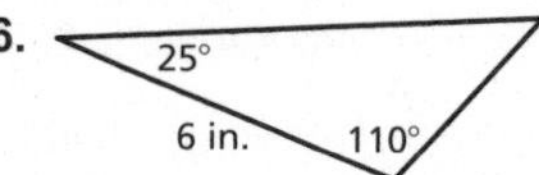

7. one triangle; Only one triangle can be drawn with a 2-inch side, 4-inch side, and 5-inch side.

8. no triangles; None of the sides of a scalene triangle are equal.

9. many triangles; The other two sides of the triangle can be many different lengths.

10. a. 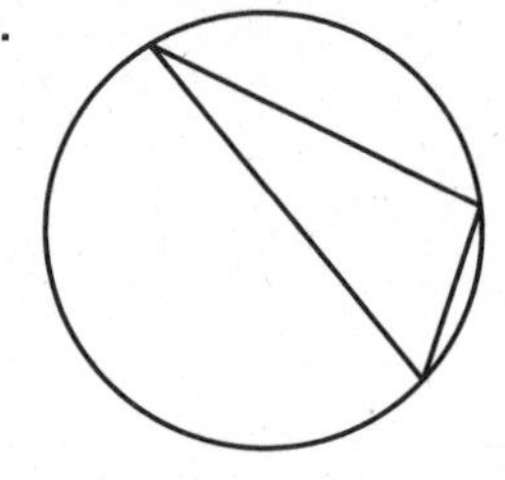

b. 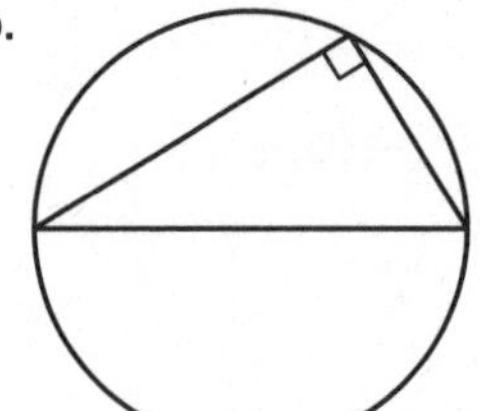

c. 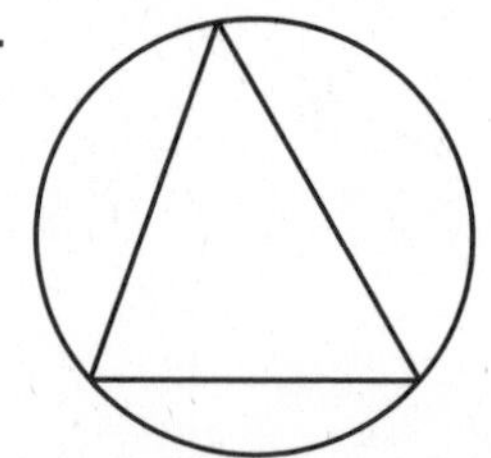

7.3 Enrichment and Extension

1. $\frac{4}{5}$
2. $\frac{12}{5}$
3. $\frac{3}{5}$
4. $\frac{4}{3}$
5. $\frac{12}{13}$
6. $\frac{5}{13}$

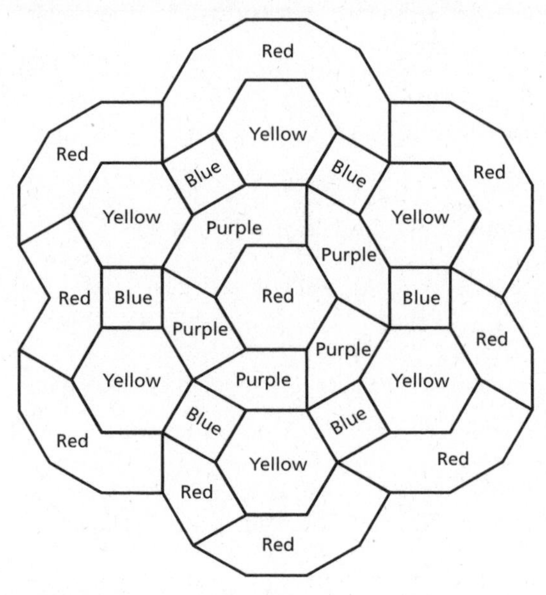

7.3 Puzzle Time

SHE WANTED TO LEARN TO READ BETWEEN THE LIONS

Extension 7.3 Start Thinking!

For use before Extension 7.3

Sample answer: construction, bridge building

Extension 7.3 Warm Up

For use before Extension 7.3

1. acute, isosceles
2. right, isosceles
3. obtuse, scalene
4. obtuse, isosceles
5. right, scalene
6. equilateral, equiangular, acute

Extension 7.3 Practice

1. $x = 102$; obtuse, scalene
2. $x = 60$; equiangular, equilateral, acute
3. $x = 28$; isosceles, obtuse
4. $x = 12$; right, scalene
5. $x = 45$; right, isosceles
6. $x = 22.5$; isosceles, obtuse
7. yes

Answers

8. no; 24.9°, 121.4°, 33.7°

9. **a.** 60°, 60°, 60°

b. equiangular, equilateral, acute

c. *Sample answer:* 110°, 35°, 35°; isosceles, obtuse

d. *Sample answer:* 40°, 70°, 70°; isosceles, acute

e. Two of its angles are equal.

7.4 Start Thinking!

For use before Activity 7.4

Sample answer: quadrennial (occurring every 4 years), quadriceps (4 part muscle at the front of the thigh), quadrilingual (can speak 4 languages), quadruped (animal with 4 feet), quadruple (4 times as great), quadruplets (4 children born of one pregnancy), quad (4-wheeled all-terrain vehicle); quad- means four.

7.4 Warm Up

For use before Activity 7.4

1. trapezoid **2.** parallelogram **3.** square

4. rectangle **5.** triangle **6.** pentagon

7.4 Start Thinking!

For use before Lesson 7.4

Sample answer: square, rectangle, rhombus, parallelogram, trapezoid, kite; Check students' sketches.

7.4 Warm Up

For use before Lesson 7.4

1. rhombus **2.** square **3.** rectangle

4. parallelogram **5.** trapezoid **6.** kite

7.4 Practice A

1. kite **2.** rectangle **3.** trapezoid

4. parallelogram **5.** 56°

6. 110° **7.** always

8. sometimes **9.** never **10.** never

11. **a.** not possible; There would have to be another right angle because two sides are parallel.

b.

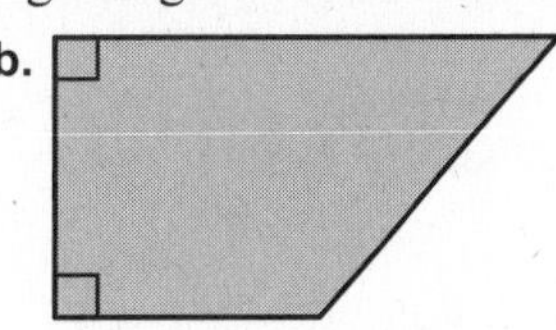

c. not possible; The sum of the measure of the angles is 360°, so the fourth angle would be a right angle. So it would be a square, not a trapezoid.

d. not possible; A trapezoid has two sides that are not parallel.

7.4 Practice B

1. rhombus **2.** trapezoid **3.** 112°

4. 125° **5.** sometimes **6.** always

7. never **8.** sometimes

9. **a.** false; The width needs to be half the length.

b. true; Each square has side length 10 inches.

c. true; Each rhombus has side length 3 feet.

d. true; *Sample answer:*

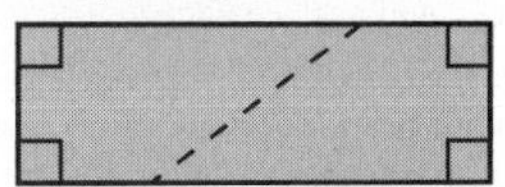

e. true; *Sample answer:*

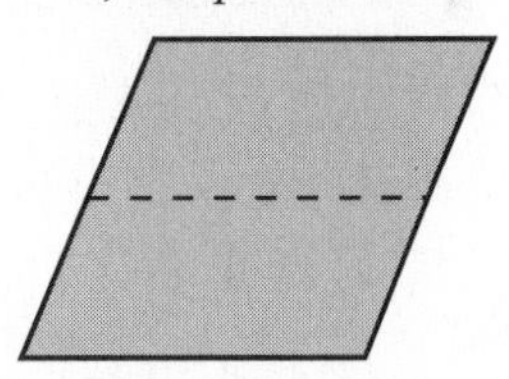

7.4 Enrichment and Extension

1. triangle; 60° **2.** octagon; 135°

3. pentagon; 108° **4.** hexagon; 120°

5. nonagon; 140° **6.** decagon; 144°

7.4 Puzzle Time

VITAMIN BEE

7.5 Start Thinking!

For use before Activity 7.5

Sample answer: The architect drew the building to scale. If the contractor does not build to the same scale as the architect used, the building will not go together correctly.

Answers

7.5 Warm Up

For use before Activity 7.5

1. $x = 24$ **2.** $x = 10$ **3.** $x = 20$

4. $x = 28$ **5.** $x = 80$ **6.** $x = 81$

7.5 Start Thinking!

For use before Lesson 7.5

Sample answer: You must first determine the scale. You will also need to use a measuring device to measure the distance you will be traveling. Then you need to convert the scale distance to actual distance.

7.5 Warm Up

For use before Lesson 7.5

1. 28 ft **2.** 42 ft **3.** 120 ft

4. 7.6 ft **5.** 160 ft **6.** 1700 ft

7.5 Practice A

1. a. 24 ft **b.** 8 ft by 4 ft **c.** 8 ft by 8 ft

d. $33.\overline{3}\%$ **e.** 3 : 4 **f.** 1 : 2 **g.** no **h.** $33.\overline{3}\%$

2. 15 ft **3.** 35 m **4.** 4 yd **5.** 2.5 cm

6. a. 1 in. : 0.5 ft **b.** $\frac{1}{6}$

7.5 Practice B

1. a. 12 ft by 20 ft **b.** 8 ft by 6 ft **c.** 14 ft **d.** 8 : 7

e. 1 : 1; They both have the same number of squares.

f. closet **g.** both the same **h.** 144 ft^2

2. 25 km **3.** 12.5 in. **4.** 9.6 ft **5.** 13 m

6. should be model : actual; $\frac{1}{8} = \frac{x \text{ ft}}{48 \text{ ft}}$

$x = 6$ ft

7.5 Enrichment and Extension

Answer should include, but is not limited to: The drawing should fit on the grid and a scale should be included. All items should be drawn to scale, and their lengths should be labeled.

7.5 Puzzle Time

MICE CUBES

Technology Connection

1. *Sample answer:*

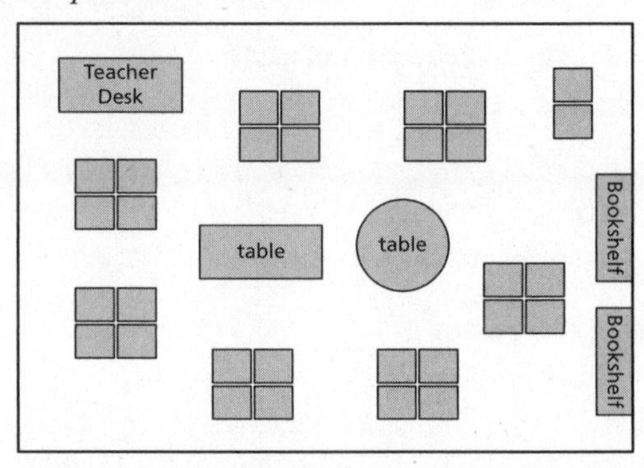

2. *Answer should include, but it not limited to:* aspects of design such as sufficient walking space, teacher accessibility to all desks, ability or inability to see a particular chalkboard, ease of collaboration among peers, etc.

Chapter 8

8.1 Start Thinking!

For use before Activity 8.1

Check students' piems.

8.1 Warm Up

For use before Activity 8.1

1. 24 m **2.** 40 cm **3.** 36 yd

4. 120 ft **5.** 16 in. **6.** 60 mm

8.1 Start Thinking!

For use before Lesson 8.1

Sample answer: The circumference of a tree trunk is easier to measure. Unless the tree is cut down, you do not have access to the circular cross section in order to measure the diameter. The diameter of a quarter is easier to measure. You only need a ruler to measure across the quarter. In general, it is usually easier to measure the diameter of a circular object. The circumference is easier to measure only when you cannot access the diameter to measure it.

8.1 Warm Up

For use before Lesson 8.1

1. about 37.68 in. **2.** about 12.56 ft

3. about 22 cm **4.** about 62.8 ft

5. about 44 mm **6.** about 3.14 in.

8.1 Practice A

1. 30 ft **2.** 4 m **3.** 32 mm

4. 5 cm **5.** 12 in. **6.** 3.5 yd

7. about 53.38 m
8. about 18.84 ft
9. about 88 in.
10. about 56.54 mm
11. about 20.56 in.
12. about 25.7 yd
13. a. 8 mm; 24 mm b. 3 times greater
14. $131.56
15. about 5.14 ft

8.1 Practice B

1. about 81.64 m
2. about 37.68 ft
3. about 264 in.
4. about 72 yd
5. about 31.05 ft
6. about 28.27 cm
7.

Circle	*A*	*B*	*C*	*D*
Radius	2.5 ft	1 ft	32 ft	3.5 ft
Diameter	60 in.	24 in.	768 in.	84 in.

8. about 9.42 in.
9. yes; The diameter of each of the small circles in diagram B is 2 feet. Because there are 5 circles along each side of the square, the length of each side is $5 \bullet 2 = 10$ feet.
10. about 9.6 times

8.1 Enrichment and Extension

1. 25.12 in., 50.24 in.
2. 43.96 ft, 87.92 ft
3. 47.1 cm, 94.2 cm
4. The circumference is multiplied by 2.
5. The circumference is multiplied by n.
6. 12.85 mm, 6.425 mm
7. 5.14 in., 2.57 in.
8. 51.4 m, 25.7 m
9. The perimeter is multiplied by $\frac{1}{2}$.
10. The perimeter is multiplied by n.

8.1 Puzzle Time

HE HAD TOO MUCH TIME ON HIS HANDS

8.2 Start Thinking!
For use before Activity 8.2

Sample answer: To find the perimeter of an irregular shape, you can add all of the different straight sections (or circular sections). To find a distance on a map, assuming that the route traveled is not along a straight line, you must add all of the straight sections to estimate the total distance.

8.2 Warm Up
For use before Activity 8.2

1. 16 cm
2. 17 ft
3. 52 in.
4. 25 m
5. about 56.52 in.
6. about 62.8 ft

8.2 Start Thinking!
For use before Lesson 8.2

Sample answer: yes; It is made up of a rectangle and two semicircles. Some examples: windows (rectangle and semicircle), some company and organization logos, a first place ribbon (circle and rectangles), some flowers

8.2 Warm Up
For use before Lesson 8.2

1. 24 units
2. 24 units
3. about 15 units
4. about 13.71 units
5. about 20.5 units
6. about 22.28 units

8.2 Practice A

1. about 24 in.
2. about 28 in.
3. about 21.42 in.
4. 34 yd
5. 24 in.
6. 60 mm
7. 31 m
8. 24 ft
9. 64 cm
10. about 23.42 ft
11. $3685.90

8.2 Practice B

1. about 22 in.
2. 28 in.
3. about 22 in.
4. 27 m
5. about 140.76 cm
6. 28 in.
7. The perimeter calculation included the circumference of the circle instead of the perimeter of the semicircle.

$$\text{Perimeter} \approx 2 + 7 + 2 + 10.99$$
$$= 21.99 \text{ ft}$$

8. about $211.31
9. *Sample answer:*

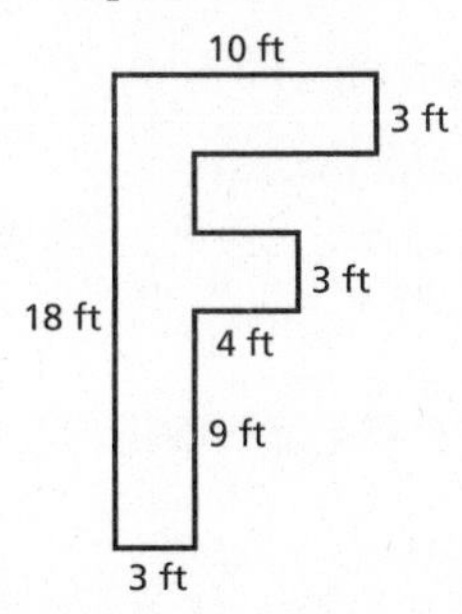

yes; For example, if the 11 ft by 3 ft rectangle on the bottom of the F is shortened to be a 6 ft by 3 ft rectangle, the perimeter decreases by 10 ft.

Answers

8.2 Enrichment and Extension

1. about 25.12 in. **2.** about 18.84 m

3. about 34.26 ft **4.** about 30.26 yd

5. about 55.4 cm **6.** about 50.24 mm

8.2 Puzzle Time

LIBRARY

8.3 Start Thinking!

For use before Activity 8.3

Sample answer: **1.** You can divide it into several triangles each with a vertex at the center of the circle. **2.** You can put it on grid paper and estimate the number of squares that are covered. **3.** You can divide it into tall, narrow rectangles.

8.3 Warm Up

For use before Activity 8.3

1. 63 in.^{2} **2.** 14 m^2 **3.** 66 yd^2

4. 50 cm^2 **5.** 22.4 m^2 **6.** 1 mm^2

8.3 Start Thinking!

For use before Lesson 8.3

Sample answer: It is easier to use $\frac{22}{7}$ if the radius is a multiple of 7. It is easier to use 3.14 in all other instances. Check students' word problems.

8.3 Warm Up

For use before Lesson 8.3

1. about 50.24 in.^{2} **2.** about 154 ft^2

3. about 3.14 cm^2

8.3 Practice A

1. about 78.5 m^2 **2.** about 7850 mm^2

3. about 1256 in.^{2} **4.** about 38.5 ft^2

5. about 1386 mm^2 **6.** about 3850 cm^2

7. about 50.24 ft^2 **8.** about 15,400 cm^2

9. about 113.04 in.^{2} **10.** about 28.26 in.^{2}

11. $\frac{99}{224}$ in.^{2} **12.** about 153.86 in.^{2}

8.3 Practice B

1. about 962.5 in.^{2} **2.** about 19.625 m^2

3. about 7234.56 mm^2 **4.** about 157 in.^{2}

5. about 100.48 m^2 **6.** about 693 ft^2

7. about 2464 ft^2 **8.** about 28.26 ft^2

9. a. $A = \pi r^2$ **b.** $A = \pi\left(\frac{r}{2}\right)^2$

c. The circle's area is one fourth of the area of the circle whose radius is twice as large; The area formula for the smaller circle can be rewritten as $A = \pi\left(\frac{r}{2}\right)^2 = \frac{1}{4}\pi r^2$; this is one fourth of the area of the larger circle.

10. The radius is 2 units; Set the circumference equal to the area to get $2\pi r = \pi r^2$. So $\pi \bullet r \bullet 2 = \pi \bullet r \bullet r$; the value of r must be 2.

8.3 Enrichment and Extension

1. 1.5625 **2.** 2.25 **3.** 16% **4.** $\frac{1}{9}$

5. a. $3.\overline{5}\%$

b.

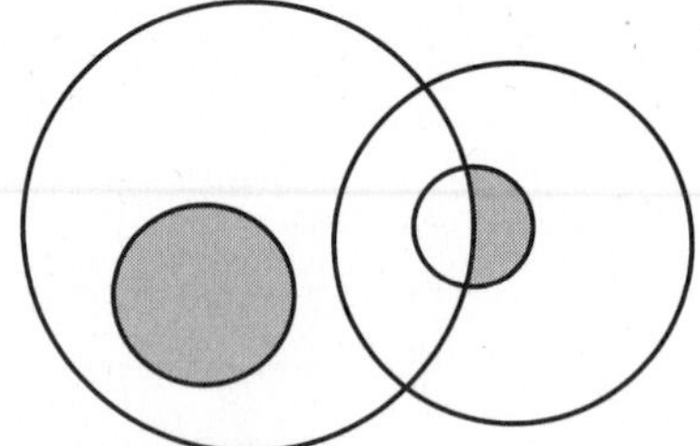

13,816 mi^2

c.

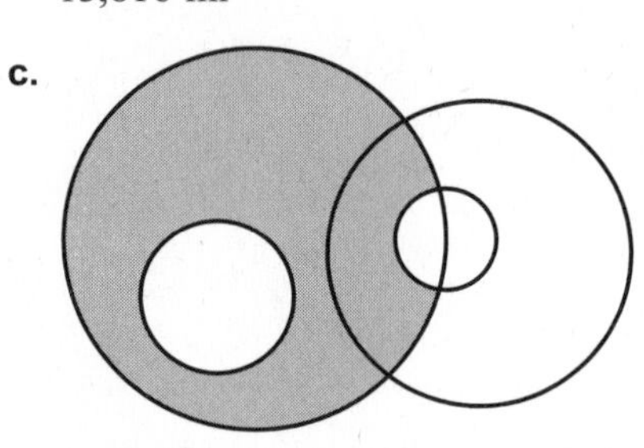

56,834 mi^2

8.3 Puzzle Time

HAMWORK

8.4 Start Thinking!

For use before Activity 8.4

Answers will vary. *Sample answer:* If you are planning on painting a room, you will need to calculate the area of each section of wall to figure out how much paint to buy. The windows and doorways make some walls composite figures.

Answers

8.4 Warm Up

For use before Activity 8.4

1. 100 ft^2
2. 256 $in.^2$
3. 150 m^2
4. 30 cm^2
5. about 314 ft^2
6. about 7850 yd^2

8.4 Start Thinking!

For use before Lesson 8.4

Check students' drawings and calculations.

8.4 Warm Up

For use before Lesson 8.4

1. 21 $units^2$
2. 20 $units^2$
3. 12 $units^2$
4. about 12.53 $units^2$
5. 18 $units^2$
6. about 17.72 $units^2$

8.4 Practice A

1. 30 $units^2$
2. 33 $units^2$
3. 16 $units^2$
4. 48 $in.^2$
5. about 178.5 mm^2
6. 402 ft^2
7. 210 cm^2
8. 60 yd^2
9. 40 m^2
10. perimeter: about 37.85 ft; area: about 69.625 ft^2
11. **a.** 144 $in.^2$ **b.** 72 $in.^2$ **c.** $172.80

8.4 Practice B

1. perimeter: about 23.42 ft; area: about 38.13 ft^2
2. perimeter: 18 mm; area: 16 mm^2
3. perimeter: 26.6 cm; area: 42.6 cm^2
4. 27.98 $in.^2$
5. about 4.55 m^2
6. 6.63 yd^2
7. The area calculation included the area of the circle instead of the area of the semicircle.

$$\text{Shaded area} \approx (4 \bullet 4) - \left(\frac{3.14 \bullet 2^2}{2}\right)$$
$$= 16 - 6.28$$
$$= 9.72 \text{ ft}^2$$

8. **a.** 39 in. **b.** 54 $in.^2$
   c. no; The dimensions of the logo are 10.5 in. by 12 in., but the dimensions of the notebook cover are 8.5 in. by 11 in.

8.4 Enrichment and Extension

1. I: 78 m^2
2. P: about 50.7 m^2
3. M: 50 m^2
4. S: about 75.36 m^2
5. C: about 21.98 m^2
6. T: 82.775 m^2
7. O: about 58.875 m^2
8. E: 98.935 m^2
9. O: about 37.68 m^2
10. COMPOSITE

8.4 Puzzle Time

DO WE WALK OR HOP ON A DOG

Technology Connection

1. 260.62 cm
2. 260.8571429 cm
3. 260.7521902 cm
4. no; They would be 260.62, 260.86, and 260.75.
5. *Sample answer:* In this example, rounding to the nearest integer would give a consistent value.
6. 433 mi or 434 mi

Chapter 9

9.1 Start Thinking!

For use before Activity 9.1

Sample answer: Measure the length, width, and height of the box and find the sum of the areas of each side. Take the box apart and find the area of the net.

9.1 Warm Up

For use before Activity 9.1

1. 62
2. 28
3. 54
4. 142
5. 40
6. 304

9.1 Start Thinking!

For use before Lesson 9.1

Sample answer: Area is the amount of space a two-dimensional object takes up. Surface area is the sum of the areas of each surface of a three-dimensional object. Surface area is measured in square units.

9.1 Warm Up

For use before Lesson 9.1

1. 32 $in.^2$; 
2. 22 $in.^2$; 
3. 52 $in.^2$; 
4. 46 $in.^2$; 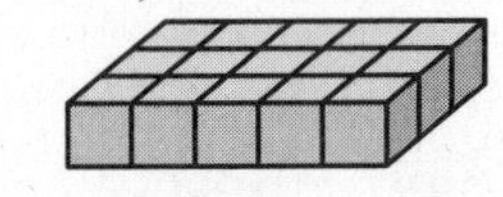

Answers

9.1 Practice A

1. 232 cm^2 2. 336 $in.^2$ 3. 360 ft^2

4. 672 m^2 5. 2640 $in.^2$

6. 384 cm^2 7. 171 $in.^2$

8. *Sample answer:*

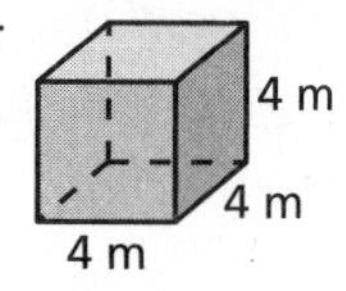

9.1 Practice B

1. 2520 cm^2 2. 136 ft^2 3. 85.6 $in.^2$

4. 1544 m^2 5. 600 cm^2 6. 456 ft^2

7. a. 482 cm^2 b. 452.75 cm^2

8. 5 in.

9.1 Enrichment and Extension

1. a. *Sample answer:*

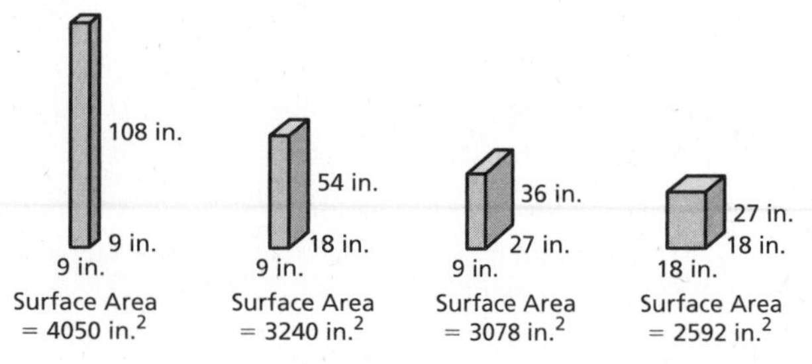

b. *Sample answer:*
28.125 ft^2; 22.5 ft^2; 21.375 ft^2; 18 ft^2

c. *Sample answer:* 1012.5 ft^2

2. a. Original box: 283.12 $in.^2$; 532.8 $in.^2$; 238.74 $in.^2$; 231.78 $in.^2$

b. 356.5 ft^2

c. *Sample answer:* The design with the least surface area will be the cheapest to make and produce the least waste. But, a cube is inconvenient to store in a cupboard. It would take up the same amount of space, but because it is wider and shorter, you would have to stack them on top of and in front of each other in order to best use the cupboard space. The original design is still probably best for storage reasons because it is tall and skinny, but not really tall like one of the new designs.

3. a cube

9.1 Puzzle Time

A BIG WHEEL

9.2 Start Thinking!

For use before Activity 9.2

Sample answer: Yes, because a pyramid goes to a point. No, the base can be any polygon.

9.2 Warm Up

For use before Activity 9.2

1. 24 cm^2 2. 21 $in.^2$

3. 171 ft^2 4. 195.5 cm^2

9.2 Start Thinking!

For use before Lesson 9.2

Sample answer: To make sure your neighbor buys enough roofing materials.

9.2 Warm Up

For use before Lesson 9.2

1. 297 $in.^2$ 2. 148.3 ft^2

9.2 Practice A

1. 37 cm^2 2. 7 ft^2

3. 33 m^2 4. 210.6 $in.^2$

5. a. 30 ft^2 b. 30 ft^2 c. 30.4 ft^2

d. yes; The lateral surface areas are almost the same, so you will get the same amount of coverage from each.

9.2 Practice B

1. 70.09 cm^2 2. 525.7 ft^2

3. a. 52 ft b. 3640 ft^2 c. $12,740.00

4. 7.8 m 5. 7 yd

9.2 Enrichment and Extension

1. *Sample answer:* If you use 1 m = 1 cm as your scale, then the model of the Cheops Pyramid would have a side length of 230 centimeters, which is more than 7.5 feet. A square that is 7.5 feet by 7.5 feet cannot be cut from a sheet of plywood that is 4 feet by 8 feet. If the scale is 1 m = 0.5 cm, then the model's side length will be half as long, which is less than 4 feet and can be cut from the sheet of plywood.

Answers

2.

Pyramid	Model side length (cm)	Model slant height (cm)
Cheops Pyramid in Egypt	115	93
Muttart Conservatory in Edmonton	13	13.5
Louvre Pyramid in Paris	17.5	14
Pyramid of Caius Cestius in Rome	11	14.5

3. 36,371.25 cm^2; 2 sheets of plywood

4. a. yes; A full sheet and a half sheet have an area of about 44,593.5 cm^2, which is greater than the amount of plywood needed for the models.

   b. You cannot use a half sheet. With one full sheet and one half sheet, you can cut out everything except one triangle needed for the sides of the Cheops Pyramid. So, you need 2 full sheets of plywood.

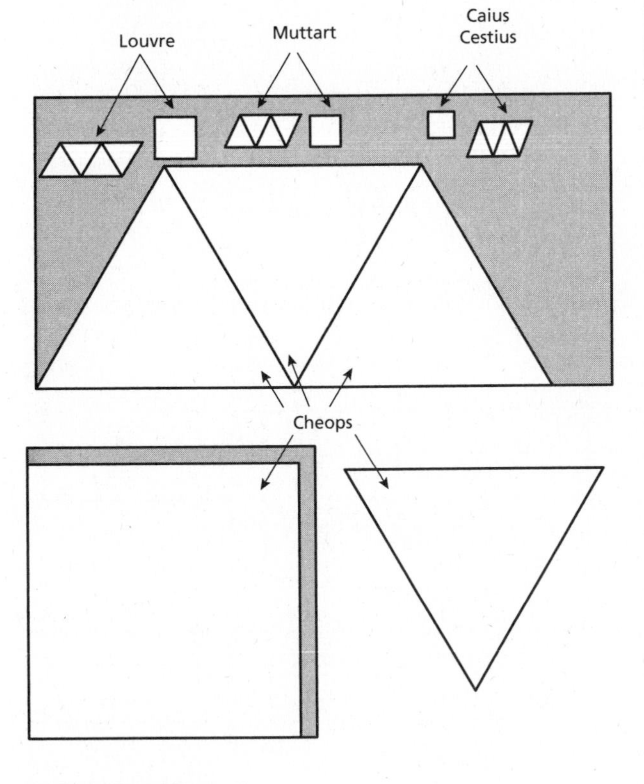

9.2 Puzzle Time

THE INFANTRY

9.3 Start Thinking!

For use before Activity 9.3

Sample answer: recovering a cylindrical ottoman with new fabric

9.3 Warm Up

For use before Activity 9.3

1. 28.26 $in.^2$
2. 153.86 ft^2
3. 314 cm^2
4. 50.24 cm^2

9.3 Start Thinking!

For use before Lesson 9.3

The cylinder with a radius of 10 centimeters and height of 4 centimeters has a greater surface area (approximately 879.2 square centimeters compared to approximately 351.68 square centimeters).

9.3 Warm Up

For use before Lesson 9.3

1. surface area: $36\pi \approx 113.0$ $in.^2$

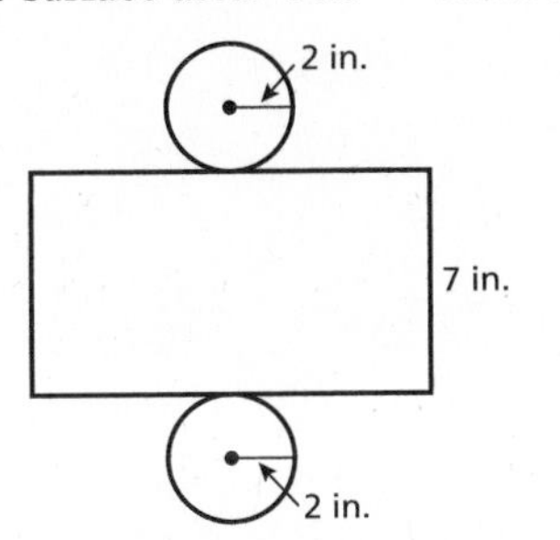

2. surface area: $42\pi \approx 131.9$ $in.^2$

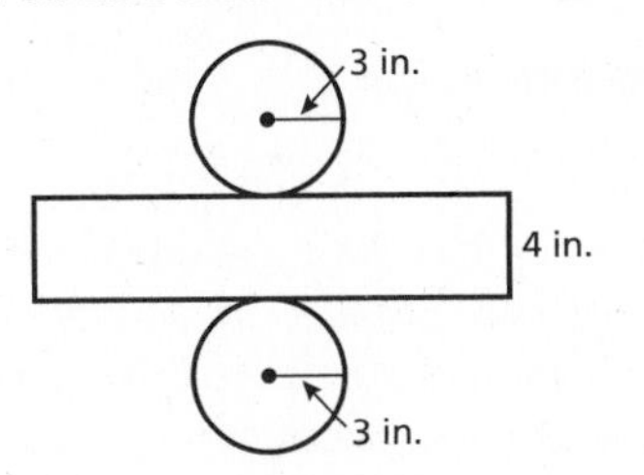

3. surface area: $12\pi \approx 37.7$ ft^2

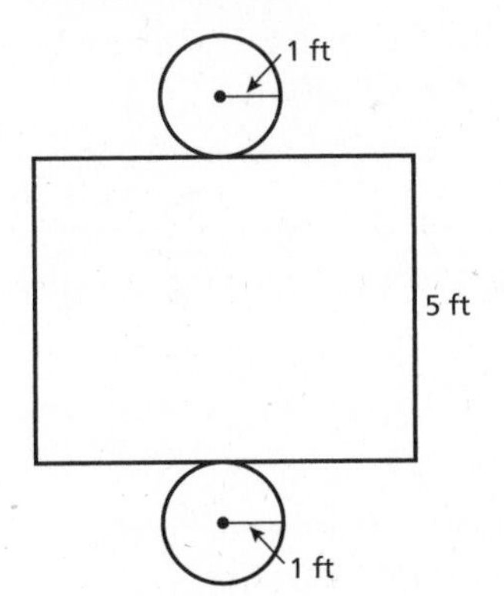

Answers

4. surface area: $32\pi \approx 100.5 \text{ ft}^2$

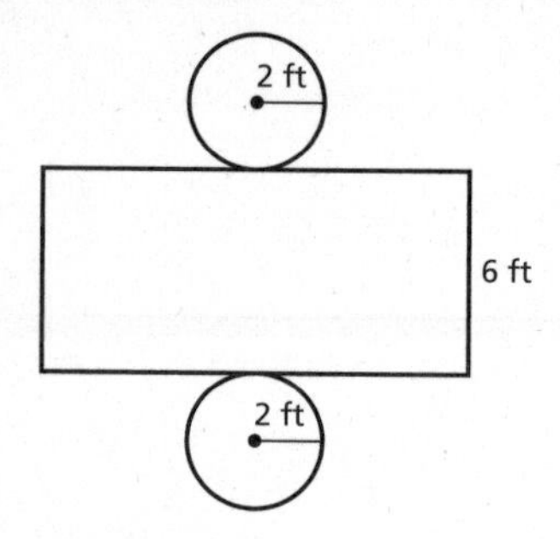

9.3 Practice A

1. $48\pi \approx 150.7 \text{ in.}^2$
2. $60\pi \approx 188.4 \text{ m}^2$
3. $128\pi \approx 401.9 \text{ ft}^2$
4. $176\pi \approx 552.6 \text{ mm}^2$
5. $18\pi \approx 56.5 \text{ cm}^2$
6. $18\pi \approx 56.5 \text{ yd}^2$
7. $84\pi \approx 263.8 \text{ in.}^2$

9.3 Practice B

1. $408\pi \approx 1281.1 \text{ ft}^2$
2. $20\pi \approx 62.8 \text{ cm}^2$
3. $260\pi \approx 816.4 \text{ m}^2$
4. 552.6 in.^2
5. a. $\frac{255}{512}\pi \approx 1.56 \text{ in.}^2$ b. $\frac{189}{256}\pi \approx 2.32 \text{ in.}^2$

   c. $\frac{\$0.25}{1.56 \text{ in.}^2} \neq \frac{\$0.50}{2.32 \text{ in.}^2}$; $0.58 \neq 0.78$

   d. $\frac{\$0.25}{1.56 \text{ in.}^2} = \frac{\$0.50}{x \text{ in.}^2}$; $x = 3.12 \text{ in.}^2$

9.3 Enrichment and Extension

1. a. 169.6 in.^2 b. 304.6 in.^2 c. 34.8%

   d. *Sample answer:* After the cake is cut into pieces, it has more surface area. Because more of the cake is exposed to air, more moisture can escape and evaporate. So, the cake does not stay as moist.

2. a. 12.7 oz

   b. He will have to buy 2 containers, and he will have 3.3 ounces of icing left over.

9.3 Puzzle Time

THE MAN WHO BOUGHT A NEW BOOMERANG AND SPENT A WEEK TRYING TO THROW HIS OLD ONE AWAY

9.4 Start Thinking!

For use before Activity 9.4

No, two-dimensional figures do not have volume because they do not have depth (height). Yes, three-dimensional figures do have volume because they have depth (height).

9.4 Warm Up

For use before Activity 9.4

1. 280
2. 672
3. 910
4. 495
5. 1120
6. 4032

9.4 Start Thinking!

For use before Lesson 9.4

Sample answer: Find the volume of the sand and the box. If the volume of the box is greater than the volume of the sand, then the sand will fit in the box.

9.4 Warm Up

For use before Lesson 9.4

1. 24 cm^3
2. 168 ft^3
3. 1120 in.^3
4. 126 m^3

9.4 Practice A

1. 36 in.^3
2. 135 m^3
3. 120 cm^3
4. 16 yd^3
5. 15 in.^3
6. 300 m^3
7. 8 in.^3
8. 5280 in.^3
9. 8.7 gal

9.4 Practice B

1. $12{,}000 \text{ cm}^3$
2. 1875 m^3
3. 63.7 in.^3
4. 22.5 cm^3
5. 10.8 yd^3
6. 1290 m^3
7. 936 in.^3
8. 0.85 L
9. 27 ft^3

9.4 Enrichment and Extension

1. a–b.

Length (in.)	Width (in.)	Height (in.)	Volume (in.^3)	Surface Area (in.^2)
5	5	5	125	150
5	5	10	250	250
5	5	15	375	350
5	10	10	500	400
5	10	15	750	550
10	10	10	1000	600
5	15	15	1125	750
10	10	15	1500	800
10	15	15	2250	1050
15	15	15	3375	1350

Answers

2. a–b.

Length (in.)	Width (in.)	Height (in.)	Volume ($in.^3$)	Surface Area ($in.^2$)
5	5	5	125	150
5	5	8	200	210
5	8	8	320	288
5	5	15	375	350
8	8	8	512	384
5	8	15	600	470
8	8	15	960	608
5	15	15	1125	750
5	5	50	1250	1050
8	15	15	1800	930
5	8	50	2000	1380
8	8	50	3200	1728
15	15	15	3375	1350
5	15	50	3750	2150
8	15	50	6000	2540
15	15	50	11,250	3450
5	50	50	12,500	6000
8	50	50	20,000	6600
15	50	50	37,500	8000
50	50	50	125,000	15,000

3. The surface area usually increases. There are a few times in the second table where the surface area will decrease.

4. In order to maximize volume and minimize surface area, the dimensions should be close to the same value. A cube is the rectangular prism with the least surface area for its volume.

9.4 Puzzle Time

TIC TAC DOUGH

9.5 Start Thinking!

For use before Activity 9.5

Sample answer: Slant height deals with the height of the triangle on the outside of the pyramid and is used to find the surface area of a pyramid. The height of the pyramid is the distance from the top of the pyramid straight down to the base and is used to find the volume of the pyramid.

9.5 Warm Up

For use before Activity 9.5

1. 10 **2.** 6 **3.** $4\frac{1}{5}$

4. 6 **5.** $16\frac{2}{3}$ **6.** $22\frac{2}{13}$

9.5 Start Thinking!

For use before Lesson 9.5

Sample answer: A company is making pyramid-shaped paperweights that will be filled with a liquid. The company will need to know how much liquid to fill the paperweights, so they will need to know how much each paperweight can hold (or the volume).

9.5 Warm Up

For use before Lesson 9.5

1. 45 $in.^3$ **2.** 2700 ft^3

3. 28 m^3 **4.** 180 cm^3

9.5 Practice A

1. 4 cm^3 **2.** 84 $in.^3$ **3.** 50 m^3

4. 120 ft^3 **5.** 320 ft^3 **6.** $84

7. a. 72 cm^3 **b.** 24 cm

8. The volume is halved.

9.5 Practice B

1. 600 mm^3 **2.** 140 ft^3

3. 1375 cm^3 **4.** 200 $in.^3$

5. a. 400 ft^3 **b.** 400 ft^3 **c.** same as

9.5 Enrichment and Extension

1. The sand will spill over. About 10.7 cubic inches of sand will spill out.

2. The sand will not spill over. The sand will be about 3.8 inches high in the cylindrical bucket.

3. 2 in.

4. This cylindrical bucket has a smaller diameter than the cylindrical bucket from Exercise 2. This cylindrical bucket is holding 64 cubic inches of water because that is the volume of the cube bucket. If the cylindrical bucket from Exercise 2 was filled to a height of 7 inches, it would hold about 197.9 cubic inches of water. Because the cylindrical bucket would hold less with the same height, the diameter must be smaller.

5. *Answer should include, but is not limited to:* A picture must be included in which the shapes are labeled with their names, dimensions, and volumes. The total volume of the shapes must be less than or equal to 360 in.^3

9.5 Puzzle Time

FRUIT SALAD

Extension 9.5 Start Thinking!

For use before Extension 9.5

Check students' work.

Extension 9.5 Warm Up

For use before Extension 9.5

1. cylinder
2. rectangular prism
3. pyramid
4. sphere

Extension 9.5 Practice

1. square
2. triangle
3. rectangle
4. triangle
5. point
6. triangle
7. circle
8. circle

Technology Connection

1. The surface area is multiplied by a factor of 4.
2. The surface area is multiplied by a factor of 9.
3. The surface area is multiplied by a factor of 16.
4. The surface area is multiplied by a factor of n^2. The formula for the surface area is $S = 6s^2$, so when s is multiplied by n, the formula becomes $S = 6(ns)^2 = n^2 \bullet 6s^2$.

Chapter 10

10.1 Start Thinking!

For use before Activity 10.1

2; 4; 4

10.1 Warm Up

For use before Activity 10.1

1. $\frac{6}{25}$
2. $\frac{1}{2}$
3. $\frac{4}{5}$
4. $\frac{1}{5}$
5. $\frac{3}{5}$
6. $\frac{4}{7}$

10.1 Start Thinking!

For use before Lesson 10.1

Answers will vary. Check students' work.

10.1 Warm Up

For use before Lesson 10.1

1. 8
2. 4; 4

10.1 Practice A

1. Choosing 4
2. Choosing 2, 4, 6, or 8
3. Choosing 1
4. Choosing 7 or 9
5. Choosing 2, 4, 6, or 8
6. no favorable outcomes
7. **a.** 3 **b.** Choosing any 1 of the 3 triangles
8. **a.** 1 **b.** Choosing a star
9. **a.** 6
   b. Choosing a star, choosing any 1 of the 2 circles or 3 triangles
10. **a.** 5
    b. Choosing a star, a square, or any 1 of the 3 triangles
11. **a.** 22 **b.** 24
12. **a.** 3 **b.** 2

10.1 Practice B

1. Choosing 8
2. Choosing 2, 4, or 6
3. Choosing 5 or 7
4. no favorable outcomes
5. Choosing 2 or 3
6. Choosing 3, 6, 8 or 9
7. **a.** 1 **b.** Choosing a triangle
8. **a.** 4 **b.** Choosing any 1 of the 4 stars

Answers

9. a. 8

b. Choosing any 1 of the 4 stars or 3 circles or a triangle

10. a. 8

b. Choosing a triangle or any 1 of the 4 stars or 3 squares

11. a. 12 **b.** 7 **c.** 6 **d.** 5

e. Choosing any 1 of the 4 remaining dogs

10.1 Enrichment and Extension

1. 28 cards

30 stones: 1 by 30, 2 by 15, 3 by 10, 5 by 6
31 stones: 1 by 31
32 stones: 1 by 32, 2 by 16, 4 by 8
33 stones: 1 by 33, 3 by 11
34 stones: 1 by 34, 2 by 17
35 stones: 1 by 35, 5 by 7
36 stones: 1 by 36, 2 by 18, 3 by 12, 4 by 9, 6 by 6
37 stones: 1 by 37
38 stones: 1 by 38, 2 by 19
39 stones: 1 by 39, 3 by 13
40 stones: 1 by 40, 2 by 20, 4 by 10, 5 by 8

2. a. It is more likely that the ruined card has an even number of square stones in the design because 20 of the cards have an even number of square stones, and only 8 of the cards have an odd number of square stones.

b. It is more likely that the ruined card has more than 35 square stones in the design because 14 of the cards have more than 35 stones, and only 12 of the cards have less than 35 stones.

3. 36 stones; There are 5 different rectangular designs that can be made with 36 stones. This is the most options for 30 to 40 stones.

4. He should buy at least 33 stones. His six options to consider would be 2 by 15, 3 by 10, 5 by 6, 2 by 16, 4 by 8, and 3 by 11.

5. a.

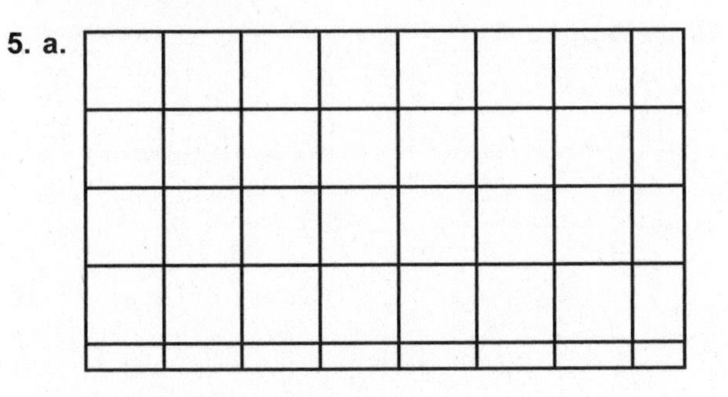

b. He will need to buy 34 stones.

c. He will use 28 whole stones. He will need to cut four of the stones so that each is two pieces, one that is 3 inches wide $\left(\frac{1}{3}\text{ of the length}\right)$ and one that is 6 inches wide $\left(\frac{2}{3}\text{ the length}\right)$. One stone will need to be cut into three pieces that are each 3 inches wide $\left(\frac{1}{3}\text{ the length}\right)$. The last stone will need to be cut into a piece that is 6 inches long $\left(\frac{2}{3}\text{ the length}\right)$ and 3 inches wide $\left(\frac{1}{3}\text{ the length}\right)$.
The only waste will be two blocks cut from this last block, one that is 6 inches by 9 inches and one that is 3 inches by 3 inches.

10.1 Puzzle Time

A VERY LOST CAMEL

10.2 Start Thinking!

For use before Activity 10.2

Your friend has a better chance of rolling a four because he has two dice.

10.2 Warm Up

For use before Activity 10.2

1. no; $\frac{3}{7}$ **2.** no; $\frac{4}{9}$ **3.** yes

4. yes **5.** yes **6.** no; $\frac{1}{2}$

10.2 Start Thinking!

For use before Lesson 10.2

Sample answer: A weather forecaster uses probability when stating the chance of precipitation for each day.

10.2 Warm Up

For use before Lesson 10.2

1. Spinner A is a better choice because you have 2 chances of getting an "Up."

2. No, it does not matter which spinner you use because both spinners have "Reverse" listed only once.

Answers

10.2 Practice A

1. Spinner A; Probability of red is $\frac{1}{2}$ versus $\frac{1}{3}$.

2. Spinner B; Probability of yellow is $\frac{1}{3}$ versus $\frac{1}{6}$.

3. no; Probability of blue is $\frac{1}{3}$ for both spinners.

4. Snow today is impossible.

5. You are likely to make a free throw.

6. It is unlikely that your band marches in the parade.

7. $\frac{1}{5}$ 8. $\frac{1}{10}$ 9. $\frac{3}{5}$ 10. 0

11. 12 games

12. a. 11 b. about equally likely

13. a. $\frac{4}{5}$; likely to happen

b. $\frac{1}{2}$; equally likely to happen

c. $\frac{9}{10}$; almost certain to happen

10.2 Practice B

1. unlikely to arrive late

2. certain to rain during a hurricane

3. likely you will go to the concert

4. $\frac{3}{16}$ 5. $\frac{1}{8}$ 6. $\frac{3}{4}$ 7. 1

8. a. 0.50 b. 0.50 c. 1 prize

9. a. $\frac{1}{2}$ before; $\frac{3}{5}$ after; probability increased

b. $\frac{3}{10}$ before; $\frac{1}{5}$ after; probability decreased

c. $\frac{1}{5}$ before; $\frac{1}{5}$ after; probability stayed the same

10.2 Enrichment and Extension

1. $\frac{1}{4}$ 2. $\frac{3}{4}$ 3. $\frac{16}{81}$

4. $\frac{65}{81}$ 5. $\frac{16}{49}$ 6. $\frac{33}{49}$

10.2 Puzzle Time

ROCK STAR WHO WENT TO THE COMPUTER STORE SO HE COULD GET A GIG

10.3 Start Thinking!

For use before Activity 10.3

Sample answer: There are a total of 18 marbles in your bag. Six of the marbles are red. The theoretical probability of pulling a red marble is $\frac{6}{18}$ or $\frac{1}{3}$.

10.3 Warm Up

For use before Activity 10.3

1. $\frac{1}{6}$ 2. $\frac{1}{2}$ 3. $\frac{1}{3}$

4. $\frac{1}{6}$ 5. $\frac{1}{2}$ 6. $\frac{1}{2}$

10.3 Start Thinking!

For use before Lesson 10.3

Sample answer: When you conduct an experiment, the relative frequency of an event is the fraction or percent of the time that the event occurs.

$$\text{relative frequency} = \frac{\text{number of times the event occurs}}{\text{total number of times you conduct the experiment}}$$

10.3 Warm Up

For use before Lesson 10.3

1. $\frac{2}{10}$, 20% 2. 0%

3. $\frac{4}{10}$, 40% 4. $\frac{8}{10}$, 80%

10.3 Practice A

1. $\frac{3}{10}$ 2. $\frac{3}{5}$ 3. $\frac{7}{10}$

4. a. $\frac{2}{5}$ b. $\frac{3}{5}$ 5. a. $\frac{5}{16}$ b. 45

Answers

6.

Outcome	Experimental Probability
2 Heads	$\frac{1}{6}$
1 Head, 1 Tail	$\frac{7}{12}$
2 Tails	$\frac{1}{4}$

7. 1 head, 1 tail
8. $\frac{1}{2}$

9. One tail had the highest probability in both Exercises 7 and 8.

10.3 Practice B

1. $\frac{7}{20}$
2. $\frac{9}{20}$
3. $\frac{4}{5}$
4. $\frac{13}{20}$

5. a. $\frac{7}{30}$ b. 28
6. 400 packages

7. a. 50 times b. 6 times c. 3 tails
   d. $\frac{3}{25}$ e. 24 times

10.3 Enrichment and Extension

1.

Outcome	Tally	Frequency
0		
1		
2		
3		
4		
5		
6		
7		
8		
9		
10		
11		
12		

2. *Sample answer:* The outcomes will not be equally likely. The smaller numbers will be more likely than the larger numbers. There is only one way to get an outcome of 12, and that is to pick a King and an Ace. There are 4 of each in the deck, but it is still much less likely than choosing two cards that are only one apart, which can happen in a lot more ways (Ace and 2, 2 and 3, 3 and 4, 4 and 5, etc). As the possible outcome gets bigger, there are less ways to get that outcome, which makes it less likely to occur.

3. *Answer should include, but is not limited to:* Make sure students perform the experiment at least 60 times and that they use the absolute value of the difference.

4. *Answer should include, but is not limited to:* Make sure bar graph is created correctly. Students should have similar results.

5. *Sample answer:* The most common outcome is 1. As the outcomes get bigger, each one is slightly less likely than the one before it. So, the least common outcome is 12. However, zero does not follow the pattern. The likelihood of getting a difference of zero is somewhere in the middle.

6. *Sample answer:* Because you are choosing from 52 different cards, there are too many possible outcomes to list and count.

7. Because there are a large number of items to choose from and a fairly larger number of possible outcomes, it is necessary to do a lot of trials in order for the results to more accurately coincide with the theoretical probability.

8. The possible outcomes would range from -12 to 12. The most common outcomes are -1 and 1. The least common outcomes are -12 and 12.

9. This game is not fair because the smaller outcomes are much more likely than the larger outcomes. *Sample answer:* The rules would be more fair if they said: Player 1 gets a point if the positive difference is 1, 4, 5, 8, 9, or 12, and Player 2 gets a point if the positive difference is 2, 3, 6, 7, 10, or 11.

10.3 Puzzle Time

HUCKLEBERRY FAN

Answers

10.4 Start Thinking!

For use before Activity 10.4

8; *Sample answer:* chocolate with sprinkles, chocolate with peanuts, chocolate with caramel, chocolate with whipped cream, vanilla with sprinkles, vanilla with peanuts, vanilla with caramel, vanilla with whipped cream

10.4 Warm Up

For use before Activity 10.4

1. 60 **2.** 126 **3.** 100

4. 1080 **5.** 1350 **6.** 504

10.4 Start Thinking!

For use before Lesson 10.4

60; *Sample answer:* Multiply 3, 4, and 5 together.

10.4 Warm Up

For use before Lesson 10.4

1. 1000

10.4 Practice A

1.

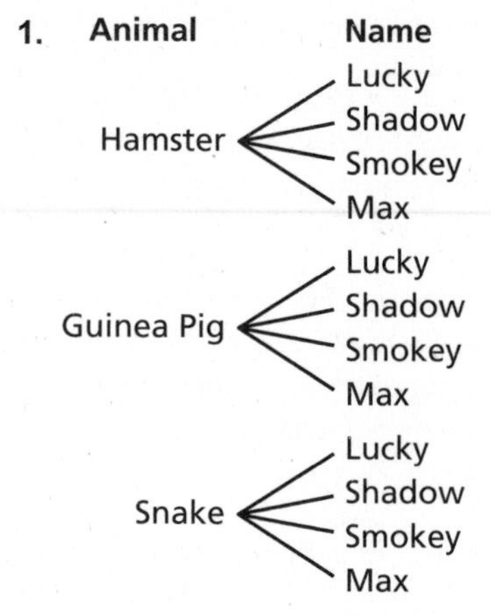

12; 12

2.

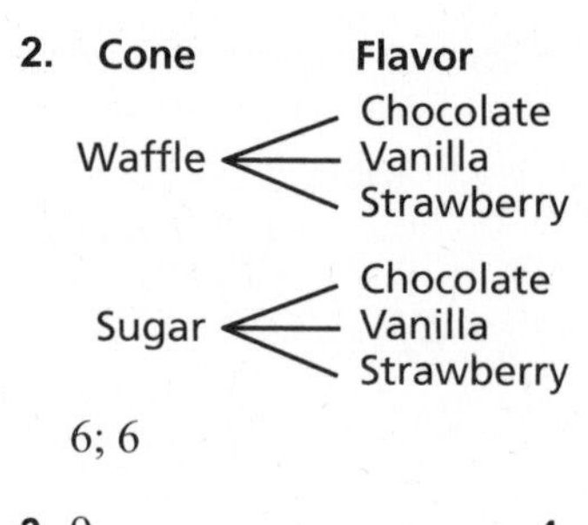

6; 6

3. 9 **4.** 16

5. 1024 **6.** 18,137,088

10.4 Practice B

1.

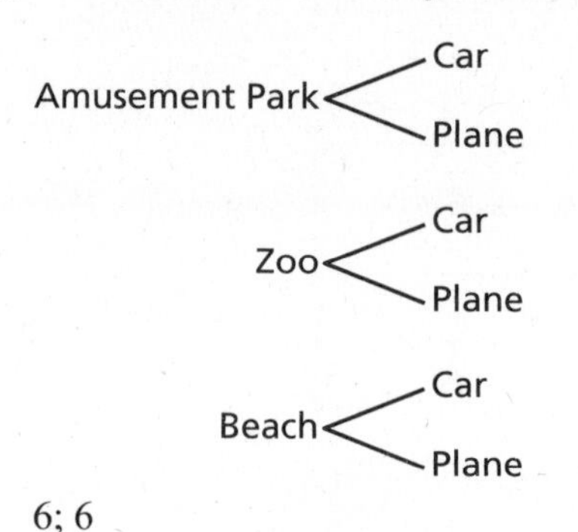

6; 6

2.

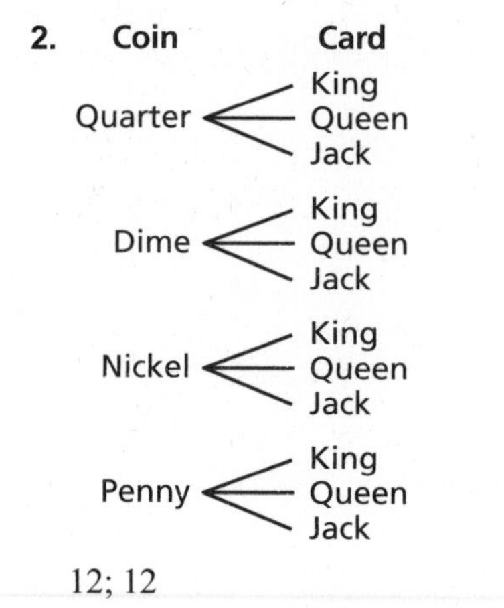

12; 12

3. 8 **4.** 18

5. a. 5040 **b.** 720

6. 6,760,000; 60,840,000

10.4 Enrichment and Extension

1. 256 sandwiches **2.** 64 sandwiches

3. a. There are 3 choices for sides which means the total combinations from Exercise 1 needs to be multiplied by 3.

b. 768 sandwich platters

Answers

4.

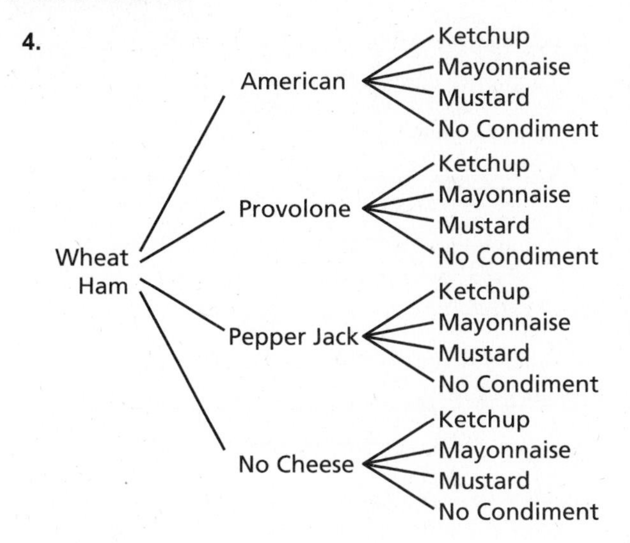

5. $\frac{1}{256}$

10.4 Puzzle Time

A TYRANT

10.5 Start Thinking!

For use before Activity 10.5

Sample answer: Independent means not relying on others; dependent means relying on others; I did my homework independently, but was dependent on my group members during our project.

10.5 Warm Up

For use before Activity 10.5

1. 5	**2.** 2	**3.** 2
4. 1	**5.** 5	**6.** 8

10.5 Start Thinking!

For use before Lesson 10.5

No, your friend is not correct. The events are dependent because the marble you did not replace affects the outcome of the second event.

10.5 Warm Up

For use before Lesson 10.5

1. Independent because the outcome of rolling the cube the first time does not affect rolling the cube the second time.
2. Independent because the outcome of flipping the coin the first time does not affect flipping the coin the second time.
3. Dependent because one of the marbles is removed from the bag and set aside so it cannot be drawn the second time.
4. Independent because the first marble is placed back in the bag and can be drawn the second time.

10.5 Practice A

1. independent; The second spin is not affected by the first spin.
2. dependent; One person cannot be both president and vice president.

3. $\frac{3}{10}$ **4.** $\frac{2}{5}$ **5.** $\frac{2}{21}$ **6.** $\frac{2}{7}$

7. $\frac{1}{15}$ **8.** $\frac{1}{36}$ **9.** $\frac{1}{4}$

10.5 Practice B

1. dependent; There are 10 pins on first throw and 4 pins on second throw.
2. independent; The second roll is not affected by the first roll.

3. $\frac{1}{5}$ **4.** $\frac{3}{10}$ **5.** $\frac{1}{21}$ **6.** $\frac{1}{21}$

7. $\frac{3}{51}$ **8.** $\frac{1}{1000}$ **9.** $\frac{1}{8}$

10.5 Enrichment and Extension

1. $\frac{1}{20}$ **2.** 3; $\frac{3}{20}$ **3.** 16; $\frac{4}{5}$

4. a. \$55, \$60, \$65 **b.** $\frac{3}{20}$

5. a. \$30, \$35, \$40 **b.** $\frac{3}{20}$

6. The probabilities are the same. The winning values change from the first spin to the second spin, but the number of winning sections does not change because you need to get a total of \$90, \$95, or \$100 to win. So the probability does not change.

7. $\frac{3}{20}$

8. a. $\frac{1}{4}$ **b.** $\frac{7}{40}$ **9. a.** $\frac{11}{20}$ **b.** $\frac{11}{50}$

10. a. $\frac{1}{10}$ **b.** $\frac{17}{200}$ **11. a.** $\frac{3}{4}$ **b.** $\frac{3}{20}$

10.5 Puzzle Time

A SHEEPDOG

Answers

Extension 10.5 Start Thinking!

For use before Extension 10.5

Sample answer: Find the probability of the event.

$P(\text{mistakes}) = \frac{2}{50} = \frac{1}{25}$

To make the prediction, multiply the probability of mistakes by the number of papers, $\frac{1}{25} \bullet 1000 = 40.$

You can predict that there will be 40 mistakes.

Extension 10.5 Warm Up

For use before Extension 10.5

1. $\frac{1}{10}$, 10% **2.** $\frac{4}{10}$, 40% **3.** $\frac{1}{10}$, 10% **4.** $\frac{8}{10}$, 80%

Extension 10.5 Practice

1. 13 **2.** 13%

3. 12%; increasing the number of trials would bring the experimental probability closer to 12%

10.6 Start Thinking!

For use before Activity 10.6

Sample answer: Events are independent if one occurrence of one event does not affect the likelihood that the other event(s) will occur. For example, you spin a 2 on a spinner and roll a 2 on a number cube. Events are dependent if the occurrence of one event does affect the likelihood that the other event(s) will occur. For example, you randomly choose a tile from a bag and without replacing the tile, you choose another tile.

10.6 Warm Up

For use before Activity 10.6

1. $\frac{1}{12}$, $8\frac{1}{3}\%$ **2.** $\frac{1}{4}$, 25%

3. $\frac{1}{4}$, 25% **4.** $\frac{1}{3}$, $33\frac{1}{3}\%$

10.6 Start Thinking!

For use before Lesson 10.6

Sample answer: No, because the sample is too small.

10.6 Warm Up

For use before Lesson 10.6

1. population; sample **2.** sample; population

3. population; sample **4.** sample; population

10.6 Practice A

1. population: All students in a school;
sample: 30 students in a school

2. population: All the strawberries in the field;
sample: 75 strawberries in the field

3. a. all the students in your school

b. 30 random students that you meet in the hallway between classes

c. unbiased; You are surveying at different times of the day and in the hallway rather than in your classrooms

4. sample A; surveyed the county, rather than just one city

5. sample A; larger sample size

6. population; You have access to all of the members of your family.

7. sample; It would not be easy to contact or visit every grocery store in the state of Florida.

8. 840 students

10.6 Practice B

1. a. fans at the Miami Dolphins and Dallas Cowboys game

b. 50 fans with season tickets for the Dolphins

c. not reasonable; Dolphins fans are more likely to say that the Dolphins will win. You must also survey the Cowboys fans.

2. sample B; surveyed your town, rather than just your neighborhood

3. sample; It would not be easy to survey every student at your school.

4. population; It is possible for you to ask all of the students in your history class.

5. 12 students

6. a. teenagers; *Sample answer:* It is more likely that teenagers favor rap music.

b. *Sample answer:* circle graph, bar graph

c. yes; *Sample answer:* When people in their 70s were younger, there was no rap music.

10.6 Enrichment and Extension

1. No, the Electoral College votes represent the majority of the state. Even if a candidate gets just a few more votes than their opponent does, the candidate will most likely earn all of the Electoral College votes for that state.

Answers

2. California: 134 students; District of Columbia: 7 students; Florida: 66 students; Louisiana: 22 students; Montana: 7 students

3. *Sample answer:* The Electoral College votes are not proportional to the state's popular vote. So, a presidential candidate can have big wins in some states and narrowly miss in others resulting in more popular votes and less Electoral College votes than their opponent has.
   Sample answer: In 2000, Al Gore won the popular vote by over 500,000 votes, but earned fewer Electoral College votes than George W. Bush. The Florida votes had to be recounted by hand in many precincts. Before the recount was complete, the U.S. Supreme Court ruled that the most recent data would stand, which gave Bush the win by less than 1000 votes. So, Bush earned all of the state's Electoral votes, which gave him the narrowest win in the Electoral College since the 1876 election.

4. *Sample answer:* Television stations work with other organizations to gather data from pre-election polls, exit polls, and early vote reporting. The pre-election polls give them an idea of who might win the state before the election. Then they collect data the day of the election to verify this prediction before making a projection. Precincts from across the state are chosen at random so that the sample of precincts will be representative of the state's population. Within the sample precincts, people take exit polls randomly (from every third or fifth person leaving the polls, for example). They ask voters about themselves and who they voted for. Some of the actual votes from these precincts are reported early as well. The results from these sample precincts are the most important predictor for making a projection for the population of the state because they tend to be more accurate than pre-election polls. In states where the polls are close, however, a projection is not made until more actual votes are reported and there is more certainty about who will win. Some states, such as California, traditionally vote for the same party one election after another. When the sample polling confirms that the vast majority of the state is in favor of one candidate, the television station can project the state with a high degree of certainty very early on. For states such as Florida, the margin of victory is often much closer and harder to determine without more official votes reported.

10.6 Puzzle Time

GO FISH

Extension 10.6 Start Thinking!

For use before Extension 10.6

Sample answer: Step 1: Order the data. Find the median and the quartiles; Step 2: Draw a number line that includes the least and greatest values. Graph points above the number line that represent the five-number summary; Step 3: Draw a box between the quartiles. Draw a line through the median. Draw whiskers from the box to the least and greatest values.

Extension 10.6 Warm Up

For use before Extension 10.6

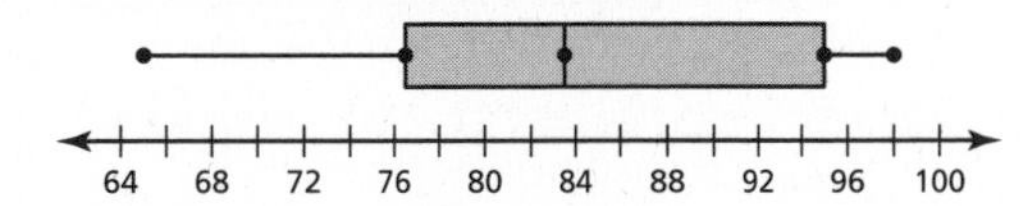

Extension 10.6 Practice

1. **a.** Check students' work.
   b. Check students' work.
   c. Check students' work. *Sample answer:* yes; Increase the number of random samples.

2. **a.** Check students' work.
   b. Check students' work.
   c. Check students' work.
   d. Check students' work.

10.7 Start Thinking!

For use before Activity 10.7

Sample answer: Given the data 2, 3, 4, 4, 6, 8, 8; The mean of a data set is the sum of the data divided by the number values. So, the mean is

$$\frac{2 + 3 + 4 + 4 + 6 + 8 + 8}{7} = \frac{35}{7} = 5.$$

The median is found by first ordering the data and then finding the middle value. So, the median is 4.

10.7 Warm Up

For use before Activity 10.7

1. 12	**2.** 62	**3.** 37
4. 78.5	**5.** 38	**6.** 65.5

10.7 Start Thinking!

For use before Lesson 10.7

Sample answer: Compare the heights of girls in 6th and 8th grades. You can compare the measures of center, the measures of variation, the shape of the distribution, and the overlap of the two distributions.

Answers

10.7 Warm Up

For use before Lesson 10.7

1. Team 1: mean: 57; median: 56; mode: 56; range: 25; IQR: 12; MAD: 5.8

   Team 2: mean: 58; median: 57; mode: 49; range: 23; IQR: 11; MAD: 5.6

2. Team 2 had a greater mean and median. Team 1 has a greater mode, range, and measures of variation.

10.7 Practice A

1. **a.** Varsity Team: mean: 17.5; median: 18; mode: 18; range: 3; IQR: 1; MAD: 0.75

   Junior Varsity Team: mean: 16.25; median: 16.5; mode: 17; range: 4; IQR: 1.5; MAD: 0.92

   b. The Varsity Team has greater measures of central tendency because the mean, median, and mode are greater. The Junior Varsity Team has greater measures of variation because the range, interquartile range, and mean absolute deviation are greater.

   c. mean and MAD; Both distributions are approximately symmetric.

   d. The difference in the means is about 1.36 to 1.67 times the MAD.

2. **a.** City A: median: 3; IQR: 4

   City B: median: 6; IQR: 4

   The variation in the number of inches of snow is the same, but City B had 3 more inches than City A.

   b. The difference in the medians is 0.75 times the IQR.

10.7 Practice B

1. **a.** Football:
   mean: 189, median: 178, mode: 178, range: 158, IQR: 28, MAD: 22.89

   Basketball:
   mean: 199, median: 194, no mode, range: 145, IQR: 47, MAD: 31.67

   b. The basketball pep rallies have greater measures of central tendency because the mean and median are greater. Football has greater range, but basketball has greater interquartile range and mean absolute deviation.

   c. The median and the IQR; both distributions are skewed.

   d. The difference in the medians is about 0.34 to 0.57 times the IQR.

2. **a.** Garden A: median: 54; IQR: 18

   Garden B: median: 42; IQR: 18

   The variation in the height of the corn stalks is the same, but Garden A had 12 more inches than Garden B.

   b. The difference in the medians is about 0.67 times the IQR.

10.7 Enrichment and Extension

1. expected value: 6; Daulton should pass because the expected value is more than 4.

2. expected value: 5.25; Ally should be confident because the expected value is greater than the value of her friend's card.

3. expected value: 6; Paxton should spin because the expected value is equal to the lowest number he can get to win.

4. $P(\text{Exercise 1}) = \frac{1}{3}$

   $P(\text{Exercise 2}) = \frac{1}{2}$

   $P(\text{Exercise 3}) = \frac{3}{5}$

   The advice would be the same for Exercises 1 and 3, but not Exercise 2. Because the probability for Exercise 2 is equal to $\frac{1}{2}$, Ally should not be confident she will win.

5. Expected value and probability are not the same thing. *Sample answer:* Probability is better because the expected value may not be a possible outcome.

10.7 Puzzle Time

LIGHTHOUSE

Technology Connection

1. The results of the two cases should be similar. As long as the areas of the two outcomes are equal, regardless of whether the sections are adjacent or not, the results should be fairly close.

2. The spinner should be labeled with sections from 2 to 12. The weights of the sections should be the following: 2 and 12 set to 1; 3 and 11 set to 2; 4 and 10 set to 3; 5 and 9 set to 4; 6 and 8 set to 5; and 7 set to 6. The theoretical probability of rolling each number in 1000 trials is approximately the following: 2 or 12: 28 times; 3 or 11: 56 times; 4 or 10: 83 times; 5 or 9: 111 times; 6 or 8: 139 times; 7: 167 times